# Structural Geology
## An Introduction

## Frontispiece

Leonardo da Vinci's *A Sketch from Nature*. For the first time, and well ahead of his time, an artist truly sees the fabric of rocks. (From an original drawing in the Royal Library, Windsor Castle. Copyright reserved. Reproduced by gracious permission of Her Majesty Queen Elizabeth II.)

# Structural Geology
## An Introduction

John G. Dennis
California State University
Long Beach

**wcb**
Wm. C. Brown Publishers
Dubuque, Iowa

Cover Image

The cover photo is of the San Andreas Fault along the western foothills of the Temblor range, California, some 25 miles west of Taft. The view looks northwest from an altitude of 3,000 feet. The western strand of the fault, currently active, cuts off ridge spurs. An older strand to the east has beer deeply eroded. It has not been active in recent times. The flat area at the western (left) side of the picture is the eastern edge of the Carrizo plains. Photo by Barrie Rokeach/The Image Bank.

Book Team

Editor *Edward G. Jaffe*
Developmental Editor *Lynne M. Meyers*
Designer *Mary K. Sailer*
Production Editor *Michelle M. Kiefer*
Photo Research Editor *Shirley Charley*
Permissions Editor *Mavis M. Oeth*


**wcb group**

Wm. C. Brown *Chairman of the Board*
Mark C. Falb *President and Chief Executive Officer*


**wcb**

Wm. C. Brown Publishers, College Division

G. Franklin Lewis *Executive Vice-President, General Manager*
E. F. Jogerst *Vice-President, Cost Analyst*
George Wm. Bergquist *Editor in Chief*
Edward G. Jaffe *Executive Editor*
Beverly Kolz *Director of Production*
Chris C. Guzzardo *Vice-President, Director of Sales and Marketing*
Bob McLaughlin *National Sales Manager*
Craig S. Marty *Manager, Marketing Research*
Julie A. Kennedy *Production Editorial Manager*
Marilyn A. Phelps *Manager of Design*
Faye M. Schilling *Photo Research Manager*

# Contents

# Preface

This book introduces the principles of structural geology for undergraduate students of geology. It also provides the basic knowledge of structural geology needed for professionals in related disciplines of science and engineering, and is designed to be equally suitable as a text for classroom instruction and for self study. Prior knowledge of elementary geology, physics, and algebra is assumed. This book succeeds my earlier *Structural Geology* (Ronald Press 1972, later John Wiley & Sons), incorporating many of its proven features. The organization and treatment reflect many years of experience in the classroom and in the field, with feedback from both students and colleagues that it would be impossible to acknowledge here appropriately.

The book approaches concepts from first principles. Descriptive matter has been separated from interpretation and theory to the greatest possible extent, and the student is led from observation to inference. I have taken particular care to identify as clearly as possible areas of doubt and uncertainty, where better understanding can be expected in the years to come.

I have tried to make the subject accessible to students who traditionally approach it with apprehension, and to show that structural geology is not a difficult subject, even though there is much meat in it. It is, above all, a fascinating subject.

The book is divided into four parts. In Part I, Fundamentals, the first chapter provides an overview of plate tectonics as a setting for examples of structures illustrated throughout the book; the remaining chapters of this part cover such fundamentals as geometrical principles, stress, strain, and flow of rocks. Part II, Continuous Structures, includes folds and fabrics. Part III, Discontinuous Structures, deals with fractures and faults. Part IV, Structures of Igneous Rocks, consists of a single chapter on the subject.

Appendix A introduces the student to graphic solutions of problems that arise in the course of geological field work and in the preparation of reports and maps. The exercises develop basic skills and good three-dimensional perception. At California State University, Long Beach, we have found them very successful in preparing students for our geological mapping courses. I have used a slightly modified version of the Laboratory Exercises in Charles Nevin's classical text (Nevin 1949). I gratefully acknowledge permission by Brian Nevin on behalf of the Nevin family. The remaining appendices contain additional material and techniques that might have encumbered the main body of the text. More advanced geometrical techniques for specialists will be found in some of the excellent manuals now available.

The book may be readily adapted to both shortened and extended courses. Boxes in the text contain material which, while useful and interesting, may be omitted where time is limited. The same is true of chapters 2, 4, and 16, although a complete course should include them. Chapter 3 may be integrated in the laboratory part of the course. On the other hand, an extended, possibly two-semester or two-quarter, course may include additional material from works listed below and under "Additional Reading" at the end of each chapter, as well as from some of the references cited at the end of the book. Review questions at the end of each chapter help to focus on important concepts in the chapter. Boldface denotes terms introduced or defined for the first time. The book concludes with a glossary that gives brief definitions of some of the more important terms found in the text.

I have included a large number of illustrations on the premise that, especially for geological structures, pictures speak more eloquently than words. While geometric and mechanical principles are best illustrated diagrammatically, I have tried also to present as much as possible of the illustrative material in photographs of actual examples; this helps enormously in creating a base for true structural intuition.

I appreciate the many permissions that were freely given to reproduce instructive illustrations throughout the text. The sources are acknowledged by name, keyed to the bibliography at the end; photographs by the British Geological Survey are reproduced by permission of the Director, British Geological Survey: U. K. Crown Copyright Reserved. I am indebted to Mason Hill; C. N. Nevin; John Christie, University of California at Los Angeles; Peter W. Huntoon, University of Wyoming; Richard Sibson, University of California at Santa Barbara; Jay Zimmerman, James W. Sears, University of Montana; Jeremy Dunning, Indiana University; Clarence J. Casella, Northern Illinois University; William B. Travers, Cornell University; and many other colleagues and students for very helpful critical advice while preparing the manuscript. I also wish to record my debt to Gilbert Wilson, who first interested me in the subject, and to Eugene Wegmann and Walter Bucher, who gave direction to my further studies. Thank you to my former student Laura Krol for compiling the glossary and index for this book. The help and advice of the editorial and production staff of Wm. C. Brown Company, particularly Edward Jaffe, Lynne Meyers, and Michelle Kiefer have been invaluable.

John G. Dennis<br>
Long Beach, California

# Introduction

S tructural geology is the science that deals with the shape and internal fabric of deformed rock bodies and the processes that deform them. It is an interesting and rewarding study that forms a link between most other branches of geology.

Some basic terms need to be defined here so that we may avoid misunderstandings as we proceed. For instance, *deformation,* as used in geology, is the process that results in a change of the shape or size of a coherent rock body. Deformation may be *continuous,* that is, distributed over the whole rock body; or it may be *discontinuous,* that is, localized in fractures or in narrow zones. The *fabric* of a rock is the internal arrangement of repetitive constituent elements, such as mineral grains or planes of weakness. *Tectonics,* in modern usage, is the study and interpretation of regional structural patterns.

Early structural geologists did not distinguish clearly between cause and effect. One of the first to do so was Eduard Suess, the great Austrian geologist, in a book analyzing the structure of the Alps (Suess 1875). In 1893 G. F. Becker showed that it was necessary to analyze the geometry of deformed rocks before attempting to interpret it in terms of causes of rock deformation. In 1911 the Austrian geologist Bruno Sander initiated an approach to structural analysis which, with some modification, is still followed today. It proceeds by the three successive steps of geometric, kinematic, and dynamic analyses of structures.

*Geometric analysis* describes the external form and the internal fabric of rock bodies. A complete analysis should include available observations on all scales.

Geometric analysis is not directly concerned with deformation; thus, an inventory of geometric elements includes *primary structures,* which a rock body acquires in the process of deposition or emplacement, and *secondary structures,* which it acquires as a result of diagenesis and deformation. Geometric analysis is equally concerned with the form of rock bodies, the boundaries or contacts between them, their internal fabric, and the discontinuities that traverse them.

*Kinematic analysis* is the analysis of *displacements.* Displacements within a body that lead to changes in size and in shape of the body constitute *strains.* Strain is caused by *stress,* which is the force per unit area acting at any point of the body. These concepts will be developed in chapter 5.

*Dynamic analysis* is the attempt to find the stress configuration responsible for observed strains.

The integration of geometric, kinematic, and dynamic analysis leads to *structural synthesis.* At their most complete, structural and tectonic syntheses give a picture of the evolution in time of a deformed segment of the earth's crust. This demands a proper perspective in space and in time. The methods are both analytical and historical, and the tools come from many other branches of science. However, the synthesis is only as good as each step that has led to it.

The total structure of a deformed segment of crust gives it a certain imprint, a *style,* in the same way that the sum of structural and ornamental features gives a building its particular style. In rocks, tectonic style of a given domain reflects, above all, relative mobility, intensity of deformation, and the relative and absolute sizes of the rock units involved. This is what Maurice Lugeon, an Alpine geologist who had in his family several architects, called *tectonic style.* Thus, we may speak, for instance, of *supercrustal style, fluid style, brittle style,* and many others.

We will begin our study of structural geology with a review of the global setting of geological structures. This setting has been revealed by the most encompassing synthesis in the geological sciences: plate tectonics.

## Selected General References for Further Study

Condie, Kent C. 1982. *Plate Tectonics & Crustal Evolution.* Oxford: Pergamon. 310 pp.

Jaroszewski, W. 1984. *Fault and Fold Tectonics.* Warsaw: Polish Scientific Publishers; Chichester: Ellis Horwood. 565 pp.

Lowell, J. D. 1985. *Structural Styles in Petroleum Exploration.* Tulsa: OGCI Publications. 460 pp.

Ragan, D. M. 1985. *Structural Geology, An Introduction to Geometrical Techniques.* 3d ed. New York: Wiley. 393 pp.

Ramsay, J. G. 1967. *Folding and Fracturing of Rocks.* New York: McGraw-Hill. 568 pp.

Ramsay, J. G., and Huber, M. I. 1983. *The Techniques of Modern Structural Geology. Vol. 1, Strain Analysis.* London: Academic Press. 307 pp.

Uemura, T., and Mizutani, S., eds. 1984. *Geological Structures.* New York: Wiley. 309 pp.

# Part One
# Fundamentals

# 1

# Review of Plate Tectonics

## Figure 1.1

Block diagram illustrating schematically the configurations and roles of lithosphere, asthenosphere, and mesosphere in plate tectonics. Arrows in lithosphere indicate relative movements of adjoining plates. Arrows in asthenosphere represent possible compensating flow in response to downward movement of lithosphere. One arc–arc transform fault appears to the left, between oppositely facing subduction zones (trenches). Note the lack of relative movement between any trench and the bordering plate that is not being consumed. Therefore, between the two trenches at left, relative movement must be occurring that is taken up along the transform fault. Two ridge–ridge transform faults appear along the ocean ridge at center. (After Isacks et al., 1968. Copyright by the American Geophysical Union.)

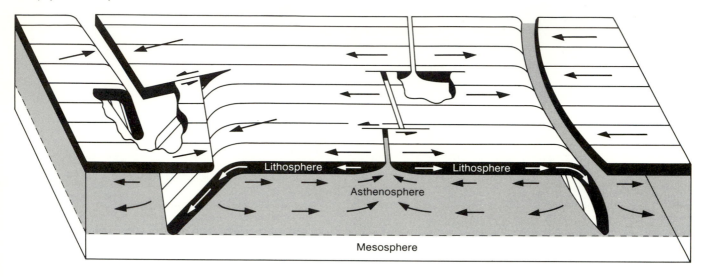

Lithosphere · Lithosphere

Asthenosphere

Mesosphere

**P**late tectonics is the name given to the theory that provides a unifying model and framework for most displacements and deformation in the earth's outer layers. Two basic assumptions underlie the theory: (1) the outermost earth shell, called the **lithosphere,** rests on and is stronger than the underlying shell, which is a weak layer called the **asthenosphere** (figure 1.1); and (2) the lithosphere is not continuous but consists of a mosaic of horizontally rigid shell segments or **plates.** The plates are in constant relative motion with one another, and their individual surface areas are constantly changing. In this the earth is unique: No other known planet appears to have an outer shell of plates that are in relative motion with one another. Most (but not all) rock deformation on earth is related to geological processes that occur along plate boundaries.

## Earth Shells

To place plate tectonic processes within the framework of the whole earth, we must first review the large-scale seismological subdivisions of the earth. The boundaries between the different subdivisions or **shells** (figure 1.2) have been determined largely by seismic methods. Seismic waves travel at different velocities through rocks of different composition, and they change in other properties as they cross boundaries between different rock shells.

The **core** of the earth is metallic and probably consists mainly of iron. It may be subdivided, on seismic evidence, into a solid inner core with a 1,200 kilometer (km) radius, and a liquid outer core 2,300 km thick.

**Figure 1.2**
Earth shells, as determined from seismic refraction studies. (From Plummer and McGeary, 1985.)

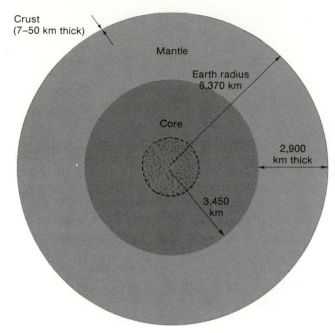

The overlying **mantle** comprises the bulk of the earth's volume and mass. It is 2,900 km thick and probably consists mainly of ultramafic rock with a composition somewhere between that of basalt and peridotite. It is solid when considered over short time spans ($n \times 10^3$ years), but it can flow in the solid state when considered over longer time spans. Some regions of the mantle flow much more readily than others. We shall reconsider this property in chapter 6.

The mantle is overlain by another solid layer, the earth's crust, which varies considerably in thickness. Oceanic crust, which is relatively dense and consists mainly of basaltic rocks, ranges in thickness from 5 to 6 km. Continental crust is less dense than oceanic crust and has an average composition of granodiorite. It varies in thickness from 20 to more than 70 km. These thickness variations cause relief, both in the lower boundary of the crust—the **Mohorovičić discontinuity**—and in its upper boundary, the earth's surface.

## Isostasy

In 1889, C. E. Dutton (1841–1912), in a classic paper, examined the conditions of equilibrium in a heterogeneous earth's crust. He showed that, as a result of gravity acting on segments of unequal density, areas underlain by rocks of low density would bulge upward, whereas areas underlain by rocks of high density would tend to be depressed. All components of the crust would tend to reach a gravitational equilibrium, which he termed **isostasy.** Dutton also pointed out that, in order to maintain isostatic equilibrium, areas being eroded would have to rise, whereas areas receiving sediments would have to sink.

Speculation on isostasy started with Leonardo da Vinci, who, believing in a perfectly rigid earth, thought that changes in relief, such as mountain building and erosion, would displace the earth's center of mass. In

**Figure 1.3**

Topographic relief by density contrast in isostatic equilibrium; less dense rocks supported by a denser substratum. (a) Model according to Pratt. The less dense columns of rock carry higher elevations. Level of compensation at depth D is 100 km. (b) Model according to Airy. Higher elevations have deeper roots.

(Source: Airy, G. B., "On the contribution of the effects of the attention of mountain-masses" in Royal Society of London Transactions, 145, pp. 101–104, 1855.)

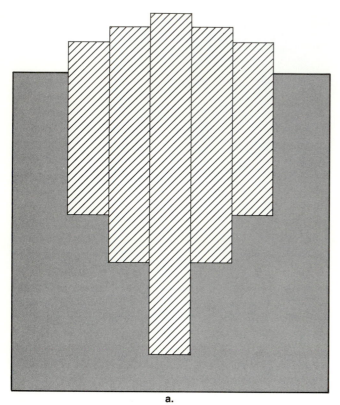

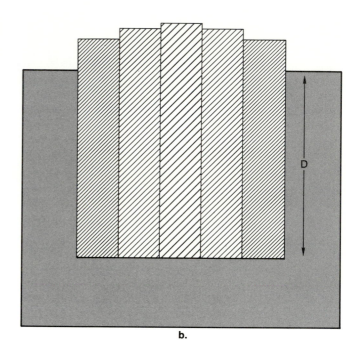

a.

b.

1700 Clüver, in *Geologia* (the first recorded use of "geology" in its present sense), considered that mountains should either displace the earth's center of mass or sink into subterranean cavities, "if these exist." However, he pointed out that these things do not happen, because mountain "columns" are made up of rocks having an average specific gravity of 2.5, "without considering the empty cavities which [the mountains] contain." Clüver thought that the inner earth had a specific gravity varying from 3 to 5, as determined from "denser rocks found at the bottom of mines."

Between 1735 and 1745, the French geodesist Pierre Bouguer measured a meridian arc in the foothills of the Peruvian Andes, using a plumb line to determine the vertical. He expected to see the plumb line deflected by the mass of the adjacent mountains. But the gravitational attraction of the Andes was much less than expected. In 1755 R. J. Boscovich explained this mass deficiency by assuming that voids existed below the mountain that would compensate for the overlying mass. Between 1774 and 1776, Maskelyne and Charles Hutton compared the specific gravity of the earth with

that of Schiehallion Mountain in the Grampians of Scotland. They did this by evaluating the deflection of a plumb line from the vertical at two stations near the foot of Schiehallion. As a result, they computed a mean earth density of 4.713. This figure is of the correct order, but it is not quite accurate, because of the heterogeneous composition of the mountain and the crudity of the measurements.

In 1854, in a paper submitted to the Royal Society of London, Archdeacon J. H. Pratt (Pratt 1855) commented on the results of Sir George Everest's recent geodetic mission to the Himalayas. He observed that an error in vertical (astronomical) angles had been introduced in Everest's geodetic measurements through a lower-than-expected deflection angle of plumb lines used to position his instruments. The Astronomer Royal of the time, G. B. Airy (1855), thereupon proposed that the anomalous deflection implied a mass deficiency at depth, and that mountains are buoyed up by light roots "floating" in a denser substratum. Pratt disagreed, and in 1858 (Pratt 1860) he proposed his own model, also based on mass deficiency. Both models are illustrated in figure 1.3.

**Figure 1.4**
Hypothetical sections through Fiji, Tonga, and Rarotonga.
Distribution of high-Q and low-Q zones. Boundaries are approximate.
(From Oliver and Isacks, 1967. Copyright by the American
Geophysical Union.)

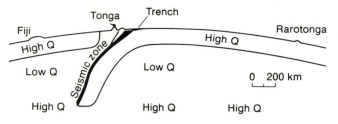

**Figure 1.5**
Lithosphere (zone of relative strength), asthenosphere (zone of
relative weakness), and mesosphere (zone of relative strength),
assuming Q correlates with strength. (From Oliver and Isacks, 1967.
Copyright by the American Geophysical Union.)

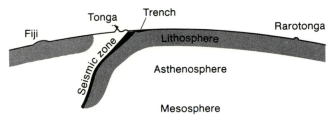

According to Airy, the extent of crust that carries continental elevations above sea level (now known as the continental crust) is of uniform density, less than that of the substratum. Hence, elevations, in accordance with Archimedes' principle, are directly proportional to depths of columns of continental crust below them (figure 1.3a).

According to Pratt, elevations at the earth's surface are compensated by columns of rock whose density is an inverse function of topographic elevation. All columns are assumed to extend downward to a uniform "level of compensation," to a depth of slightly less than 100 km (figure 1.3b). Seismic evidence favors Airy's model, but Pratt's model is used to modify it to some extent.

## Lithosphere and Asthenosphere

Based on what he believed to be the requirements of the theory of isostasy, Barrell (1914) postulated an uppermost, "strong" earth shell—the **lithosphere**—and a lower, yielding shell, the **asthenosphere.** Barrell's concept was a very useful one but, unfortunately, almost entirely hypothetical. Direct evidence for the existence of an asthenosphere was lacking, although Gutenberg

(1955) was able to show that a likely low-velocity channel in the upper mantle might be a yielding layer. Later it became possible to use direct physical measurements to determine the boundary between the lithosphere and the asthenosphere.

The physical property of rocks used in these measurements is the attenuation or damping of seismic waves. A convenient measure of damping is the ratio of energy dissipated to the total energy carried. The reciprocal of this ratio is known as Q. Anderson et al. (1965) have suggested that Q may be a measure of relative strength. Based on that assumption, Oliver and Isacks (1967) mapped high- and low-Q zones in the crust and mantle (figure 1.4), and these zones seem to agree with the lithosphere and asthenosphere as proposed by Barrell (1914) and Daly (1940). This mapping showed that the lithosphere seems to descend into the asthenosphere along zones of high seismic activity (figures 1.4 and 1.5). These zones were previously noted by Wadati, Zavaritsky, and Benioff and are called **Benioff zones** in honor of the latter. The lithosphere normally comprises both the crust and part of the upper mantle.

**Figure 1.6**
Diagrams illustrating a convection-current mechanism for
"engineering" continental drift and the development of new ocean
basins (A represents ascending currents; B and C represent
descending currents), as proposed by A. Holmes in 1928, when the
oceanic crust was thought to be a thick continuation of a continental
basaltic layer (line shading). (After Holmes, 1931; from Meyerhoff,
1968. Copyright by the American Geophysical Union.)

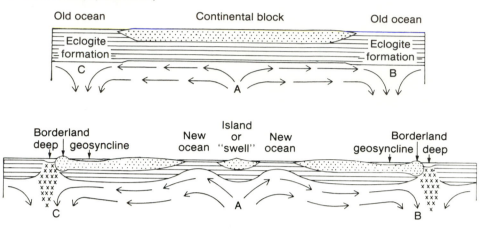

## Evolution of a Model

Until the beginning of the twentieth century, most geologists believed that the solid outer shell of the earth was stable and that rock deformation occurred because of comparatively small displacements within it. Taylor (1910) was the first to seriously argue that the continents might be in relative motion with one another. But it was Wegener (1915), Argand (1924), and du Toit (1937) who pioneered the theory of continental drift, largely on the basis of geological evidence.

Physicists were unable to accept Wegener's idea of strong continental rafts "drifting" through weak oceanic crust. Holmes (1931, 1945) pointed out that the continents could suffer mutual displacement without "drifting" through oceanic crust, for they might ride passively on a convecting lower crust and mantle. He showed how the mid-Atlantic ridge might be the origin of continental displacement in the Atlantic region, marking emergence of an upward branch of a convection system (figure 1.6).

In the early 1950s, K. Runcorn and his associates investigated rock magnetism, particularly the direction of magnetization locked in ancient rocks containing ferromagnetic minerals. Measurements of magnetization directions in ancient rocks hinged on two basic assumptions: (1) certain rocks have sufficient magnetic stability to retain the direction of permanent magnetization they acquired at their formation; and (2) the geomagnetic field, when averaged over very small time spans compared with a geological period, is symmetrical about the earth's axis of rotation and is dipolar, as at present.

As results came in and were evaluated, it became clear that, over geologic time, there had been shifts in the position of the poles relative to sites of measurement. This meant that the poles had moved or *wandered* over geologic time, that the continents had moved with respect to the poles, or perhaps both. By 1956 sufficient data had accumulated to make it clear that poles computed for North America and for Europe had each described a different path (figure 1.7). Since there was no reason to doubt reasonably persistent symmetry and dipolarity of the geomagnetic field, this meant that there had, in fact, been relative displacement between the two continents since Triassic time (Runcorn 1956). The paths plotted in figure 1.7 are known as **apparent polar wander paths.** Polar wander is apparent, but continental displacement is real, since the discrepancy in pole positions for the two continents is systematic and not random. As it became inescapably clear that relative displacement between continents had occurred, physicists who were once opponents of continental drift became serious supporters of it; a fairly rapid shift from "fixism" (a belief in fixed continents) to "mobilism" (a belief in mobile continents) began to take place among the geological establishment. But more evidence was needed before a majority was convinced.

**Figure 1.7**
Apparent polar wander paths for North America and Europe. Bold numbers show radiometric ages of rocks used, in hundreds of millions of years. If the Atlantic Ocean had not opened about 200 million years ago, the two paths would nearly have coincided. (From Peterson et al., 1980.)

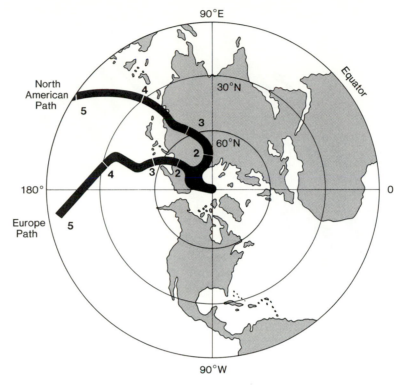

Intensive exploration of the seafloor had by now yielded important new information. This led H. Hess (1962) to conclude, like Holmes before him, that the continents were riding passively on convecting mantle. In Hess's more explicit model, seafloor was created at thermally and seismically active oceanic rifts and was destroyed at seismically active oceanic trenches. R. Dietz (1961) called this phenomenon **seafloor spreading.**

Egyed (1957), Carey (1958), and Heezen and Tharp (1965) sought to explain continental drift by postulating a gradually expanding earth. Expansion could take place by light fractions of mantle material continually rising at oceanic ridges from below. But this hypothesis could not explain conditions at deep-sea trenches, and it is also incompatible with observed rates of plate motion (see p. 13), which are an order of magnitude too high to accommodate earth expansion, and are also too irregular in time and place. (Earth expansion at a rate substantially slower than current plate motions cannot be disproved, but it cannot account for continental drift.)

At about the same time, marine magnetic surveys (Mason 1958) revealed an interesting magnetic anomaly pattern on the flanks of some known active oceanic ridges (e.g., figure 1.8). In each basin, strips of anomalies of alternating high and low magnetic intensities are oriented parallel to the rift at the crest of active ridges. The pattern of the anomalies is astonishingly symmetrical about the rift axis. Vine and Matthews (1963) explained this symmetry by hypothesizing that new material, constantly added at ridge crests, spreads outward, away from the ridges. With each magnetic field reversal, lavas emplaced at the rift acquire corresponding reversed polarity of magnetization, and their continuing outward migration results in strips of alternating polarity that register as high- and low-intensity anomalies. This occurs because "normal" polarity magnetization reinforces the present-day field, whereas "reverse" polarity magnetization opposes it. The assumption is that lavas that erupt at the rift retain the magnetic polarity of the time of eruption as they are carried outward by the spreading ocean floor.

**Figure 1.8**
Magnetic anomaly pattern over Reykjanes Ridge, southwest of
Iceland. Straight lines at ends of pattern indicate the axis of the
ridge and the central positive anomaly. (From Vine, copyright 1966
by the American Association for the Advancement of Science.)

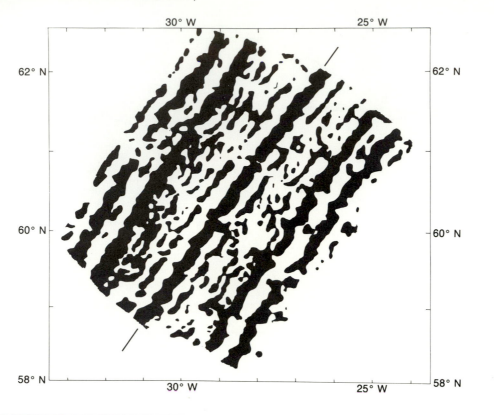

**Figure 1.9**
Paleomagnetic stratigraphy showing epochs of normal and reversed
geomagnetic field to 4 million years B.P., based on pole reversal and
potassium-argon (K-Ar) dating of lava flows. (From Vine, 1966; based
on Cox et al., 1964. Copyright 1966 by the American Association for
the Advancement of Science.)

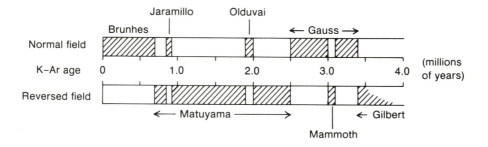

In confirmation, Vine (1966) matched the seafloor magnetic anomaly pattern (e.g., figure 1.8) to the magnetic reversal chronology then known to occur on land (figure 1.9), assuming a best-fit velocity of spreading (figure 1.10). Vine showed that it is possible to make such a match on all active ridges for which the magnetic anomaly pattern is known, assuming a characteristic spreading velocity in each case. This mechanism appears to have created the oceanic crust, accounting for the displacement of continents in a physically documented manner.

J. T. Wilson, an early opponent of continental drift, soon recognized the evidence in favor of it and became a pioneer of plate tectonics. He showed that certain chains of volcanic islands in the Pacific Ocean increased in age westward, as if the ocean floor had moved

**Figure 1.10**
Inferred normal-reverse magnetic boundaries within the crust plotted
against the reversal time scale of figure 1.9, for two different active
ridges. Note the similar deviations from linearity for the East Pacific
Rise and the Juan de Fuca Ridge. Both are in the eastern Pacific
and are believed to be branches of one spreading system that
became obliterated by the North American continent. (From Vine,
copyright 1966 by the American Association for the Advancement of
Science.)

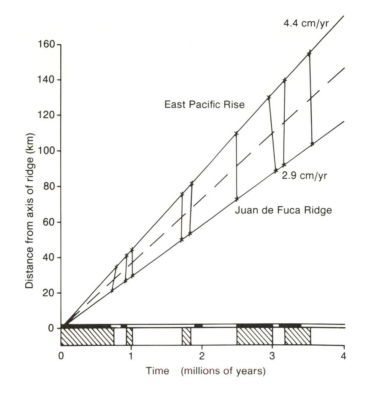

relatively east over a fixed magma supply. More important (though few recognized it at the time) was his
demonstration of transform faults (J. T. Wilson 1965)
linking rifts that create lithosphere and trenches that
consume it, and also transform faults that link rifts with
rifts and trenches with trenches (figures 1.1 and 1.11).
L. R. Sykes (1967) measured earthquake first motions
and demonstrated that motion along plate boundaries
actually occurs in the directions predicted by the seafloor spreading model. He found that along ridges
seismic first motions indicate extension; whereas along
transform faults, they indicate strike-slip. After the
Alaskan earthquake in 1964, Plafker (1965) found that
surface displacement associated with the earthquake
could be explained only in terms of the continent overriding the ocean.

**Figure 1.11**
Displacement along a ridge–ridge transform fault. (From Dennis,
1967.)

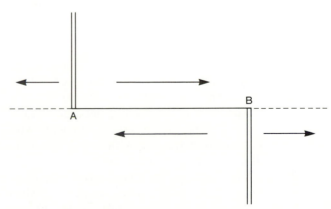

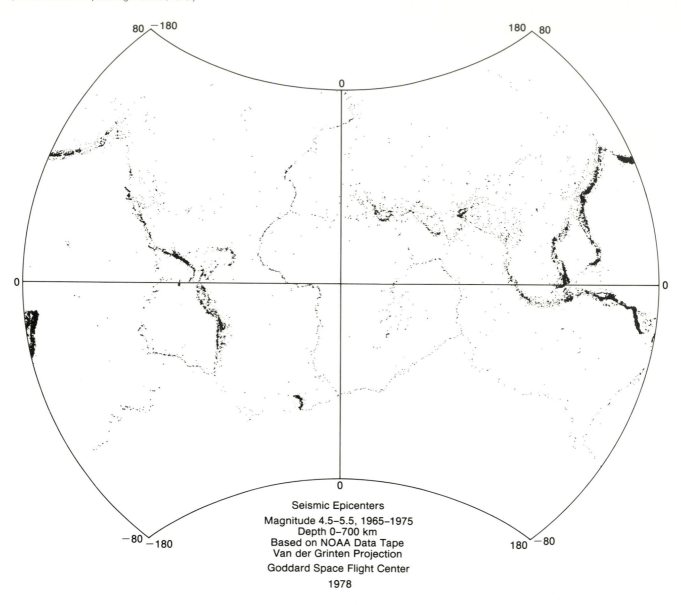

Seismic Epicenters
Magnitude 4.5–5.5, 1965–1975
Depth 0–700 km
Based on NOAA Data Tape
Van der Grinten Projection

Goddard Space Flight Center

1978

At about this same time, observational seismology was advancing rapidly. The distribution of epicenters over the whole earth was revealing a pattern of narrow seismic zones surrounding relatively quiet areas (figure 1.12). The seismic zones, it turned out, coincide with oceanic ridges and trenches and with active mountain belts, as well as with certain great faults.

The stage was now set for a grand synthesis of all the evidence gathered thus far. This was accomplished by Morgan (1968), Le Pichon (1968), Isacks, Oliver, and Sykes (1968), and Heirtzler et al. (1968). The new model that emerged has become known as plate tectonics.

## Kinematics of Plate Motion

All lithosphere plates are in continual relative motion. It is possible to describe these motions and to set up something like laws of motion that reflect the natural constraints on plate movements.

We assume that the total area of all plates remains constant over a significant time span and that horizontal distortion within plates is insignificant away from plate margins. We may thus consider plates to be rigid. Since displacement along transform faults is strike-slip, it follows that plate motion must be parallel to contemporaneous transform faults. Because all this takes place on the surface of a sphere, each plate segment describes

**Figure 1.13**
On a sphere, the motion of plate 2 relative to plate 1 must be a
rotation about some pole. All faults on the boundary between plates
1 and 2 must be small circles concentric about pole *A*. (From
Morgan, 1968. Copyright by the American Geophysical Union.)

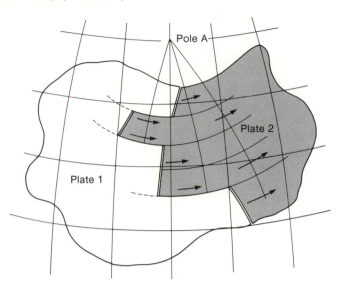

a path that is equivalent to a rotation on the spherical
surface of the earth. Plate rotation occurs about an
imaginary pivot or **pole** (figure 1.13); each pair of plates
has its own pole of rotation whose location can be constructed from known transform motion between the two
plates. The sum of the angular velocities for all pairs
of plates is zero at any instant, assuming the earth's
surface area remains constant.

Plate motions measured over a finite time interval,
such as a geological epoch or period, are averages for
that time interval. Similarly, locations of the pivots or
poles of rotation for any pair of plates over a stated time
interval are average positions. Thus, we must distinguish between **instantaneous poles of rotation,** valid for
an instant of time, and **poles of finite rotation,** for rotations averaged over a stated geologic time interval.

It is not difficult to see that *linear* relative velocities
between plates increase from zero at the poles to a
maximum at an "equator" 90° from the poles. This
maximum may exceed 15 centimeters (cm) per year.
However, few plates, if any, have boundaries extending
to a pole of rotation, so that in practice linear relative
velocities are at least 1 cm per year.

## Rates of Plate Motion

It is comparatively simple to determine average rates
of spreading at rifts. Seafloor magnetic anomalies can
be identified and dated (figure 1.10), and the distance
of the anomaly from its spreading axis is then the only
other measurement needed to obtain the average linear
spreading rate at any rotation latitude. For instance, in
figure 1.10, the distance of anomalies from their
spreading axis is plotted against their age for two different spreading axes. From this plot we can determine
that the East Pacific Rise spreads at 4 cm per year, while
the Juan de Fuca Ridge spreads at 2.9 cm per year. We
can now obtain the angular spreading rate, and hence
the linear spreading rate at any other point along the
boundary concerned.

*Spreading rates* at ridges are usually given as half-
rates, that is, the rate at which one lithosphere plate
moves away from the ridge. The full spreading rate is
the velocity differential between the two plates that
originate at the ridge. Plate velocities at trenches are
full rates, because one lithospheric plate does not move
with respect to the trench. All rates are measured relative to a plate boundary. In the pattern of motion of
plates and plate boundaries, nothing is fixed; all velocities are relative.

*Rates of convergence* between plates at trenches
and orogenic belts can be computed by vector addition
of known plate rotations.

*Rates of slip along transform faults* are easily obtained as soon as average rates of plate rotation are
known (see Le Pichon et al. 1976).

**Figure 1.14**
Major plates of the present. Divergent plate boundaries are shown as double lines; convergent plate boundaries are marked by barbed lines, with barbs on the overriding plates; transform boundaries are shown as single lines. Dashed lines indicate weak or uncertain boundaries. Arrows show direction of plate motion with respect to Africa. Only major plates are named. (From Petersen et al., 1980.)

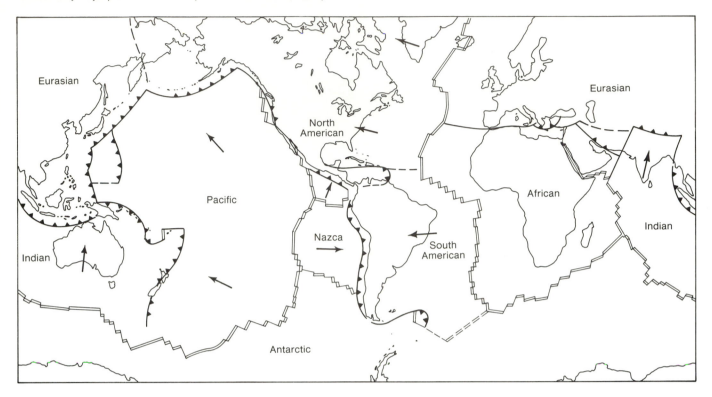

## Plate Boundaries

Most tectonic activity appears to take place through the influence of processes along the boundaries between plates. Let us review some characteristics of plate boundaries.

Plate boundaries are of three kinds: divergent, convergent, and transform. Along **divergent** plate boundaries, mantle-derived material rises and spreads outward from a rift boundary, forming new seafloor. Along **convergent** plate boundaries, one of the two converging lithosphere plates descends into the asthenosphere and is consumed by it. Along **transform** boundaries, the two adjoining lithosphere plates slip by one another, ideally without creation or destruction of lithosphere (figure 1.1).

Plate boundaries are manifested at the earth's surface by great faults or fault zones, accompanied by zones of seismic activity. We shall deal with such faults in chapters 12–15. For now, a **fault** is defined as a fracture or zone in rocks along which displacement has taken place. If the displacement is predominantly along the fault dip, the fault is a *dip-slip fault*. If displacement is predominantly along its strike, it is a *strike-slip fault*. A **rift** is a fault zone in which a central strip has subsided along faults, forming a topographic valley

(e.g., figure 14.17). The locations of plate boundaries are not fixed, either in space or in time. They are in constant relative motion, and their lengths change, just as the plates themselves are in constant relative motion and change in area over time. Present boundaries are shown in figure 1.14. In the ideal plate tectonic model, plates do not suffer horizontal distortion, except in narrow zones along some boundaries.

Some present boundaries between major plates are lined with zones of much smaller plates or "microplates," each an individually evolving segment of the lithosphere. Each is in relative motion with respect to all the others, including the major plates, and this makes detailed analysis of individual plate motions very complicated. Examples occur in the Mediterranean region between Africa and Europe, and in the Scotia Sea. These belts of microplates constitute a region of adjustment between major plates, usually where continental lithosphere interferes with an "orderly" evolution of plate boundaries. Plate boundaries that lie within continents tend to be more diffuse than those in oceanic regions; this becomes evident by a spreading out of plate boundary seismic zones (figure 1.12). Examples include the western margin of North America and the region north of the Himalayas.

**Figure 1.15**
Subduction as expressed in forearc structure, and relevant terminology. (From Dickinson and Seely, 1979.)

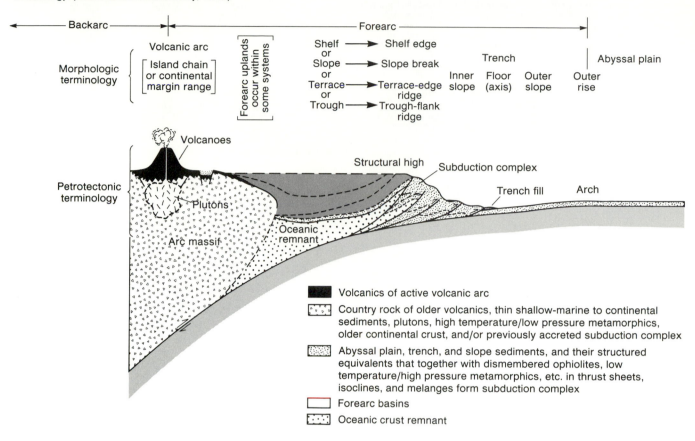

## Former Plate Boundaries

Present-day plate boundaries have been traced by seismic activity and by ocean floor topography. It is evidently difficult, if not impossible, to use the same methods for plate boundaries of the geologic past; oceanic lithosphere is destroyed after a comparatively short life cycle. The oldest known oceanic lithosphere is of Jurassic age; hence pre-Jurassic plate boundaries can be found only on the continents, greatly displaced with respect to one another since they were active. One of the most useful tools for tracing the migration of different segments of lithosphere over geological time is paleomagnetics, the use of apparent polar wander paths (see pp. 8, 9). This is not a perfect method, but it does show which segments of continental lithosphere have had independent histories and what changes in latitude and orientation they have undergone. Another clue is provided by comparing stratigraphic sequences. Some migrated continental and quasi-continental fragments of the past, now "docked" along continental margins, have been identified by these means. The west coast of North America is lined with such accreted **terranes** (see

review in Nur 1983), which migrated varying distances northward and eastward until colliding with and being incorporated into the western margin of the North American plate.

## Convergent Plate Boundaries

Along convergent plate boundaries, the plates on either side have appreciable components of motion toward one another. Such convergence could be accommodated by two possible mechanisms: (1) one or both plates might thicken and deform internally to take up the converging motion; or (2) one plate could escape downward into the asthenosphere, letting the other plate ride over it. It is the second of these mechanisms, known as **subduction,** that operates along all active convergent plate boundaries. However, where continents are involved, the first mechanism may contribute to some extent.

Subduction seems a fairly simple process—at least kinematically. The geological consequences of subduction, however, are far-reaching and complex (figure 1.15), and not all of them are as yet fully understood.

**Figure 1.16**
Seismic reflection profile across the Japan Trench extending easterly from point M, near Japan, to point N, offshore. Vertical scale represents two-way reflection time in seconds (e.g., 1 sec = 1 km of penetration for a velocity of 2 km/sec). Note block faulting along seaward slope of trench, demonstrating extension in crust and inclusion of sediments in basement rocks. Also note shoaling of oceanic basement on approaching trench. (From Isacks et al., 1968. Copyright by the American Geophysical Union.)

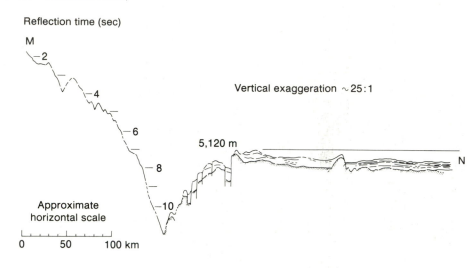

In normal plate convergence, the surface trace of the plate boundary is deeply depressed topographically, forming the well-known deep-sea trenches, some as deep as 10 km (figures 1.16 and 1.17). Some trenches are filled with young sediments that tend to obscure the structural depression. Where the downgoing slab enters the asthenosphere, magma—usually of andesitic composition—accumulates over it and forms a reservoir that feeds intrusion and extrusion immediately above. The mechanism of magma generation in this site is still controversial, but the results are conspicuous. They include volcanic island arcs in oceanic crust, and volcanoes, large batholiths, and mountain ranges in continental crust, generally a few hundred kilometers landward of the associated trenches.

Convergent plate boundaries, like all active plate boundaries, are seismically active. Earthquake foci cluster in a shallow zone near the trench, but also along the downgoing slab (figure 1.5). The zone of earthquake foci in the slab, the Wadati-Benioff zone, extends as deep as 700 km in some places but has never been found to be deeper. The slab has been outlined by plotting high Q-values (figure 1.4).

Convergent plate boundaries seem to be initiated close to continental margins in previously unbroken oceanic lithosphere. This is suggested by the distribution of most present-day active island arcs close to continental margins. Some island arcs, which now are separated from the nearest continent by small ocean basins, seem to have been formed closer to the continental margin than they are presently. The floor of the basin between them and the continent—known as a marginal or **backarc basin**—is much younger than the open ocean floor on the far side of the trench, indicating comparatively recent opening of the backarc basin that now separates the arc from the continent. Thus, backarc basins form by a kind of ocean floor spreading (plate growth) that entails oceanward migration of an arc-trench system.

## Collision

If a subducting plate carries a continent or even a small portion of continental crust, such continental crust cannot normally be subducted when it reaches the trench, because its low density does not permit it to sink into the asthenosphere. Hence, further convergence is impeded. If the overriding plate margin is oceanic, the sense of subduction may now be reversed (it may "flip"), and the previously overriding plate begins to be subducted. The abandoned slab becomes absorbed into the asthenosphere. If, however, the overriding margin is also continental (or carries only vestiges of oceanic crust), arrival of continental lithosphere, however small, at the subducting margin causes **collision** when the two segments of continental lithosphere meet, because low-density continental lithosphere cannot escape downward. Where only a comparatively small continental

**Figure 1.17**
Trenches, Benioff zones, and volcanicity for the western Pacific. Trends of trenches are shown by heavy black lines; depth to Benioff zone is shown by 50-km contours; volcanoes are represented by black dots. (From Oxburgh and Turcotte, 1970.)

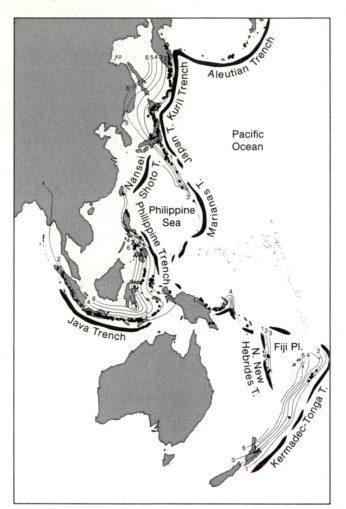

fragment (such as a microcontinent or an arc) collides, subduction may be relayed oceanward, behind the newly arrived piece of continental lithosphere; the continental fragment has "docked," and subduction continues behind it. No overall change in plate motions results. However, if both colliding continents are sizable, subduction will eventually cease. So the convergence vector across the boundary becomes zero, and the motion must be taken up elsewhere to preserve the zero sum of all plate-motion vectors on the earth's surface. Rearrangement of the global spreading and subduction pattern results.

Nevertheless, plates can continue to converge for a limited time after continents first collide. The Indian and Eurasian plates have converged for several hundred kilometers since the two continents collided. There are two possible explanations for this: The subducting plate may carry the continent along the lithosphere-asthenosphere boundary, "underplating" the overriding plate. Or the continental margin zone of one of the plates (normally the overriding one) is sufficiently yielding to take up convergence by distortion, both horizontally (along strike-slip faults) and vertically (by thickening).

When two continents collide, the intervening ocean vanishes, leaving a **suture** between the collided continents. Clues for the identification of sutures include (*a*) strings of mafic and ultramafic rocks in a characteristic association called **ophiolites** (which have been interpreted as vestiges of former oceanic lithosphere that became beached [**obducted**] on continental crust); (*b*) obvious facies contrasts (which suggest widely separated original sites of deposition); and (*c*) contrasting paleomagnetic histories.

Convergent plate boundaries are preferred sites of some of the most important geological processes, especially those linked with mountain building (**orogeny**): intense rock deformation; magmatism, metamorphism, and ore deposition; and relatively rapid uplift leading to intense erosion and consequent copious sedimentation, not infrequently linked with hydrocarbon accumulation.

## Divergent Plate Boundaries

Oceanic lithosphere is constantly being created along a global network of rifts totaling some 70,000 km in length (figure 1.14). Lithosphere plates diverge on either side of the rifts and therefore constitute *divergent plate boundaries*. Most of these rifts are on the crests of great submarine ridges. Some—notably the mid-Atlantic ridge—are midway between continents. But others, such as the East Pacific Rise, come close to continents and may even enter them. Therefore the commonly used name "mid-ocean ridge" is misleading.

Ridge elevation above adjacent ocean bottoms (abyssal plains) is normally between 1,000 and 3,000 meters (m); a representative width is around 1,000 km. The associated rift is in a valley whose central axis may be as much as 2,000 m below the crest of the ridge. The rifts are over 10 km wide and as wide as 30–32 km in the central Atlantic. The fast-spreading East Pacific Rise carries no morphological rift. In a few places, divergent plate boundaries are more diffuse; they may rise above sea level, as in Iceland, the Azores, and in the Afar Triangle (figure 1.18).

No divergent plate boundary has a smooth, continuous trace: all are offset by transform faults (figure 1.14). Occasionally, a portion of a spreading axis will "jump" to a new location, abandoning its old trace. Once initiated, spreading along a divergent plate boundary may change rate or direction of spreading, usually in response to global changes in plate motions. As far as is known, spreading along any rift can only be stopped by subduction of that rift.

As lava cools in the rift, it acquires magnetization. Consequently, in the course of spreading, magnetized seafloor moves laterally and symmetrically away from the spreading axis. Each newly formed element acquires the direction of magnetization of the prevailing magnetic field. Periodic geomagnetic field reversals result in the magnetic anomaly strips discussed on pages 9 and 10.

Clearly, magma production at rifts causes abnormally high heat flow. Fracturing and the circulation of seawater through the fractured hot seafloor result in hydrothermal processes and in chemical reactions that concentrate some metallic elements, such as copper, nickel, and cobalt. Fracturing also generates seismic activity. First motions of rift earthquakes indicate dip-slip displacement. Originally, divergent plate boundaries were located by topographic mapping of the seafloor, using depth-sounding techniques (Heezen, Tharp, and Ewing 1959).

### *Rifts on Land*

New rifts may start in oceanic lithosphere, dividing existing oceanic plates, as in the case of the Cocos ridge off Central America; or, more commonly, they may start in continental lithosphere, dividing continents and creating new oceans. The Red Sea rift (figure 1.18) has just begun to create a new ocean, and the East Pacific Rise is creating a new arm of the Pacific Ocean in the Gulf of California. Some continental rifts do not evolve into true oceans. (We shall discuss these under Triple Junctions, pp. 20, 21.)

## Transform Plate Boundaries

Along transform plate boundaries two plates slip by each other, ideally without either divergence or convergence. Since plate motions are around pivots (poles), as illustrated in figure 1.13, transform boundaries should follow small circles around the pole for the pair of plates involved. This is indeed the case for transform boundaries in oceanic lithosphere, where such boundaries are relatively sharp, narrow fault zones, the transform faults. Note (figure 1.11) that the offset between rift segments is not due to faulting: it was there from the beginning of rifting, and fault motion has not changed it. Also, movement on the fault is in a sense opposite to that of the ridge offset.

All oceanic rifts are segmented by relatively short transform faults. Since plates cannot distort horizontally, there is no other way for divergent plate boundaries to curve while maintaining a constant spreading direction. The whole system of rifts and transform faults

**Figure 1.18**
Rift zones in the Middle East. (From Gass and Gibson, 1969.
Reprinted by permission from *Nature,* copyright © 1969 Macmillan
Journals Limited.)

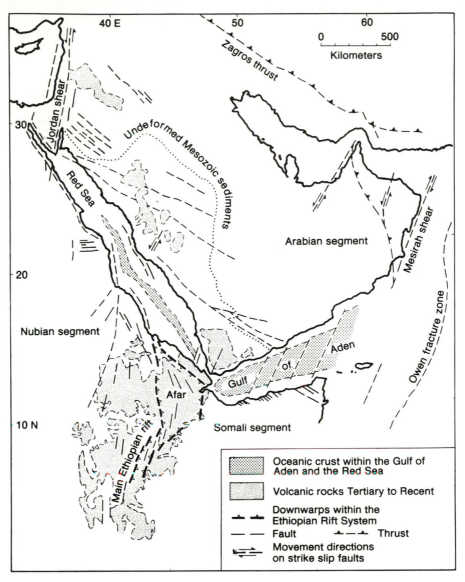

form one *spreading system:* in some segments, divergence predominates, as in most of the Atlantic Ocean. In others, transform motion predominates, as in the Gulf of California.

In continental lithosphere, transform boundaries are not so clean and uncomplicated as in oceanic crust. They do tend to follow, more or less, small circles around the poles of two adjoining plates. But displacement does not seem to propagate to the top of the crust in an orderly manner. Above transform boundaries in continental crust, transform motion is taken up by strike-slip faults that are not entirely in the theoretical location of small-circle transform boundaries. Changes in fault strike result in converging or diverging components of motion, and this in turn results in structural complications in the rocks involved as described on pages 272–276.

## Triple Junctions and the Genesis of the San Andreas Fault

There are, evidently, a number of points where three plates meet. These are of considerable interest because they are not fixed in place or time, and because their evolution determines the structure of adjacent portions of the crust. Let us look first at a simple example involving only two plates. In figure 1.19a, plate *X* is consumed at the trench between *b* and *c,* and plate *Y* is consumed between *a* and *b*. This is not a stable situation, and it will result in strike-slip faulting at *b,* giving rise to a new transform fault, *bb'* (figure 1.19b). At triple junctions, the situation is somewhat more complex. McKenzie and Morgan (1969) list a total of 16 possible triple junctions, using all possible combinations of trenches, ridges, and transform faults. They make an important distinction between stable and unstable triple junctions, the last being destined to evolve to a more stable configuration.

One important example of the evolution of an unstable triple junction is the genesis of the San Andreas fault (figure 1.20). There is now little doubt that, until Oligocene time, the Farallon plate was moving toward North America and was being consumed at a trench along its west coast faster than the spreading rate at the East Pacific Rise to the west. Consequently, the East Pacific Rise eventually collided with the trench (figure 1.20b). The first collision was at the junction between the Mendocino transform and the eastern segment of the rift, resulting in a momentary quadruple junction between the Pacific, the American, and the two parts of the now-divided Farallon plate. This junction, as a result of continuing plate movement, immediately split into two triple junctions, giving rise to a right-slip transform between the two. As triple junctions 1 and 2 moved apart, they generated the San Andreas fault. All but one of the facts thus far collected appear to fit this reconstruction: if the terminal triple junctions first formed in Oligocene time, some 32 million years ago, as postulated by McKenzie and Morgan, displacement along the transform fault joining them should be about 2,000 km. The greatest postulated strike-slip along the San Andreas fault since Oligocene time is only 350 km (Hill and Dibblee 1953). Atwater (1970) suggests that transform motion at continental plate margins may not be confined to a narrow zone as in the oceans, but may be taken up by a fairly wide marginal zone. Thus, slip along the San Andreas fault would indicate only part of the total motion between the two adjoining plates, and a "soft" continental border zone would take up the rest.

**Figure 1.19**
Evolution of a trench. Arrows show the relative motion vector and are on the plates being consumed. Thus, plate *Y* in (a) and (b) is consumed between *a* and *b*, and plate *X* is consumed between *b* and *c*. A trench evolves to form two trenches joined by a transform fault. (c), a sketch of New Zealand, suggests that the Alpine fault might be, or might be evolving into, a trench–trench transform fault of the type in (b). (After McKenzie and Morgan, 1969. Reprinted by permission from *Nature,* copyright © 1969 Macmillan Journals Limited.)

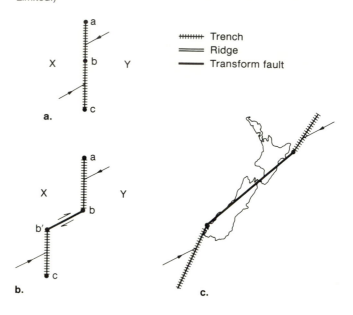

Triple junctions are commonly labeled according to the types of plate boundaries that meet at them. McKenzie and Morgan (1969) used the symbols R (ridge or rift) for divergent boundaries, T (trench) for convergent boundaries, and F (fault) for transform boundaries. Thus, the Mendocino triple junction (figure 1.20) is an FFT triple junction.

Rift-rift-rift (RRR) triple junctions are of particular interest, for they represent triple divergence. Evidence suggests that many RRR triple junctions form over upwelling mantle currents or **plumes,** which heat the lithosphere and thus cause **hot spots** above them. Because of continuous relative motion of plate boundaries, an RRR triple junction, once formed, will move away from its generating plume, for plumes are assumed to be fixed over long time spans. Rift systems may be regarded as strings of RRR incipient triple junctions joined by rifts (see figure 1.21). When continents break up by rifting, the RRR triple junctions play an important role. Two of the arms of the triple junctions evolve into oceans, while the third or "failed" arm remains an intracontinental rift, called an **aulacogen.** It is typically located at a reentrant of the rifted continental margin, as becomes clear from figure 1.21.

**Figure 1.20**
Evolution of triple junctions off western North America. (a) Simplified
geometry of the northeast Pacific before collision of the Pacific plate
with the North American plate. All fracture zones except the
Mendocino and the Murray have been omitted for simplicity.
(b) Stable triple junctions after the eastern Pacific plate met the
trench off western North America. Double-headed arrows show the
motion of junctions (1) and (2) relative to American plate A. (c) is a
vector velocity diagram for junction (1) and shows that the junction
will move northwestward with the main Pacific plate. (d) is a similar
diagram for (2). Dashed lines *ab, ac, ad, bc, cd,* in the vector
velocity diagram join points whose vector velocities leave the
geometry of *AB, AC, AD, BC, CD,* respectively, unchanged. Vector
*AB* shows the direction and rate of consumption of the Farallon plate
by the American plate. (See legend of figure 1.19.) (From McKenzie
and Morgan, 1969.)

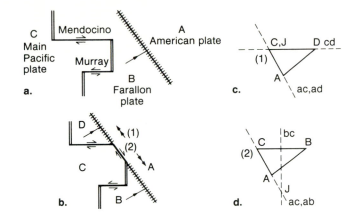

**Figure 1.21**
Sketch maps representing early stages of a Wilson cycle, showing
role of hot spots, and evolution of aulacogens from failed arms of
triple junctions. (From Dewey and Burke, 1974.)

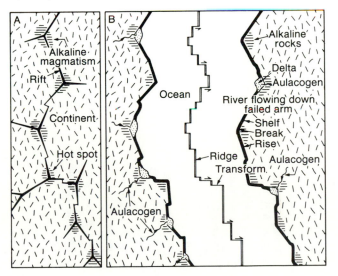

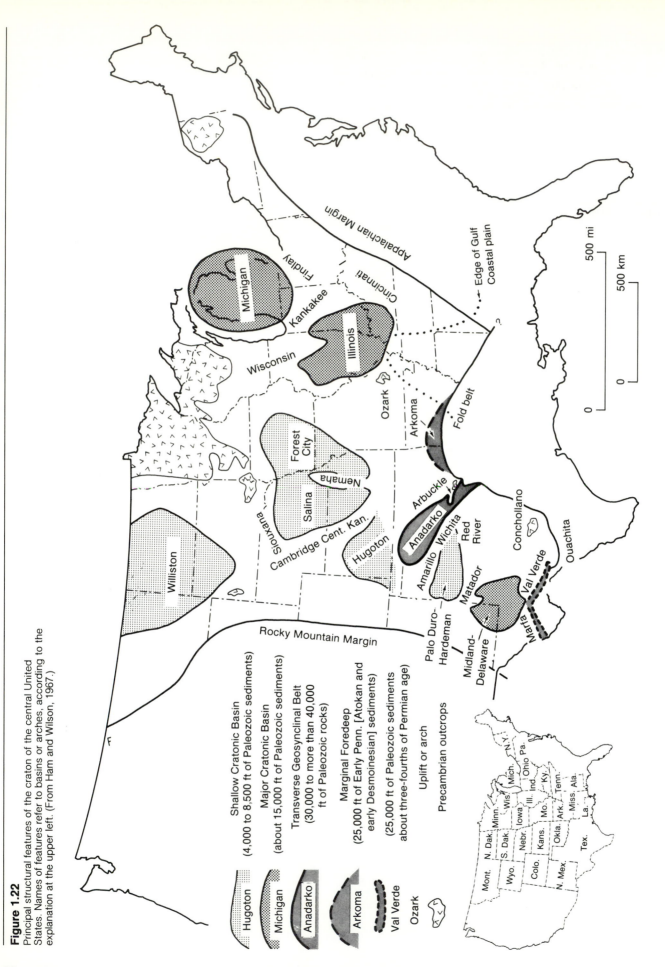

**Figure 1.22**
Principal structural features of the craton of the central United States. Names of features refer to basins or arches, according to the explanation at the upper left. (From Ham and Wilson, 1967.)

Shallow Cratonic Basin
(4,000 to 8,500 ft of Paleozoic sediments)

Major Cratonic Basin
(about 15,000 ft of Paleozoic sediments)

Transverse Geosynclinal Belt
(30,000 to more than 40,000 ft of Paleozoic rocks)

Marginal Foredeep
(25,000 ft of Early Penn. [Atokan and early Desmoinesian] sediments)

(25,000 ft of Paleozoic sediments about three-fourths of Permian age)

Uplift or arch

Precambrian outcrops

Hugoton

Michigan

Anadarko

Arkoma

Val Verde

Ozark

## The Role of Continents

### The Wilson Cycle

Oceans are normally destined to go through a cycle of rifting, seafloor spreading, net convergence of bordering continents and, finally, continent–continent collision. This cycle is known as the **Wilson cycle,** after J. T. Wilson, who first proposed it. Intracontinental rifting marks the geological manifestation of the beginning of this cycle, continental margin sedimentation accompanies its mature evolution, and mountain belts formed by collision *(collision orogens)* mark its end.

### Tectonic Elements of Continents

The main regional structures on continents are cratons, rifts, and orogenic belts. They are sometimes referred to as tectonic or geotectonic elements. Cratons are relatively stable areas, rifts are subsidence zones along faults, and orogenic belts are zones of intense deformation associated with mountain building. We shall briefly discuss each in turn.

#### *Cratons*

A craton is a relatively stable part of the continental crust that undergoes broad, reversible, vertical deformation only. Such deformation, called **epeirogenic,** forms regional basins, domes, and arches, hundreds of kilometers across and thousands of meters deep (figure 1.22). Deformation is recorded mostly by basin stratigraphy. Cratons are commonly subdivided into **platforms** and **shields** (figure 1.23). Platforms consist of a **basement** and its **cover.** The basement is a complex of metamorphic and igneous rocks underlying a major unconformity on which rests the platform cover of relatively little-deformed sedimentary rocks. Shields are areas in which the basement is wholly exposed and in which there are no cover rocks.

#### *Rifts*

Rifts in continental crust form along narrow zones of extension. They may be defined as zones of subsidence along faults, over an extending portion of the lithosphere. Rifts will be further discussed in chapter 14.

#### *Orogenic Belts*

Orogenic belts (also known as orogens, orogenes, mountain belts, or fold belts) are elongated zones of intensely deformed continental crust generally associated with igneous activity and metamorphism. The processes that combine to form orogenic belts are collectively known as **orogeny.** Orogenic belts originate along converging plate boundaries. They were first integrated in plate tectonics by Dewey and Bird (1970). There are two main types of orogenic belts: *marginal* belts, formed along the margins of an overriding plate above a subducting slab of lithosphere; and *collision* or

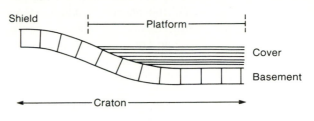

**Figure 1.23**
Generalized model of structural elements of a craton. (From Dennis et al., 1979.)

*alpine-style* belts, formed along the suture zone of two continents or fragments of continental crust that have collided. A subsidiary type—a *transform* orogenic belt—results from a convergent component of motion along a continental transform fault.

*Marginal orogenic belts* are of two types: arc and Andean style. Arc orogens, along active island arcs, are separated from nearby continents by a backarc basin. The islands of Japan are a good example of an arc orogen. Andean-style orogens are formed along a continental margin over a subducting slab. The central Andes of South America are a good example of this type of orogen. Characteristic features of arc orogens are chains of volcanoes with predominantly andesitic to andesitic basalt volcanism, some intrusive activity, some metamorphism (which may be high temperature-low pressure or high pressure-low temperature), intense rock deformation along relatively narrow zones, and predominantly normal faulting. Andean orogens share many of the same features, but they are built on continental margins, have more silicic igneous activity, and more of it is intrusive than along arcs. Many (but not all) have thrust belts along their continental flanks (see chapter 15), and many are lined with accreted terranes (see p. 15). Both styles of marginal orogens are rich in mineralization.

Collision orogenic belts are characterized by intense deformation. Igneous activity is subordinate, but regional metamorphism is prominent. Along the suture zone, former oceanic crust and mantle may be beached in the form of ophiolites (see p. 18).

Large volumes of sedimentary rocks are deposited before and deformed during an orogeny along zones of subsidence, which precede and accompany orogeny and which, in the past, were called **geosynclines.** These zones of thick sedimentation develop from continental shelves and backarc basins. **Thrusting** (chapter 15) is prominent in collision orogens, with one direction of thrusting usually far more prominent than the other. Ore deposits are of minor importance. Collision may be between two major continents, but the results of collision are not much different where a continent collides with an arc or with a microcontinent.

## Time Frame

All these geotectonic elements of the continental crust have limited lifespans that are dependent on the plate tectonic evolution of the lithosphere as a whole. Hence, an orogen or a craton must always be referred to a time interval, e.g., the Paleozoic craton of North America, or the Hercynian orogen of western Europe. An orogen or a rift can always become part of a later craton, and any cratonic part of the lithosphere may turn into an orogen or be rifted, depending on the evolution of the plate tectonic setting. In addition, orogenic belts may incorporate rifts, and collision orogens may incorporate marginal orogens.

## Conclusion

The crust and the upper mantle of the earth are in a state of continuing unrest, and this unrest is responsible for the tectonic structures observable today. Two basic tectonic processes affect the earth's crust: the motion of lithospheric plates (which is linked to orogeny and rifting) and epeirogeny (the slow up-and-down movements of isostatic adjustments). No evidence for plate motions has been observed on the moon or on other planets, although tectonic structures provide evidence of limited movements there. What makes the earth different?

Plate motion must be maintained by an evolving source of energy. Clearly, the spreading centers are heat sources, and the trenches are heat sinks. The prime source of energy in the earth is doubtless thermal; the earth is a heat engine. However, plate motions do not reflect convection currents in a conventional sense. The picture is more complex than that, for the budget along rifts and trenches does not balance within domains that would be conventional convection cells. Certainly, the overall picture is one of convection in the mantle. But it is superimposed on some other phenomenon that we cannot yet recognize. The question must be asked: Is this other phenomenon cause or effect?

The movement of lithospheric plates triggers a chain of tectonic and magmatic events. Tectonic energy becomes available to the extent that the upper mantle—in fact, the whole earth—is still evolving toward thermal and chemical equilibrium. Concentrations of radioactive nuclides in the crust understandably influence this evolution.

To some extent, the changes involved are cyclic, as suggested by the cyclic nature of Phanerozoic and earlier geological history. But, to some extent, the changes are evolutionary. A declining source of energy does not allow the earth to return to a previous state, even though its manifestations may be repetitive, and this imposes limits on the actualistic approach.

Another important geological difference between the earth on the one hand, and our moon and other planets on the other, is the presence on the earth's surface of liquid water. As a result of running water, large volumes of rock are continually being transferred along the earth's surface. The effects on isostasy are profound, and they significantly influence tectonic processes. Furthermore, surface weathering changes the mineralogy of rocks; for instance, a great deal of free quartz is produced. As the products of weathering are eventually incorporated in the rock cycle and carried downward, the addition of new mineral phases significantly influence mantle processes. No such changes are possible on the other planets or on the moon, and nothing indicates that they may have taken place there in the past.

Although we still know rather little about the ultimate driving forces, new observations and new interpretations are accumulating fast. As research proceeds and new methods are developed, the limits for speculation will narrow. Reliable, objective structural studies will be basic guides in the search. They provide the only evidence that can be seen and measured directly.

## Review Questions

1. Differentiate between geometric, kinematic, and dynamic analyses of rock structures.
2. Draw a diagram showing the concentric shells into which the earth has been subdivided on seismological evidence.
3. Explain the concept of isostasy.
4. How do Airy's and Platt's models of mountain roots differ?
5. How has Q been used to differentiate between the lithosphere and the asthenosphere?
6. Explain the concept of apparent polar wander paths.
7. How does the magnetic anomaly pattern document the spreading of the seafloor?
8. What assumptions are made when deriving rates of relative motion between lithospheric plates?
9. Find a published map that gives the chronological age of seafloor magnetic anomalies. Calculate the minimum linear distance of Pacific plate that has been subducted at the Lau trench since the beginning of Cretaceous time.
10. Define "pole of rotation" between plates for both instantaneous and finite rotations.
11. What geological features characterize each of the three types of plate boundaries: divergent, convergent, and transform?
12. What evidence may serve to identify present sutures between formerly separate lithospheric plates?
13. Define "collision" in the context of plate tectonics. How does this concept differ from "convergence"?
14. What is the "Wilson cycle"?
15. Name and describe the tectonic elements of the continents.

## Additional Reading

Bally, A. W. 1975. A geodynamic scenario for hydrocarbon occurrences. *World Pet. Congr. Tokyo,* Paper PD-1: 33–44.

Beloussov, V. V. 1979. Why do I not accept plate tectonics? *EOS Am. Geophys. Union, Transact.* 60: 207–211.

Condie, Kent C. 1982. *Plate Tectonics and Crustal Evolution.* Pergamon International Library: Elmsford, New York.

Hager, B. H. 1981. A simple global model of plate dynamics and mantle convection. *J. Geophys. Res.* 86: 4843–4867.

Helwig, J. 1974. Eugeosynclinal basement and a collage concept of orogenic belts. *SEPM Spec. Publ.* No. 19: 359–376.

Khain, V. Y. 1978. From plate tectonics to a more general theory of global tectonics. *Geotectonics* 12(3): 163–176.

Miyashiro, A., Aki, K., and Şengör, A. M. C. 1982. *Orogeny.* New York: Wiley.

Morel, P., and Irving, E. 1981. Paleomagnetism and the evolution of Pangea. *J. Geophys. Res.* 86: 1858–1872.

Park, J. K. 1983. Paleomagnetism for geologists: *Geoscience Can.* 10: 180–188.

Richardson, R. M., Solomon, S. C., and Sleep, N. H. 1979. Tectonic stress in the plates. *Rev. Geophys. Space Phys.* 17(5): 981–1019.

Schermer, E. R., Howell, D. G., and Jones, D. L. 1984. The origin of allochthonous terranes: Perspectives on the growth and shaping of continents. *Annu. Rev. Earth Planet. Sci.* 12: 107–131.

Şengör, A. M. C., and Burke, K. 1979. Comments on: Why do I not accept plate tectonics. *EOS Am. Geophys. Union Trans.* 60: 207–210.

Sleep, N. H., and Windley, B. F. 1982. Archean plate tectonics: Constraints and inferences. *J. Geol.* 90: 363–379.

Van der Voo, R. 1980. Paleomagnetism in orogenic belts. *Rev. Geophys. Space Phys.* 18(2): 455–481.

# 2

# Present and Recent Tectonic Movements

**Figure 2.1**

Horizontal displacements of primary triangulation points in Japan, in the period between the surveys of 1883–1909 and 1948–1958. These displacements include those resulting from slow creep as well as those resulting from earthquakes. (From Inoue, 1960.)

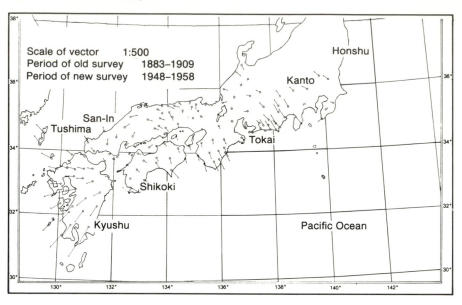

Most tectonic deformation is so slow that chances to observe it directly are few. We are, in fact, unable to follow by direct observation the complete movement history of a deforming rock body from its original emplacement (sedimentary or igneous) to the form and position in which we now see it. Our clues to its history come from very small displacements, discernible at or near the surface of the earth. These clues are of immense importance when they are interpreted in the light of the principle of uniformitarianism.

We can detect three basic kinds of small displacements: relative vertical displacements, relative horizontal displacements, and tilting. Several methods of detection are available. We can make direct measurements of very small ongoing changes by using known reference points such as benchmarks. Or we can use relatively recent historical and geological clues. The geological record allows us to assume that any movement that can be dated as Holocene (less than 11,000 years old) might recur in the same sense in the future.

## Directly Measurable Changes

Precise geodetic surveys in tectonically active areas, repeated at intervals of several decades, may reveal systematic topographic changes. Some interesting examples can be found in Japan, where Inoue (1960) compared the topographic surveys of 1883–1909 and 1948–1958. Figure 2.1 shows relative horizontal displacements on the island of Honshu detected by these surveys. Figure 2.2 shows vertical displacements, which are in remarkable harmony with present topography and with geological structure. It is clear from the diagrams that movement is not uniform over the whole island. Different structural blocks have different contemporaneous movement patterns. All measurements are with respect to assumed fixed reference points, and any possible "drift" of the island of Honshu as a whole is eliminated in the computations.

**Figure 2.2**
Vertical displacements of bench marks in Japan, in the course of approximately 60 years, by recent releveling. Displacements are systematic and in harmony with topography, except in areas near epicenters of recent earthquakes. (From Inoue, 1960.)

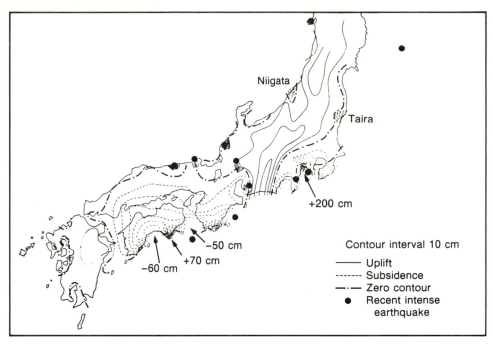

## Tilting of Crustal Blocks

Rates of crustal movements vary sharply across some faults but remain fairly constant within blocks bounded by these faults. Thus, in some places the crust, or at least its upper region, appears to form a mosaic of rather rigid blocks of the order of tens of kilometers in extent.

Current tilting of crustal blocks can be measured most easily by determining elevation changes of two or more points that are a known distance apart. In any given block, there will be one direction in which the tilt angle is greatest. This direction and angle define the tilt direction and the true tilt for that block. All other directions yield only apparent tilt angles, which are smaller than the true tilt angle. The most sophisticated method of measuring tilt is by means of a tiltmeter. This instrument can measure very small angles of tilt directly and continuously, to an accuracy of $10^{-9}$ radians per year or better. However, because most tilts are uneven on a small scale, and tiltmeter measurements are

valid for one location only, they are useful mainly for engineering geological purposes. A tilt history over an extended period of time can sometimes be reconstructed from a study of shoreline displacements; we shall consider these later.

## Fault Creep

Along some active sections of the San Andreas fault, crustal blocks are now slipping past each other at an average rate of 5 cm per year (figure 2.3), with maxima up to 10 cm per year. Movement along many segments of active faults is associated with earthquakes, but along others it proceeds slowly, almost imperceptibly, in what is known as **fault creep.**

Horizontal displacement along faults can be measured by satellite laser ranging. A pulse of light travels from a station on earth to a near-earth satellite and is bounced back to another station. The distance between

**Figure 2.3**

Annual movement along some California faults, averaged between 1959 and 1966. Lengths of arrows represent movement vectors, according to the scale given. (Base on a map by the California Department of Water Resources; courtesy of R. B. Hofmann.)

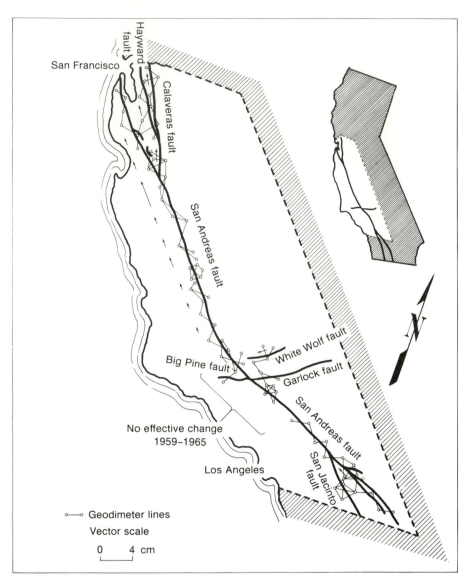

the two stations can then be computed. Such measurements reveal small changes in distance between the two earth-based stations. In order to measure horizontal displacement along the San Andreas fault (figure 2.4), stations were established about 900 km apart, 150 and 270 km, respectively, away from the fault on each side. Observations over four years showed that the stations were approaching one another at an average rate exceeding the average rate of displacement measured at the fault; it implies that part of the displacement is widely distributed on either side of the San Andreas fault zone, and that the blocks are not completely rigid near the fault. The displacement rates measured by this method are a truer reflection of relative plate velocities than are slip rates along the fault.

## Changes Associated with Earthquakes

Strong earthquakes result from sudden slips of 1 m or more on faults that may break the earth's surface to form scarps (steps in the topography), or horizontal displacements of topographic and cultural features, or both. Horizontal displacement along parts of the San Andreas fault in California associated with the 1906 earthquake was as much as 4.5 m, but vertical slip was rarely more than 30 cm. Direct surface indications of contemporary horizontal displacements usually disappear quickly unless man-made structures are involved, but vertical displacements leave scarps. In the Owens Valley earthquake of 1872, well-defined scarps formed along the active fault. At the same time, displaced

**Figure 2.4**

Laser tracking sites used to establish relative motion between the North American and the Pacific plates along the San Andreas fault. (From Smith et al., 1979.)

**Figure 2.5**

Columns of the Serapis "temple" in Pozzuoli, near Naples. Borings of intertidal organisms are visible about 5 m above present sea level. (From Lyell, 1872.)

fences revealed horizontal displacement ranging to more than 4 m. The San Fernando earthquake of February, 1971, resulted in low discontinuous scarp formation extending over several kilometers. This was most apparent in areas where man-made structures were involved.

Earthquakes are the result of sudden release of accumulated energy. In any one place they punctuate long periods of quiescence. Thus, the amount of movement averaged over a long period of time is comparatively small, only a few centimeters per year or less. Nevertheless, such movements will add up to large-scale tectonic displacement in time.

## Historical Clues to Recent Tectonic Movements

Changes recorded in human history provide some outstanding examples of crustal movements. The most striking instances involve local changes in sea level.

The British mathematician Charles Babbage (1792–1871) made some significant observations at the Serapis "temple" (actually, marketplace) in Pozzuoli, near Naples (figure 2.5). The borings of intertidal marine organisms on columns in the marketplace were then about 5 m above sea level, but the pediments of the columns were under water. Babbage made a detailed study of all geological, archeological, and physiographic features, and concluded that changes in relative elevation

of land and sea at Pozzuoli were caused principally by changes in the elevation of the land. Apparently the land had sunk, the columns had been bored, and then the whole had been re-elevated, but not quite to the original height. Babbage explained that this was the effect of volcanic heat acting on underlying rocks, which would cause contraction (subsidence) and expansion (elevation). Other geologists considered the volcanic effect purely local, confined to the general region of Mount Vesuvius, while Babbage and his friend, the astronomer Sir John Herschel, believed that the thermal expansion and contraction effects were fundamental and of global importance. Babbage further pointed out that thick sedimentation could be accommodated only by subsidence of the seafloor. He finally concluded that the earth's surface is in a continual, slow up-and-down motion, and that stationary portions of the crust are indeed few. The record of "dead" rocks is a record of long-continued movements.

Present and Recent Tectonic Movement    31

## Figure 2.6

Deformation of the Lake Bonneville shoreline. The many inlets and islands of the former lake provide an unusually wide coverage of shoreline points on which present levels could be measured. The resulting contour map faithfully reflects warping of the former lake level, a direct result of isostatic adjustment following disappearance of the lake water. (Source: Critten, D. "New data on the isostatic deformation of Lake Bonneville" in *U.S. Geological Survey Professional Paper* 454–E, 76, p. 31.)

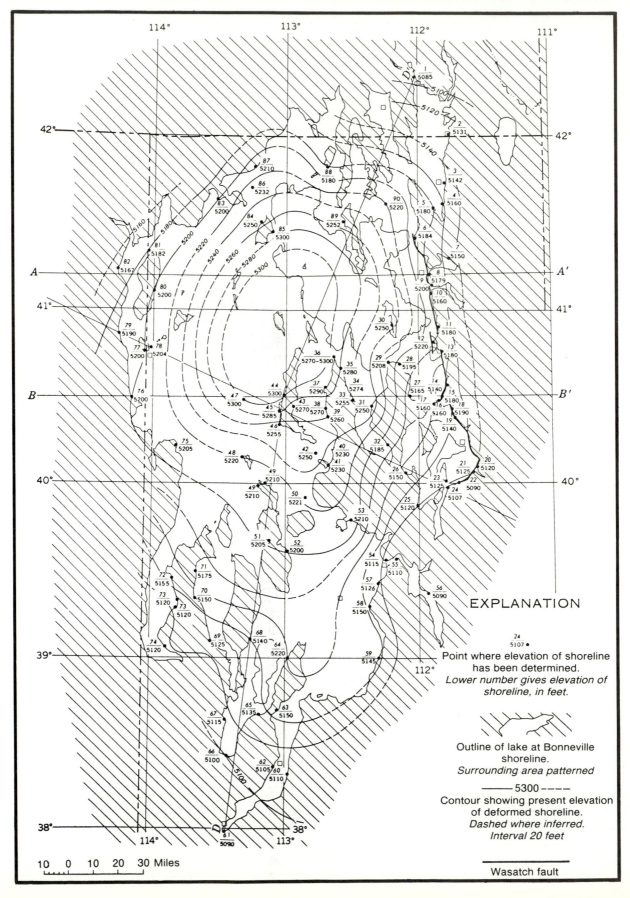

EXPLANATION

24
5107•

Point where elevation of shoreline
has been determined.
*Lower number gives elevation of
shoreline, in feet.*

Outline of lake at Bonneville
shoreline.
*Surrounding area patterned*

———— 5300 - - - -
Contour showing present elevation
of deformed shoreline.
*Dashed where inferred.
Interval 20 feet*

Wasatch fault

10  0   10   20   30 Miles

## Geological Clues to Recent Tectonic Movements

Geological methods may reveal crustal displacements over a time span which, though recent, exceeds historical time. The displacements may be continuous (warping, tilting) or discontinuous (faulting, rifting).

### Warping and Tilting

Shorelines, being horizontal originally, are good markers of crustal displacement. However, several nontectonic factors influence the displacement of shorelines, and we must eliminate these factors before attempting any tectonic interpretation. Changes in sea level that can be shown to be worldwide are *eustatic*. These movements certainly have no direct tectonic significance. Lake levels are very sensitive to climatic change, to erosion, and to sedimentation. The most significant observations are those showing *differential* shoreline displacements; if an ancient shoreline is now at different elevations in different places, it must have undergone warping or tilting. However, we must be sure that we are measuring the same shoreline throughout, and that there has been no differential compaction of underlying rocks.

Much useful work has been done in Scandinavia, where the retreat of the last Pleistocene ice sheet has provided excellent chronological and sedimentological markers. Scandinavian geologists have used both annual varves and recessional moraines as well as the dispersion of the shorelines over time and place to separate the relative effects of postglacial isostatic uplift and postglacial eustatic rise in sea level. Warping of comparatively recent horizontal features such as terraces and shorelines is well documented. It is this kind of slow vertical crustal movement that Gilbert (1890) called *epeirogenic*. Crittenden (1963) extended Gilbert's data, and drew a contour map showing warping of the formerly level Lake Bonneville shore (figure 2.6). Warping may also interfere with an established drainage pattern. Raised marine terraces (figure 2.7) always indicate some crustal movement, since their rise must have exceeded any postglacial eustatic rise of sea level.

**Figure 2.7**
Several levels of raised beaches on Palos Verdes Peninsula, California. The former beaches cannot be correlated far laterally, an indication that vertical displacement at Palos Verdes occurred in an individualized tectonic unit. Other such segments of raised beaches occur along much of the coastline of the northern Pacific Ocean. (Photograph by John S. Shelton.)

**Figure 2.8**
Deformed beach ridges across the active Whareama Syncline, New
Zealand. Fine lines show calculated heights; black dots show
observed heights for five beach ridges. Datum is normalized value
for ridge A. (From Wellman, 1971.)

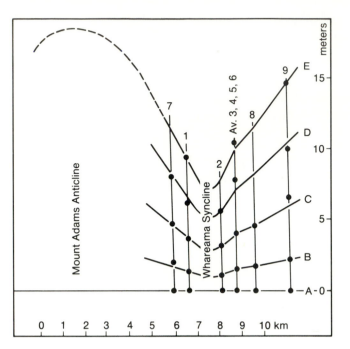

By recording the tilts of different shorelines of known ages, it is possible to determine the history of tilting over a given time span. H. Wellman (1971) measured deformation of a series of shorelines across the Whareama Syncline in New Zealand. Figure 2.8 shows the results. Clearly, the shorelines are being warped in a pattern that reflects the growing syncline. Wellman determined the rate of tilting of the southeast flank of the syncline to be about 0.016° per thousand years. At the same time, the inclination of the synclinal axis is increasing by about 0.002° per thousand years.

Zones of "live" tectonic uplift have been relatively less exposed to erosion (or, under water, to deposition) than their surroundings. For instance, Ewing and Ewing (1962) found that domes in the Sigsbee abyssal plain have kept their tops above continuing contemporary sedimentation for at least the last 200,000 years. The rate of dome rise calculated from their observations averages about 1 m per 1,000 years. Te Punga (1957) observed that the crests of some gentle anticlines in a peneplained terrain in New Zealand coincide with zones of no dissection. These anticlines are evidently recent structures and have controlled a very recent drainage pattern on their flanks.

Figure 2.9 shows tilted alluvium near Twizel, New Zealand. Here the history of tilting can be observed in a striking natural exposure: the dip increases from the upper to the lower layers; an unconformity documents an interval of erosion. Clearly, the lower (older) the beds, the greater the tilt.

Pávai Vajna (1926), in a classic study, found that recent differential changes in surface level in the Hungarian plain were related to structure. In the 25 to 30 years following a survey in 1890, some areas rose as much as 105 millimeters (mm), while some subsided as much as 222 mm (although subsidences of 100 mm or less were more common). The rising zones are over anticlines in late Cenozoic sediments; the subsiding zones are over synclines. Furthermore, the river Tisza is eroding over topographically barely perceptible anticlines, while it is alluviating over synclines. The anticlines are associated with gravity highs, the synclines with gravity lows. (Where the succession includes rock salt, however, gravity lows coincide with anticlines.)

An interesting example of recent crustal movement, dating from historic times but based on geological and geodetic methods, is the current tilting of the

**Figure 2.9**
Tilted alluvium near Twizel, New Zealand. Tilting was active during deposition, so that the older layers are more steeply tilted than the younger ones.

floor of Death Valley (Greene and Hunt 1960). Archeological and geological studies have shown that the eastern shoreline of a shallow lake that flooded the area just before the introduction of the bow and arrow, is now 7 m lower than the western shoreline. Part of the tilt can be explained by the presence of a recent fault scarp with 3 m of vertical displacement at the foot of the Black Mountains; the balance appears to be the result of slow surface deformation. Using a tiltmeter, Greene and Hunt were able to demonstrate that the tilting is still continuing. The direction and amount of tilting measured seem to agree with the known geological structure and with movements in the recent geological past.

Recent and current sedimentation is capable of registering recent and current earth movements. Some of the best examples can be found in Holland. Depth contours to the base of the Pleistocene there reveal a basin shape. The lower Pleistocene is marine, and there is gradual transition to fresh water sediments toward the top. The depression is clearly the result of Pleistocene and post-Pleistocene subsidence. Comparison of two precision levelings, 1875–1887 and 1926–1940, shows that relative subsidence of northern Holland is still going on, while the southern part of the country is rising with respect to sea level.

**Figure 2.10**
Stream offset approximately 400 m by the San Andreas fault, Carrizo Plains, California. (Photograph by R. E. Wallace and P. D. Snavely, Jr., U.S. Geological Survey.)

## Faulting and Rifting

Currently active faults may leave a number of clues of recent activity. There are obvious indications such as displacement of recent sediments and topographic (figure 2.10) and cultural features (figure 2.11) and recent sediments or soil. But one of the most striking indications consists of "triangular facets" along active fault scarps (figure 2.12). Normally, side valleys emerging in a main valley are separated by ridge spurs which, more or less attenuated, come down to the main valley floor. Where recent fault scarps form main valley flanks, spurs that separate tributary valleys are truncated by the fault scarp along the main valley. The fault scarp is a youthful form, usually interrupting a more maturely dissected landscape. Of course, triangular facets may also form as a result of differential stream erosion, as in the "flatiron" effect (figure 7.12), or by glacial erosion.

**Figure 2.11**
Deformed drainage ditch at the Almaden winery, Hollister, California.
The ditch was constructed in 1949, the photo taken in 1964.
Displacement is by slow creep along a fault of the San Andreas
system. (Northwest is at left.)

## Figure 2.12

Fault scarp of Paleozoic basement rising above a trough filled with Cenozoic sediments. Ivanpah area, Mojave Desert, California, looking westward. Note triangular facets on truncated intercanyon spurs, characteristic of active fault scarps.

## Figure 2.13

Displaced alluvial terraces on Wairau fault, New Zealand. The older the terrace, the greater the displacement. (From Stevens, 1975.)

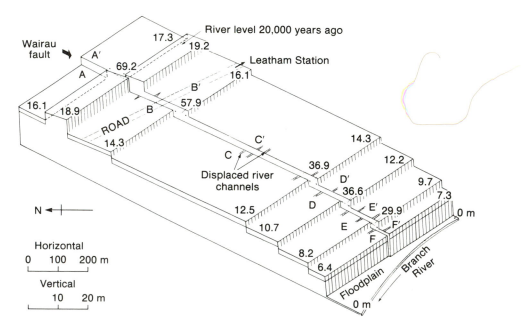

**Figure 2.14**
Cross section of Almannagja, west of Thingvalla Lake, Iceland. Horizontal scale = vertical scale, 1:2,700. Figures indicate widths of *gjar* in meters. The slope *XY* is the surface of a tilted block, following downfaulting of the graben to the east. (Modified after Bernauer, 1943.)

**Figure 2.15**
Aerial view of Almannagja (looking northward).

Stream terraces displaced by the Wairau fault in New Zealand document the fault's recent history (figure 2.13). The higher (older) terraces show greater horizontal displacement than the younger ones. If the ages of the terraces are known, displacement rates can be established. G. J. Lensen (1968) of the New Zealand Geological Survey concluded that in the past 20,000 years, horizontal movements of 3–6 m along the fault have occurred every 500–900 years. This represents an average rate of about 3.4 mm per year.

Bernauer (1939, 1943) described and measured postglacial rifting in Iceland (figure 2.14). The lack of erosion of these rifts or *gjar* (figure 2.15) makes it clear that the rifts are postglacial. According to Walker (1965), opening of the *gjar* indicates that crustal extension in Iceland is occurring at an average rate of 5 mm per year. This is a manifestation of crustal spreading along the mid-Atlantic ridge.

Since 1964, Möller and Ritter (1980) have made geodetic measurements in the active volcanic zone of northeast Iceland, essentially continuing work that had been begun in 1938. They found that for the first few years, contraction occurred across the rift zone. In 1971 expansion began which reached 1 m per kilometer per year by 1977. An interesting feature came to light: while this expansion took place across the inner rift zone, actual local contraction occurred across the outer zones bordering the rift, so that net overall expansion amounted to only 40 cm between 1971 and 1977.

Near Houston, Texas, a number of currently active faults have become apparent by small scarps in road pavement. Displacement ranges from 15 to 40 mm per year, with the amount decreasing toward the Gulf of Mexico. The faults appear to be related to an extensive slow creep of large parts of the western Gulf coast toward the Gulf of Mexico. However, current movement may be related to groundwater withdrawal.

**Figure 2.16**
Generalized topographic contour map of California. Within the Coast Ranges, a granitic basement lies between the two dotted lines; a Franciscan basement lies to the east and west. Compare this map with figure 2.17. (From Christensen, 1965.)

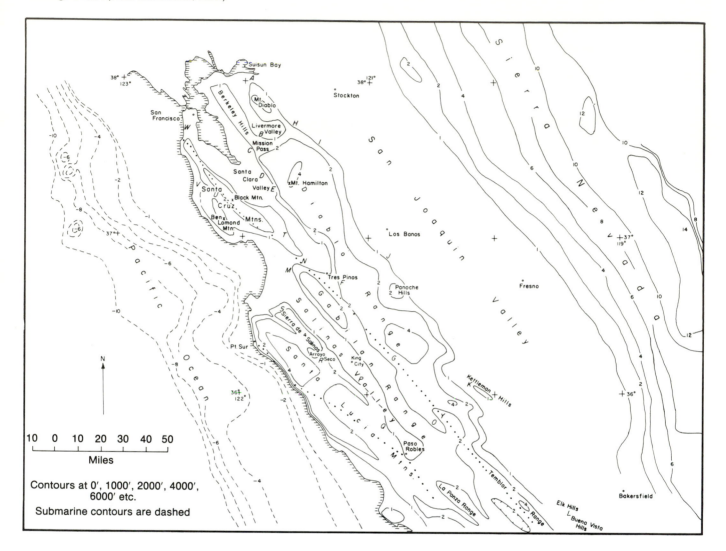

# Neotectonics

In central California, Christensen (1965) analyzed deformed sedimentary rocks of Pliocene age. Figures 2.16 and 2.17 show that such late Tertiary vertical deformation is in reasonably good harmony with present-day topography.

A distinct branch of tectonics, **neotectonics,** is the study of recent crustal movements beginning when the geological record first began to show a coherent and continuous tectonic history. For example, one neotectonic map shows that, since Miocene time, the entire area of the USSR has experienced movement either up or down. Nothing has stood still (except, theoretically, the zero contours between positive and negative areas). The evidence for subsidence is fairly straightforward; it is based mainly on sedimentation. Evidence for uplift must be taken from physiographic clues, and these are sometimes ambiguous. Other neotectonic maps show a number of other active processes, such as faulting, tilting, and seismicity.

In contrast to neotectonics, *recent* (or *active*) tectonics is confined to currently observed tectonic movements.

## Figure 2.17

Contour map of the present configuration of a surface that was the geoid approximately 3 milion years ago. Contour interval=1,000 feet. Dotted lines represent supplementary contours. (From Christensen, 1965.)

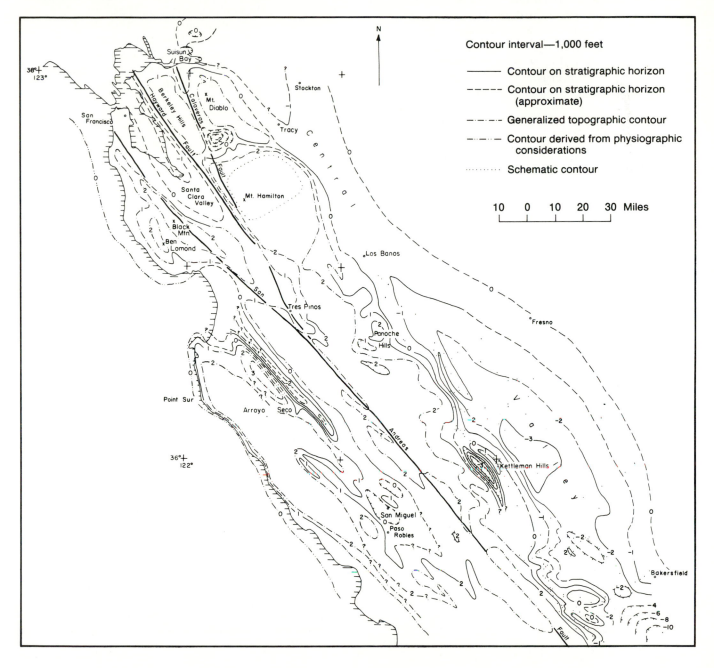

N

Contour interval—1,000 feet

——— Contour on stratigraphic horizon

- - - - Contour on stratigraphic horizon (approximate)

—·—·— Generalized topographic contour

—··—··— Contour derived from physiographic considerations

·········· Schematic contour

10   0   10   20   30   Miles

## Conclusion

The crustal movements we have discussed allow us to think of deformation in terms of the rates at which tectonic processes occur. In chapter 1, we learned that relative plate motions are of the order of 1–15 cm per year. This amount is compatible with the displacements considered in this chapter, most of which have proceeded at the rate of a few millimeters to centimeters per year. Knowing the rates of crustal deformation will help us to reconstruct what we cannot see directly.

## Review Questions

1. What is the range in magnitude of displacement of topographic features that may accompany single earthquakes?
2. What are representative rates of measurable slow creep caused by tectonic deformation?
3. List some methods used to measure rates and amounts of present-day tectonic deformation.
4. How can worldwide change in sea level be differentiated from local rise or subsidence of the land, on the basis of ancient shoreline evidence?
5. List some geological clues for recent crustal deformation.
6. What is *neotectonics?* How does this differ from *active tectonics?*
7. Discuss and interpret a segment of a neotectonic map available to you.

## Additional Reading

Adams, J. 1980. Active tilting of the United States midcontinent: geodetic and geomorphic evidence. *Geology* 8: 442–446.

Allen, C. R. 1975. Geological criteria for evaluating seismicity. *Geol. Soc. Am. Bull.* 86: 1041–1057.

———, Gillespie, A. R., Han Yuan, Sieh, K. E., Zhang Buchum, and Zhu Chengman. 1984. Red River and associated faults, Yunnan Province, China: Quaternary geology, slip rates, and seismic hazard. *Geol. Soc. Am. Bull.* 95: 686–700.

Bilham, R. G., and Beavan, R. J. 1979. Strains and tilts on crustal blocks. *Tectonophysics* 52: 121–138.

Bull, W. B. 1984. Tectonic geomorphology. *J. Geol. Educ.* 32: 310–324.

Cotton, C. 1968. Relation of the continental shelf to rising coasts. *Geogr. J.* 134(3): 382–389.

Geophysics Study Committee, National Research Council, 1986. *Active Tectonics: Studies in Geophysics.* Washington D.C.: National Academy Press. 266 pp.

Howard, J. H. 1968. Recent deformation at Buena Vista Hills, California. *Am. J. Sci.* 266: 737–757.

Kahle, J. E., Barrows, A. G., Weber, F. H., Jr., and Saul, R. B. 1971. Geologic surface effects of the San Fernando earthquake. *Calif. Geol.* 24: 75–79.

Keller, E. A., Bonkowski, M. S., Korsch, R. J., and Shlemon, R. J. 1982. Tectonic geomorphology of the San Andreas fault zone in the Indio hills, Coachella Valley, California. *Geol. Soc. Am. Bull.* 93: 46–56.

King, P. B. 1965. Tectonics of Quaternary time in middle North America. Pages 831–870 in: H. E. Wright, Jr. and D. G. Frey, eds. *The Quaternary of the United States.* Princeton, NJ: Princeton University Press.

Lees, G. M., and Falcon, N. L. 1952. The geographic history of the Mesopotamian plains. *Geogr. J.* 118: 24–39.

Mescheryakov, Y. A. 1959. Contemporary movements in the earth's crust. *Int. Geol. Rev.* 1: 40–51.

Sieh, K. E. 1978. Prehistoric large earthquakes produced by slip on the San Andreas fault at Pallett Creek, California. *J. Geophys. Res.* 83: 3907–3939.

Vyskocil, P., Wassef, A. M., and Green, R., eds. 1983. Recent crustal movements, 1982. *Tectonophysics* 97: 351.

Winslow, M. A. 1986. Neotectonics: concepts, definitions, and significance: *Neotectonics: An International Journal of Crustal Dynamics* 1: 1–5.

Yeats, R. S. 1983. Large-scale quaternary detachments in Ventura basin, Southern California. *J. Geophys. Res.* 88(B1): 569–583.

Yoshikawa, T., Kaizuka, S., and Ota, Y. 1981. *The landforms of Japan.* New York: Columbia Univ. Press.

# 3

# Geometric Representation of Rock Structures

One task of the structural geologist is to reconstruct geological structures from scattered observations and measurements made in the field. Another is to describe and to measure them. In this chapter we will discuss geometric principles that will help us in these tasks.

## Lines and Planes

The basic geometric abstractions we will use are lines and planes. They define the external form of rock bodies as well as the internal ordering of structural elements such as grains, layering, folds, and fractures. The external boundaries of rocks make up their **form.** The internal ordering of structural elements is their **fabric.**

## Some Definitions

For accurate geometric description, we need accurately defined terms. We will use no more than the minimum necessary to convey clearly what we mean. Lines and planes are the first terms we must learn. These are so basic that mathematicians do not even define them. They are idealizations of what we commonly mean by the words. A line is either straight or curved. A plane is an abstraction of a flat surface, such as a tabletop or the surface of a quiet pool.

A plane or line (or any other form) that is a boundary or that has boundaries in space is **discrete.** A tabletop, a fault, the outcrop edge of a fault, all are discrete if they can be identified separately and individually. In other words, a feature is discrete if it is a boundary or has boundaries itself. A discrete plane is commonly called a *surface;* thus, we speak of fault surfaces and bedding surfaces.

A line or plane that can be defined *throughout* a given space or entity is **penetrative.** Flow lines in a lava stream, for instance, are penetrative: they exist throughout the flowing lava. Bedding planes in a sedimentary rock are also penetrative, even though we can only see certain discrete bedding surfaces.

The **trace** of a plane or surface on a differently oriented plane or surface is its line of intersection with that plane or surface (figure 3.1). Traces of bedding planes, for instance, may be seen on rock outcrops, on fault surfaces, and on cleavage surfaces. The trace of a line on an intersecting plane is a *point.* However, in orthographic projection (p. 48), the trace of a line on a plane is also defined as the perpendicular projection of the line on that plane.

**Figure 3.1**
Traces of a plane and of a line. *XY* is the trace of *ABCD* on *MNOP;* it is also the trace of *MNOP* on *ABCD*. Point *Z* is the trace of line *JH* on plane *MNOP*.

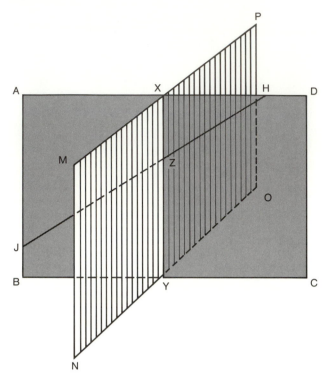

Lines and planes are uniquely fixed in three-dimensional space by their position and their attitude. **Position** is geographic location. **Attitude** is orientation in space, and it is measured by the angles made by planes or lines with a standard reference system. For the earth, this system consists of any horizontal plane (for measuring vertical angles), and any north-south line (for measuring horizontal angles) (see figures 3.2 and 3.3). The attitude of plane *ABCD* in figure 3.2a is fixed by two angular measurements: (1) the angle between it and the horizontal plane *ADN;* and (2) the angle between its trace *AD* on the horizontal plane and a reference direction (north). The angle between a given plane and a horizontal plane is the **dip** of the given plane. The direction of the intersection between a given plane and a horizontal plane is its **strike.** Dip and strike determine the attitude of the plane; they are its *coordinates of attitude.*

## Figure 3.2

Components of attitude of geological surfaces: dip and strike.

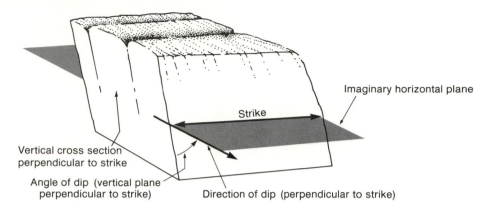

Imaginary horizontal plane

Strike

Vertical cross section perpendicular to strike

Angle of dip (vertical plane perpendicular to strike)

Direction of dip (perpendicular to strike)

## Figure 3.3

Attitude of a line. Angle *ACE* is the plunge of line *AC*; the bearing of *EC* is the trend of *AC*. Angle *BAC* is the pitch of *AC* in the surface *ABCD*. (See figure 3.4.)

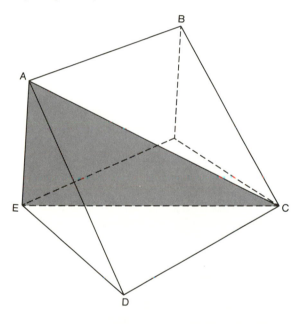

## Figure 3.4

Pitch of a line in a given surface. Angle *AOC* is the pitch of *OC* in the shaded surface. Where the angle of pitch of a line is recorded, the attitude of the reference surface must also be given.

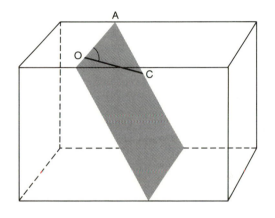

The angle between a *line* in space and a horizontal plane measured in a vertical plane is its **plunge.** The strike of a vertical plane containing the line is its **trend** (figure 3.3). Penetrative lines and planes have attitude but not unique position.

We may measure the attitude of a line or set of lines that lie within a known plane by either plunge and trend, as above, or, if more convenient, with reference to its attitude *within* the plane. The **pitch** (sometimes called the *rake*) of a line within the given plane is defined as the angle between that line and any horizontal line in the plane; that is, it is measured within the plane. It follows that pitch must always be given in terms of a plane of reference (figure 3.4).

**Figure 3.5**

Apparent dip. A strictly homoclinal sequence of beds is cut by
vertical faces along different bearings. Only the face at right angles
to the strike shows true dip. In the other faces, the trace of the beds
makes an angle (apparent dip) with the horizontal, which decreases
as the angle between the strike of the face and the strike of the
beds decreases.

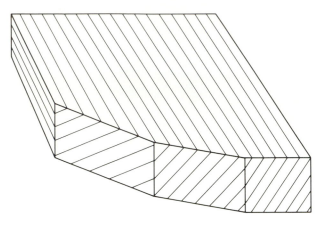

**Figure 3.6**

Apparent dip. The attitude of a plane is uniquely determined where
two different apparent dips are known.

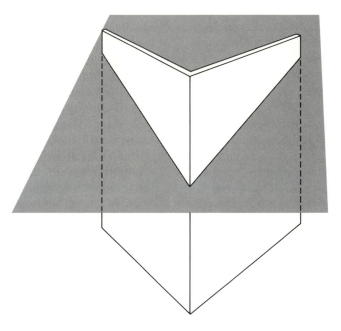

In some outcrops, the attitude of some or all of the
geological surfaces cannot be measured directly. Only
their traces on random-exposure faces may show. In that
case we can measure instead the attitudes of the traces,
also known as **dip components.** They are lines (figure
3.5), and as such they have plunge and trend. The
plunge of the trace of a surface in a given direction is
better known as the **apparent dip** of the surface it represents, in that direction (see figure 3.5). Vertical quarry
faces and all geological cross sections show apparent,
not true dips, unless they happen to be perpendicular
to the strike of the surfaces concerned.

It is important to bear in mind that apparent dips
are traces; that is, they are lines. Since two nonparallel
lines within a plane determine that plane geometrically
in space, two apparent dips or dip components are sufficient to determine the attitude of a plane (figure 3.6).

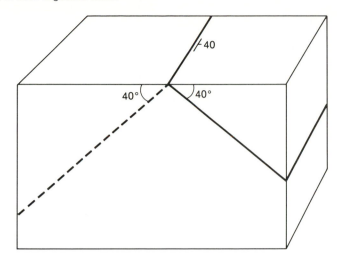

## Practical Implications and Conventions

It follows directly from the definition of dip and strike (p. 44) that the direction of dip of a plane is at right angles to its strike. It lies in the vertical plane that contains the line of greatest slope. By convention, the direction of dip is measured downard from the horizontal. Since there are two possible senses in which a plane may dip for a given strike and dip angle (figure 3.7), the *sense* of the dip must be given. Its general compass direction is usually sufficient. This requires no great precision, since we need only distinguish between two alternatives for a given strike. Actually, a less error-prone way of measuring and reporting attitude is by dip and dip direction. This eliminates the ambiguity of dip sense for a given strike and is in a form that can be directly used in statistical evaluations and in computers. Adjustable compasses can be set so as to register dip direction instead of strike.

Lines plunge along their trend. The bearing of the trend is always read in the down-plunge direction. Both strike and trend are compass bearings, that is, angular measurements taken from the north-south direction (meridian). There are two notation conventions. In one—the azimuthal notation—measurements are from true north (0°) eastward, 360° around a circle. In the other convention—the quadrant notation—common in the United States, angles are measured in quadrants, going east and west from true north, and east and west from south. The 360° convention is more straightforward and less subject to error. (In Europe, especially in France, a 400° "grad" circle is used occasionally.) With a two-circle compass (figure 3.8), dip and strike may be measured in one setting.

**Figure 3.8**
Two-circle compass. The box is leveled by means of the bubble, and the lid is adjusted so that it is parallel to the surface to be measured. The dip is measured along the small circle, and the strike is obtained from the compass needle. (Courtesy Breithaupt, Kassel.)

**Figure 3.9**

Construction principles in orthographic projection. See Appendix A. (Modified from Hammond et al., 1971.)

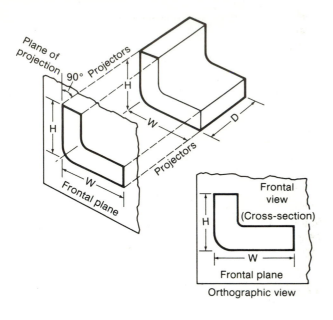

**Figure 3.10**

Three views, mutually at right angles. *H, F,* and *P,* refer to horizontal, frontal, and profile planes, terms used in engineering graphics. In geology the reference plane *(H)* need not be horizontal (as when projecting plunging folds). (Modified from Hammond et al., 1971.)

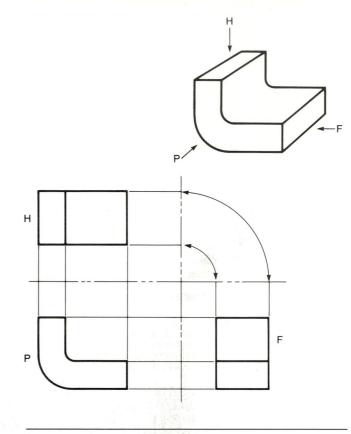

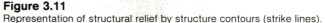

## Representation

We can represent a three-dimensional configuration (such as a deformed bed) on a two-dimensional sheet of paper by reduction in scale and by projection. Reduction in scale is straightforward; all lengths are reduced in a stated proportion, the **scale** of the representation. In **projection,** each point of a three-dimensional configuration has a corresponding point on a *plane of projection* joined to it by a *line of projection.* In *orthographic projection*—the most common—lines of projection are parallel to one another and are perpendicular to the plane of projection (figures 3.9 and 3.10). The most useful orthographic projections for solving problems in structural geology are (1) the map projection, which is along vertical lines onto a horizontal plane; (2) the projection along lines parallel to the strike of a planar structure, such as a fault or a sequence of beds; and (3) the projection parallel to the axis of a linear structure, such as a fold axis.

Geological maps show attitudes of geological lines and planes by characteristic symbols. Appendix D lists the most important of these.

### Representing Angles

In many problems of structural geology, we need to represent angles between lines and planes, regardless of their actual position. A convenient way of doing this is to enter the angular relationships on a *reference sphere,* as explained in Appendix A (pp. 364–374).

**Figure 3.11**

Representation of structural relief by structure contours (strike lines).

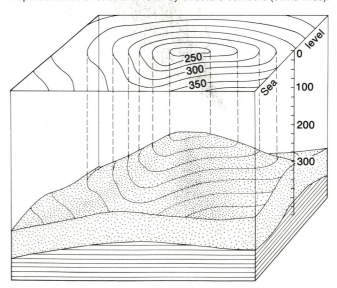

### Representing Shapes

Curved surfaces may be represented by means of **contours.** Contours are lines of equal distance from a reference plane, projected on the reference plane (figure

**Figure 3.12**
Relationship between outcrop pattern and attitude of beds. (Crown
Copyright, British Geological Survey. Reproduced by permission of
Her Britannic Majesty's Stationary Office.)

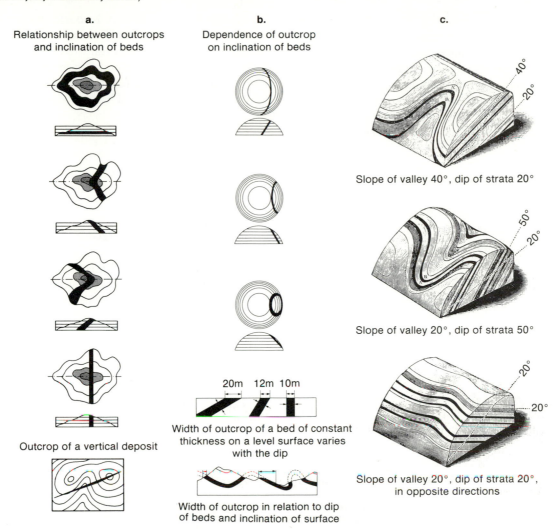

**a.**
Relationship between outcrops
and inclination of beds

Outcrop of a vertical deposit

**b.**
Dependence of outcrop
on inclination of beds

20m  12m  10m

Width of outcrop of a bed of constant
thickness on a level surface varies
with the dip

Width of outcrop in relation to dip
of beds and inclination of surface

**c.**

Slope of valley 40°, dip of strata 20°

Slope of valley 20°, dip of strata 50°

Slope of valley 20°, dip of strata 20°,
in opposite directions

3.11). Usually the reference plane is mean sea level.
Contour lines on geological surfaces are **structure contours.** (See Appendix A pp. 374–380.)

Geological surfaces may be contoured in the same
way as topographic surfaces. Structure contours over a
horizontal datum plane are lines of equal elevation; they
are strike lines. The direction of the tangent to a contour line at any point is the strike of the surface at that
point. A closed contour line on a surface that is convex
upward marks a dome or **closure** (figure 3.11). Closures may be important traps for petroleum.

### Representation on Maps

Scattered observations of attitude on discrete structures on a topographic base map, together with lithologic information, form the raw material for the
geological map.

The **outcrop** of a geological surface is where it intersects the topographic surface. Some geological maps
contain only information obtained directly from exposed outcrops. These are *outcrop maps*. More commonly, the geologist attempts to interpolate and
extrapolate field data so as to obtain a coherent *outcrop
pattern*. This is interpretation. Obviously, the geologist
must be aware of what patterns or structures are reasonably possible in a given area; he must be well
grounded in structural geology.

Figure 3.12 shows the relationship between outcrop pattern and dip of beds in areas of topographic
relief. Figure 3.12c illustrates the "rule of Vs": In a
valley, dipping beds form a V-shaped outcrop, the V
pointing in the direction of dip, unless the gradient of
the valley bottom (stream) is greater than the dip of
the beds.

## Extrapolation

To obtain a three-dimensional picture, we must extrapolate both into the earth's crust and upward into eroded space. Such extrapolations are commonly based on some of the following assumptions:

1. Strata were laid down horizontally and follow one another in chronological sequence.
2. Strata were at one time continuous parallel layers.
3. Lithologic contacts are sharp and well defined.
4. Discontinuities in structure patterns are faults or unconformities.
5. Folds are cylindrical; that is, they are generated geometrically by a line sweeping out a curved surface in space while the line remains parallel to its original attitude (see chapter 7).
6. Fold axes are parallel straight lines.
7. Separated but aligned exposures of the same rock type imply continuity of outcrop.

Many more assumptions of this kind are cited. Each one may be valid under given circumstances, but it is advisable to remember that they are, after all, mere assumptions. Whenever possible, they should be checked by first-hand observations. Unfortunately, some geologists have treated the assumptions as axioms, to the detriment of many existing maps and reports. The careful geologist will be alert for such sources of error and will make clear how many interpretations are based on direct observations and how many are merely assumptions and extrapolations.

**Cross sections** are extrapolations from surface data, both downward to subsurface and upward into eroded space. Whenever subsurface data (from drilling and from geophysical studies) are available, they must be used to constrain interpretations. Since many cross sections are across folded rocks, we will return to them in chapter 7.

## Some Field Techniques

Field observations for tectonic analysis go beyond the much-too-common random measurements of dips and strikes. First of all, we must remember that dip and strike, as well as plunge and trend, are coordinates of attitude and that they belong together. Readings for each measurement of dip and strike must be made at the same spot, for the coordinates of each attitude refer to, and are valid for, one point only.

The attitudes of lines are best taken in a vertical plane. In other words, one should read plunge rather than pitch. Only if this is impractical should pitch readings be taken.

At key exposures it may be desirable to take specimens oriented in the field. For instance, we may wish to have a record of an important reading, or to cut oriented thin sections, or to make more precise measurements later under more favorable conditions than may have existed in the field. The procedure is as follows:

1. Break off a specimen.
2. Replace it in its exact original position.
3. Mark a horizontal line on any convenient flat surface of the intended specimen. (A pen with indelible black ink is convenient for marking.)
4. Add an arrowhead.
5. Read the bearing of the marked line in the sense of the arrowhead, and record it immediately in the field notebook.
6. Add a dip barb to the line on the specimen straight down the dip if the surface faces up, or with a "reversed dip" hook if the surface faces down (figure 3.13).
7. Measure the dip and indicate direction of dip. The specimen is now fully marked for reorientation in the laboratory and for cutting oriented thin sections.

Certainly, the value of a rock sample is greatly enhanced if its original orientation in the field has been measured and recorded.

**Figure 3.13**
Markings on oriented specimens. (a) On upward-facing surface, showing strike and dip. (b) On downward-facing surface, showing strike and dip. (c) On upward-facing surface, showing dip direction and dip.

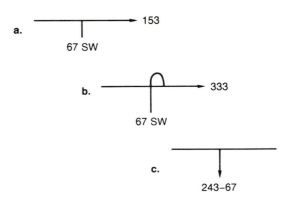

A good way to avoid unnecessary error and confusion is to enter attitude data in the notebook by map symbols in appropriate orientation (figure 3.14). Where different types of surfaces are distinguished, each should receive a distinctive symbol. Orientation of strike lines in the notebook can be approximate; since the symbols are merely guides, only the numbers need be accurate. Localities should receive numbers. Since the notebook is organized by dates, it is a good idea to enter, on the back of the field map or aerial photograph used, each locality and specimen number under the date on which they are observed. This is a valuable cross-reference and a time-saving device when trying to locate information later.

Along some slopes, bedding attitudes may be deflected by **downslope creep** of a surficial layer (figure 3.15). Hence, great care is necessary when recording attitudes where creep is apparent or suspected.

**Figure 3.14**
Notebook record of structural measurements. The arrow of the lineation symbol indicates plunge, the dashed extension, trend. The triangular plunge arrow should indicate one specific type of lineation. Others may be designated by open triangles, Vs or other symbols. In this instance, bedding (*upper left*) and slaty cleavage (*lower left*) are parallel.

**Figure 3.15**
Downslope creep of dipping beds may give erroneous indications of attitude at the surface. (a) Miocene rhyolite, Tick Canyon, California. (b) Paleozoic schists, Spessart, West Germany.

a.

b.

## Conclusion

In this chapter we have become acquainted with some of the basic geometric language used in describing most geological structures. Let us not forget that geometry deals with ideal forms, and that nature almost never provides ideal forms; even crystals are full of imperfections. This is particularly true of geological structures.

Before we move on to structures resulting from deformation—the main topic of this book—we will take a brief look at some primary, mainly depositional, structures.

## Review Questions

1. Define and explain dip and strike, using appropriate diagrams.
2. Define plunge, pitch, and trend, and explain how they are measured.
3. Explain the concepts of apparent dip and dip component, and describe how they are measured.
4. What assumptions are commonly made when extrapolating geological observations for the purpose of defining a structure?
5. How would you take an oriented specimen from an outcrop?
6. A sloping embankment exposes a straight dipping bedding outcrop. How would you measure the apparent dip of the bed in the embankment? What else would you need to know in order to obtain the true dip and dip direction? (If there is more than one possibility, give all you can think of.) Illustrate your answer, and label all geometric elements.

## Additional Reading

Bengtson, C. A. 1980. Structural uses of tangent diagrams. *Geology* 8: 599–602.

Bucher, W. H. 1944. The stereographic projection, a handy tool for the practical geologist. *J. Geol.* 52: 191–212.

Knutson, R. M. 1958. Structural sections and the third dimension. *Econ. Geol.* 53: 270–286.

McIntyre, D. B., and Weiss, L. E. 1956. Construction of block diagrams to scale in orthographic projection. *Geol. Assoc. London Proc.* 67: 145–155.

Ragan, D. M. 1985. *Structural Geology, an Introduction to Geometrical Techniques.* 3d ed. New York: Wiley.

Reeves, R. G. 1969. Structural geologic interpretations from radar imagery. *Geol. Soc. Am. Bull.* 80: 2159–2164.

Roberts, J. L. 1982. *Introduction to Geological Maps and Structures.* Oxford: Pergamon.

Secrist, M. H. 1936. Perspective block diagrams. *Econ. Geol.* 31: 867–880.

# 4

# Primary Structures

Structural geology deals principally with deformed and deforming rock bodies. Undeformed rocks would, at first sight, seem to be outside its province. But, in order to understand rock deformation, we must know something of the form and fabric of rocks before they were deformed, the original or *primary* structures.

Primary structures are acquired when a rock is emplaced. They are the result of two processes: the settling of solid particles from a fluid medium, and the crystallization of mineral grains from a liquid. Most sedimentary rocks, but not all, result from settling, and most igneous rocks result from crystallization. Here we consider mainly depositional structures and some others that are closely related to the original emplacement of rocks.

## Stratification

The fundamental sedimentary structure is stratification, or bedding, a planar arrangement of particles. Stratification provides reference horizons whose significance has been known at least since Steno's (1669) classic *Prodrome*. The apparent simplicity of the phenomenon may be misleading, which is why it needs closer study.

Stratification surfaces become apparent in two ways: through changes in texture, or composition, or both across layers (figure 4.1); and through discontinuities, which generally mark episodic nondeposition. Some discontinuities are difficult to detect, unless changes such as induration, leaching, or marked erosion have taken place along them.

In certain cases, bedding may be parallel to a sedimentary fabric such as rock fissility in shales. Pronounced orientation in sedimentary fabric may also be the result of secondary alignment of grains, as in compaction and flowage. Deposition from a moving medium (currents) also tends to orient grains.

---

**Figure 4.1**
Bedding accentuated by badlands erosion, Eocene Wasatch Formation, Bryce Canyon, Utah.

## Initial Dip

Normally, in agreement with Steno's law of original horizontality, beds are assumed to be laid down perfectly horizontally. However, some sedimentation occurs along sloping surfaces. Where the surface of accumulation is itself sloping, strata assume a primary nonhorizontal attitude: they have an **initial dip.** This may result in **discordant bedding.**

## Discordant Bedding

In discordant bedding, internal laminations within each layer are inclined to the main (true) bedding surfaces. The most common, and best-known form of discordant bedding is **cross-bedding.** It is the characteristic internal structure of deltas, dunes, and current deposits (figure 4.2). After a layer of cross-laminations has been laid down, local erosion may remove the **topset** laminations (figures 4.3, 4.4, and 4.5), truncating the **foreset** beds. This truncation, which naturally occurs only from the top down, is one of the safest top-and-bottom indicators in sedimentary sequences. In very strong "torrential" currents, however, no **bottomset** laminations may form. So, as a result of truncation, only straight foreset laminations remain. This is torrential cross-bedding, which evidently is of little use as a top-and-bottom indicator.

Discordant bedding is most common in coarse sediments. However, it does occur in fine sediments, where its presence might not normally be suspected. In bioherms, forereef talus accumulates as foreset bedding. The thickness of bedding in drill cores may be seriously overestimated if foreset beds are mistaken for originally horizontal bedding (figure 4.2).

**Figure 4.3**
Cross-bedding in unconsolidated sediments, bed of Rio Grande north of Santa Elena Canyon, Texas.

**Figure 4.4**
Dune-bedding in Triassic sandstone, Mauchline Quarries, Ayrshire, Scotland. (U.K. Crown copyright, British Geological Survey photograph.)

**Figure 4.2**
Cross-bedding in deltas, and nomenclature and relationship between true thickness ($T_1$) and apparent thickness ($T_2$). (From Hills, 1972.)

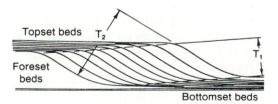

Topset beds  $T_2$

Foreset beds

Bottomset beds $T_1$

**Figure 4.5**
Overturned cross-bedding in Precambrian Grenville gneiss, Portneuf County, Province of Québec, Canada.

**Figure 4.6**
Grain relations in graded bedding. (After Kuenen, 1953.)

## Graded Bedding

In some individual beds and laminations, the grain size grades from a relatively coarse stratigraphic bottom to a finer top (figure 4.6). In fact, wherever there is gradation in grain size, it is normally upward stratigraphically, from coarse to fine. This is a standard criterion for tops of beds. However, there are exceptions, as in some deltas.

Graded bedding may reflect varying particle size of the supply, as in seasonal variations or settling of sediments from a mixed-size cloud, the coarser particles falling faster than the finer particles. This may occur either in water or in air; many tuffs are graded. It also happens in some layered igneous rocks formed by crystal settling.

In water, such mixed-size particle clouds are the second stage of **turbidity currents,** fast-moving dense suspensions of sediment in water, originating from mud slumps, submarine slides, and storm-flooded streams. The dense suspensions have an appreciably greater specific gravity than the surrounding water and will move down very gentle slopes at considerable speed, their load capacity increasing with speed. When a broadening of the channel or a decrease in slope checks the speed of turbidity currents, much of the mixed-size sediment cloud settles differentially and forms a graded bed. Each turbidity current will produce a separate graded bed. Thicknesses vary, but they rarely exceed a meter or so. The most common site of these deposits—**turbidites**—is in basins that have long and relatively steep slopes and that are tectonically unstable.

## Top and Bottom of Beds

Markings such as ripple marks, raindrops, footprints, and other impressions from organisms as well as from inorganic material may be preserved along bedding surfaces. Their orientation may serve to indicate the top of the bed (as in figure 4.7). Similarly, fossils tend to be concentrated along bedding surfaces, and they tend to lie with their convex sides up. A full account of these and other primary structures that indicate tops of beds can be found in Shrock (1948) and in Pettijohn and Potter (1964). The interpretation, oftentimes, is a matter of informed common sense.

**Figure 4.8**
Block diagram of an unconformity. Continuity of stratigraphic unit *Y* overlying older unit *O* along the discordance is a good indication that this is an unconformity, not a fault. However, the evidence is not unequivocal, since many thrust and detachment faults have continuous stratigraphic units at their base. (See chapters 14 and 15.)

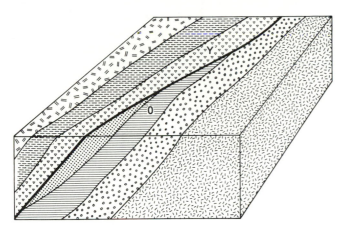

**Figure 4.9**
Unconformity with basal conglomerate over Cristianitos fault, San Onofre, California. The age of the conglomerate and that of the immediately overlying terrace formation has been determined to be at least 125,000 years. This indicates that the fault has been inactive for at least that long (an important conclusion, considering the presence of a nuclear generating station within a mile of the fault).

## Unconformities

An erosional surface beneath a sedimentary sequence is called an **unconformity.** Unconformities are structurally important, because many are clues to tectonic activity. In addition, it is sometimes necessary, and not always easy, to distinguish between those breaks or discordances that represent an unconformity, and those that represent a fault (figure 4.8).

An unconformity marks a break in the continuity of the geological record. The time interval concerned (**hiatus**) corresponds to some or all of a number of events: deposition of rocks, deformation of these rocks, emergence, erosion, and submergence. We might add an often-neglected event—nondeposition—for time elapsed without a record is time, nonetheless.

The basic event in the formation of most unconformities is erosion, which usually means emergence. A succession of unconformities in a sequence is a record of a repeated return to sea level. If the most intense tectonic upheavals occur without emergence of the affected rocks, normally no unconformity will result. On the other hand, emergence leading to unconformity may sometimes exaggerate the importance of comparatively minor warping. Submarine erosion does exist, however, and may be responsible for some unconformities.

Structurally, we differentiate among four kinds of unconformity: (1) **angular unconformity,** in which the older rock sequence has been deformed and its beds truncated so that the beds of the older sequence mark an angle with the surface of unconformity (figure 4.9); (2) **disconformity** or **parallel unconformity,** in which two rock sequences overlie one another in parallel layers, but their contact marks an erosion surface; (3) **nonconformity** or **heterolithic unconformity**

(Tomkeieff 1962), in which the younger sequence overlies eroded nonstratified rocks; and (4) **onlap unconformity** (sometimes called buttress unconformity), in which beds of the younger sequence appear truncated by the unconformity (figure 4.10).

In judging relative importance of unconformities, neither length of hiatus nor geometric relationships are good criteria; truncation is common even in cross-bedding, which is also, strictly speaking, an unconformity. On the other hand, some Pleistocene deposits overlie Paleozoic and older beds with perfect "conformity." Most geologists would agree that a "great" unconformity is one that corresponds to an important tectonic event. Thus, we would expect it to mark a radical break in the facies pattern, regardless of the time represented or the geometric discordance.

## Recognition of Unconformities: Some Criteria

### Truncation

Truncation of an older sequence, whether immediately visible or enhanced by mapping, establishes the presence of an unconformity (figure 4.8). But faults also truncate sequences. It helps to remember that the younger sequence is normally subparallel to the surface of unconformity, while most faults truncate beds on both sides. Along thrust faults (chapter 15) the sequences on either side of the fault may also be parallel to the movement surface, but the upper sequence is usually older; this identifies the surface as tectonic. As we will see later, younger rocks do overlie older rocks along some tectonic discordances, but such occurrences are few. Along some unconformities, transgressive overlap "truncates" the younger sequence at a very small angle.

**Figure 4.10**
Unconformity of Cambrian Potsdam sandstone over Precambrian Grenville gneiss, Ontario, Canada.

Transgressive relationships in such cases can be brought out by contouring bedding surfaces, using subsurface data.

### Gaps in the Succession

Identifying an unconformity often includes recognition of a time gap. In a well-established stratigraphic sequence, this may be relatively easy. But in a sequence that is not so clear, particularly in Precambrian rocks, evidence of emergence is most useful: (e.g., irregularities resulting from erosion, and chemical alteration caused by weathering). Many such criteria were summarized by Krumbein (1942).

### Basal Conglomerate

Conglomerates containing pebbles of an older sequence may lie at the base of a younger sequence (figure 4.9). This is indeed an excellent criterion for unconformity. It may not seem necessary to point out that most conglomerates are not basal, but this fact has often been overlooked.

Basal conglomerates are useful when it is not clear whether a granite contact is an intrusive contact or a heterolithic unconformity. Where a sedimentary sequence unconformably overlies a granitic basement, the basal bed of the younger sequence may be an arkose consisting largely of reworked basement rocks, grading downward into granite or gneiss. Such an unconformity has been termed a *blended unconformity*. Examples exist above the central massifs of the Alps.

## Diagenetic Structures

Diagenetic processes may form secondary structural features in sedimentary rocks, for some chemical post-deposition processes redistribute certain soluble constituents within the rock.

### Stylolites

**Stylolites** (figure 4.11) are irregular seams in soluble rocks. They result from solution along surfaces that are grossly perpendicular to contemporaneous compression: the seam contains the insoluble residue. Displaced primary structures show that as much as 40% of the rock may dissolve in this way. Commonly, the stylolitic seams are subparallel to bedding, but they may originate at an angle to it, depending on stress distribution at the time of their formation. Many are tectonic in origin (see chapter 10).

**Figure 4.11**
Different attitudes of stylolites in relation to horizontal bedding planes (corresponding to the horizontal direction in this diagram). (1) Conformable stylolites. (2) Discordant stylolites. (3) Discordant stylolites offset by conformable stylolites. (4) Perpendicular (discordant) stylolites. (5) Network stylolites. (6) Offset along discordant stylolites. In types 3a and 6a, offset might be caused by stylolitization; in types 3b and 6b, offset is caused by displacement before stylolitization of the later seam. (From Park and Schot, 1968; using a modified nomenclature.)

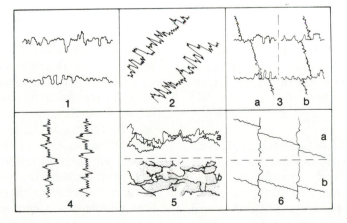

**Figure 4.12**
Liesegang banding in Silurian conglomerate, New Jersey Highlands.
Bedding traces run from top right to bottom left of the photograph.

## Liesegang Banding

Some sedimentary rocks exhibit color banding that transects bedding. This color banding is the result of diffusion in groundwater of colored impurities, usually iron oxides. Care must be taken not to confuse these **Liesegang bands** or rings with bedding (figure 4.12). The propagation of Liesegang bands appears to be little influenced by the fabric of the host rock. But fractures may impede propagation and result in "pseudofaults" (figure 4.13).

## Compaction

After deposition, newly settled sediments become compacted by expulsion of interstitial water and ordering of grains, under the load of overlying sediments. Compaction reduces pore space and can therefore be expressed in terms of relative **porosity,** the ratio of void space to total volume. For instance, a mud may have an original porosity of 80%, but in the resulting shale this may be reduced to no more than 10%—a reduction in thickness to almost one-eighth the original value. On the other hand, sands will rarely compact to less than two-thirds of their original thickness.

Structurally, compaction is important (1) because it orients platy particles, such as clay particles, and therefore imparts bedding fissility to many rocks; and (2) because it reduces the thickness of beds. Where highly compactible sediments such as clays accumulate on a floor of irregular topography (figure 4.14), the result is differential compaction of the clays around bulges in the floor. Since net compaction is proportional to original thickness, the thicker parts in the clay layer compact more, and the thinner parts, over the bulges in the floor, compact relatively less. As a result, the clay layer wraps around the floor bulges. Examples of such resistant bulges are coral reefs, sandy channels in mud, limestone lenses in clay mud, and topographic relief. Topographic relief, in particular, provides good closures for petroleum traps, and the resulting structures are known as **plains-type folds** or **compaction folds.**

**Figure 4.13**
Joint in Austin Chalk obstructs development of Liesegang bands, giving the appearance of a fault. Such relationships may also result from fracture control of metasomatism. (Photograph by Audio-Visual Department, California State University, Long Beach.)

**Figure 4.14**
Examples of differential compaction features. (From O'Connor and Gretener, 1974.)

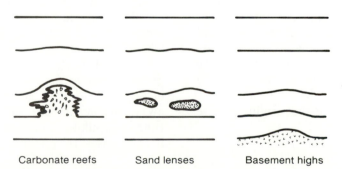

Carbonate reefs    Sand lenses    Basement highs

**Figure 4.15**
Some soft-rock deformation types. (a) Slump structures caused by horizontal movement of beds after deposition. (b) Load cast, developed by plastic deformation of a bed under a later load. (From Kuenen, 1953.)

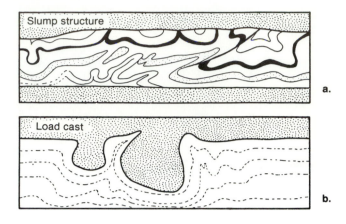

## Penecontemporaneous Deformation

Unconsolidated sediments laid down on an inclined floor may be in metastable equilibrium, ready to slide downslope if disturbed. Evidence for **penecontemporaneous deformation** resulting from subaqueous slumping is not common in the sedimentary record. Characteristically, the deformed and discordant layer lies between layers that are unaffected by the deformation (figure 4.15a). Observations on recent sediments suggest that a very small slope (perhaps 2°) may be sufficient. These structures are good "way-up" criteria if material from the original floor forms inclusions in the overlying deformed layer. Different styles of such intraformational deformation may develop, depending on the relative ductility of the layers concerned, the slope of the floor, and the extent of the deforming movement (figure 4.15a). These structures should not be confused with soft-rock deformation structures due only to load (figure 4.15b).

A submarine slide or slump may, on the other hand, disintegrate into a thick suspension, which, on an inclined floor, will initiate a turbidity current. It is not uncommon to find intraformational deformation and turbidity current sedimentation (mainly graded) associated in the same basin. In such cases, the turbidity current deposits are generally near the bottom of the original slope in the basin floor, while the intraformational structures are on those parts that originally were more inclined.

## Primary Structures of Igneous Rocks

Igneous rocks have primary structures, those structures which they acquire during emplacement. We shall discuss only a few. A fuller account can be found in some of the Additional Reading materials recommended at the end of this chapter.

Tuffs resulting from ash falls tend to acquire graded bedding in much the same way as some sedimentary rocks do. Crystallization in magma chambers may also result in layering (figure 4.16). Vesicles in lava flows tend to rise to the top of the flow, and are thus good "way up" indicators.

Many of the more mafic submarine lava flows have pillowlike internal structures, which are useful indicators of lava flow tops (figure 4.17). These are large globules of lava that appear to have rolled on top of one another in a subaqueous environment. They were soft at the time of settling into position; hence, the bottoms of the upper pillows tend to adapt themselves to irregularities in the older ones underlying them. This tendency is clearly shown in most pillow lavas, and provides excellent top-and-bottom criteria. Pillows form from tongues of lava (most commonly basaltic or andesitic) whose outside skin has solidified but whose insides are still liquid. From time to time the liquid, which is under pressure, pierces an opening in the solid skin, and blobs of lava then emerge, acquire a solid glassy skin, roll away, and accumulate downslope. The material between the pillows is derived from the pillow skin, and, in some cases, from detrital material.

**Figure 4.16**
Sedimentary structures resulting from crystal settling in layered ultramafic intrusive, Duke Island, Alaska. Note unconformity—evidence of scouring—and draping of layers around block of clinopyroxenite. (From Irvine, 1965.)

**Figure 4.17**
Pillow lavas, Point Sal ophiolite, California. Top and bottom are clearly indicated by the "points" of several pillows.

## Conclusion

A knowledge of primary structures is helpful to the structural geologist for two main reasons. Many serve as markers that reflect rock deformation, the most obvious being folded bedding. Some will indicate the upward direction in a sequence of layered rocks, or "way up," where a rock succession has been rotated out of its original horizontal attitude. Primary structures, thus, are reference structures that act as signposts for the structural geologist.

## Review Questions

1. Discuss the structural significance of layering in rocks.
2. To what extent may Steno's law of original horizontality be violated? Explain.
3. Describe primary structures that may reveal the direction in which tops of beds are facing.
4. List a sequence of geological events documented by an unconformity. Indicate which events are necessary and which are not.
5. What is the origin of stylolites? How could you estimate the loss of material along stylolite seams?
6. Explain the origin of plains-type folds.
7. Under what conditions, and in what manner, may layers of sediment be deformed a short time after they have been deposited?
8. How are pillow lavas formed? What makes them good top-and-bottom indicators of the stratigraphic sequences in which they occur?

## Additional Reading

Bailey, E. B. 1936. Sedimentation in relation to tectonics. *Geol. Soc. Am. Bull.* 47: 1713–1726.

Bayly, Brian 1986. A mechanism for the formation of stylolites. *J. Geol.* 94: 431–435.

Bishop, D. G., and Force, E. R. 1969. The reliability of graded bedding as an indication of the order of superposition. *J. Geol.* 77: 346–352.

Blackwelder, E. 1909. The valuation of unconformities. *J. Geol.* 17: 289–299.

Clark, S. K. 1932. Mechanics of plains-type folds. *J. Geol.* 40: 46–61.

Natland, M. L., and Kuenen, P. H. 1951. Sedimentary history of the Ventura Basin, California, and the action of turbidity currents. *SEPM Spec. Publ.* 2: 76–107.

# 5
# Stress and Strain

To understand rock deformation, we must become acquainted with some simple but fundamental principles of mechanics, the science concerned with the action of forces on bodies and their effects. Classical mechanics is concerned only with the action of forces on particles and rigid bodies. In rock deformation, however, we need to understand processes that result in deformation, that is, displacements *within* a body. In other words, we must know something about the relation between mechanical energy applied to a body and the response within it. Chapters 5 and 6 deal with these topics.

## Dimensions and Units

Let us begin with some basic concepts. We must first distinguish clearly between **dimension** and **unit.** To do so, we can use the simple and precise definitions given in the American Heritage Dictionary; dimension is defined as "a physical property, (such as) mass, length, time, or some combination thereof, regarded as a fundamental measure, or as one of a set of fundamental measures, of a physical quantity." Thus, mechanical properties of a material can be expressed in terms of the three fundamental dimensions of *mass, length,* and *time.* In contrast to dimension, which is a physical *property,* a unit is a physical *measurement.* The American Heritage definition of unit is "a precisely specified quantity in terms of which magnitudes of other quantities of the same kind may be stated." Thus, a meter is a unit of length; a second is a unit of time.

It is convenient to use symbols to describe dimensions. Thus, the symbol for mass is $[M]$, for length $[L]$, and for time $[T]$. Velocity combines the fundamental dimensions of length and time; it has the dimension of length divided by time. In conventional symbols this is written

$$[V] : [LT^{-1}]$$

in which the colon is read "has the dimensions of," and the brackets enclose the dimensional expression.

## Force

**Force** $[F]$, according to Newton's second law, is mass multiplied by acceleration:

$$[F] : [MLT^{-2}]$$

Force is a vector quantity. That is, it has both magnitude and direction, and it conforms to the rules of vector mathematics. Thus, any accelerating (or decelerating) body has a force acting on it.

Forces that act on the outside of a body are *applied forces* or *surface forces.* Those that result from action of a field at every point within the body are *body forces.*

Examples of applied forces include the force of a racquet on a ball, the force of expanding gases on an engine piston, and the force of the jaws of a vice on a piece of wood. Examples of body forces include the force of gravity and forces from magnetic fields. They are proportional to the mass of the body. Forces on a body may cause acceleration (equation 5.1) of the body as a whole, or deformation, or both. If some or all of the forces acting on a body are absorbed by the body (instead of moving it), the body becomes *stressed.* The forces then cause particle displacement within the body so that the body, as a result of being stressed, becomes *deformed.* In simple terms, force causes stress, and stress causes deformation.

## Stress

Stress ($\sigma$) is the concentration of force per unit area ($A$), or $\sigma = F/A$. We may visualize it as *intensity* of force. (A force acting on a small area would have a greater intensity than the same force acting on a larger area; the stress would be greater.) Since stress is force per unit area, its fundamental dimensions are

$$\left[\frac{ML}{T^2} \cdot \frac{1}{L^2}\right] \text{ or } \left[\frac{ML}{T^2L^2}\right] \text{ or } \left[ML^{-1} \cdot T^{-2}\right]$$

A component of stress in any direction is often called the stress in that direction; more correctly, it is known as the *traction* in that direction. For, while stress is a tensor, a three-dimensional entity, traction is a vector that has magnitude and direction. The stress at a small element of a body is the aggregate of all tractions (in all directions) acting on that element. Where tractions in all directions are equal in magnitude, the stress is said to be **isotropic.** Thus, in an isotropically stressed body, there is no preferred direction in which one stress component (traction) is larger or smaller than any other. Hydrostatic pressure is isotropic stress, because a fluid is normally stressed equally in all directions. We may represent isotropic stress as a sphere: all radii are equal, and each represents a traction in a different direction; (there is an infinite number of tractions through any point). Isotropy must not be confused with homogeneity, which is explained on p. 68.

If the stress is not isotropic—that is, if it is **anisotropic**—the tractions are not all equal; but they must be such that each traction is balanced by another that is equal in magnitude and of opposite sense, so that the body as a whole is in equilibrium. We may represent anisotropic stress at any point in a stressed body as an ellipsoid (figure 5.1a); then each radius vector of this **stress ellipsoid** represents a traction.

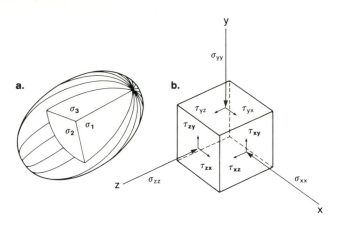

## Stress Components

Stress across any arbitrarily oriented plane can be resolved into a component normal to the plane and one parallel to it. The component normal to the plane is called the **normal stress** and is given the symbol $\sigma$ or $\sigma_n$; the component along the plane is the tangential stress, more commonly called **shear stress,** and is given the symbol $\tau$ or $\sigma_s$. Shear stress tends to move particles past each other. Most arbitrarily oriented planes within a stressed body have both normal and shear stress components acting upon them.

The state of stress at a point is three-dimensional. Let us describe it by reference to three mutually perpendicular (Cartesian) coordinate axes, $x$, $y$, and $z$. For convenience, we may consider stress at an element within a rock body as acting on an infinitesimally small cube, with sides perpendicular to each of the coordinate axes $x$, $y$, and $z$. We may then resolve the stresses acting on each face of the cube into three components (figure 5.1b). For a face normal to $x$ they are $\sigma_{xx}$, the component *normal* to that face, and $\tau_{xy}$ and $\tau_{xz}$, the two components *within* the face. The last two are shear components, each along one of the other coordinate axes $y$ and $z$, respectively. The same procedure applies for the faces normal to $y$ and $z$, so that for the three faces we obtain a total of nine stress components (figure 5.1b):

| Face normal to $x$: | $\sigma_{xx}$ | $\tau_{xy}$ | $\tau_{xz}$ |
|---|---|---|---|
| Face normal to $y$: | $\tau_{yx}$ | $\sigma_{yy}$ | $\tau_{yz}$ |
| Face normal to $z$: | $\tau_{zx}$ | $\tau_{zy}$ | $\sigma_{zz}$ |

The columns, from left to right, each represent the components in the $x$, $y$, and $z$ directions, respectively. $\sigma_{xx}$, $\sigma_{yy}$, and $\sigma_{zz}$ are normal components; the remainder are shear components. Of the six shear components, three each are equivalent to the other three ($\tau_{xy}$ to $\tau_{yx}$, $\tau_{yz}$ to $\tau_{zy}$, and $\tau_{xz}$ to $\tau_{zx}$), because they balance for a body at rest. So we are left with only six truly independent stress components acting on any arbitrary infinitesimal element in a stressed body.

We now need a sign convention. In physics and in engineering, tensile normal stress is considered positive, and compressive stress negative. In geology, it is customary to make compression positive and tension negative, because compression is dominant in most of the earth's crust.

## Principal Stresses

For any given state of stress, there are three orientations of planes, mutually at right angles, along which shear stress is zero. These planes intersect in three mutually perpendicular axes known as the **principal axes** (or directions) of stress. The stresses acting along them are the **principal stresses** for a given point or homogeneous domain within the body. Except for some special cases they are unequal, one being greatest, one least, and one intermediate in value. Each principal stress coincides with a principal axis of the stress ellipsoid (figure 5.1a). By convention, the greatest principal stress is given the symbol $\sigma_1$; the intermediate and least principal stresses acting along the other two axes are given the symbols $\sigma_2$ and $\sigma_3$, respectively.

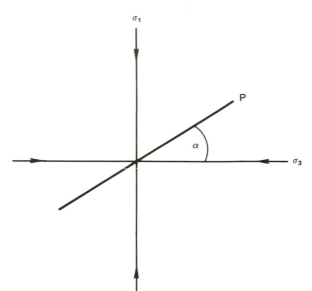

The following relationships between the principal stresses $\sigma_1$, $\sigma_2$, and $\sigma_3$ are of special interest:

| | |
|---|---|
| Uniaxial tension | $\sigma_1 = \sigma_2 = 0; \sigma_3 < 0$ |
| Uniaxial compression | $\sigma_2 = \sigma_3 = 0; \sigma_1 > 0$ |
| Biaxial (plane) stress | $\sigma_2 = 0$ |
| General triaxial stress | $\sigma_1 > \sigma_2 > \sigma_3$ |

The distribution of all the stresses in a body, or part of a body, is called the **stress field.** It can be visualized as an infinite number of local stresses distributed over the body. If the stress at each point in the field is the same in magnitude and orientation, the stress field is *homogeneous.* Thus, in a homogeneous stress field, all principal stresses everywhere point the same way.

## Relationship Between Normal Stress and Shear Stress

A common problem in rock deformation is finding the normal stress and the shear stress in an arbitrarily oriented plane, $PP'$, for a state of stress given in terms of the greatest principal stress, $\sigma_1$, and of the least principal stress, $\sigma_3$. For most practical purposes we can neglect the intermediate principal stress, $\sigma_2$, and consider only the plane containing $\sigma_1$ and $\sigma_3$. The situation is illustrated in figure 5.2. If we know the angle $\theta$ between the plane $PP'$ and one of the principal stresses (by convention, $\sigma_3$), we can determine both normal and shear stresses ($\sigma$ and $\tau$) acting on that plane. A simple method

of doing so is a construction devised by Otto Mohr (1882) and known as the **Mohr diagram** (figure 5.3). The proof of the construction is given in Appendix C.

The method of constructing the Mohr diagram is to first construct two mutually perpendicular axes $\sigma$ (abscissa) and $\tau$ (ordinate). Given a state of stress with greatest principal stress, $\sigma_1$, and least principal stress, $\sigma_3$, acting on a plane making an angle $\theta$ with the $\sigma_3$ direction (figure 5.3), we then mark off $\sigma_1$ and $\sigma_3$ along the $\sigma$-axis of the $\sigma\tau$ plot, and construct a semicircle through points $\sigma_1$ and $\sigma_3$, with $O$, the midpoint, at $\frac{1}{2}(\sigma_1 + \sigma_3)$ as center. The radius of this circle, evidently, is $\frac{1}{2}(\sigma_1 + \sigma_3)$. Next, we draw a radius $ON$ from $O$ such that angle $NO\sigma_1$ is equal to $2\theta$. The Mohr diagram is now complete, and we can read off the value of $\sigma$ along the $\sigma$-axis, and the value of $\tau$ along the $\tau$-axis, as shown in figure 5.3. From figure 5.3, we see that

$$\sigma = \frac{\sigma_1 + \sigma_3}{2} + \frac{\sigma_1 - \sigma_3}{2} \cdot \cos 2\theta$$

$$\text{and } \tau = \frac{\sigma_1 - \sigma_3}{2} \cdot \sin 2\theta \tag{5.1}$$

Thus, for each different attitude of the plane in figure 5.2 as defined by $\theta$, there is a corresponding point on the circle, the coordinates of which represent the normal and shear stresses on that plane. For example, when $\theta = 0$, $N$ coincides with $\sigma_1$, giving $\sigma_n = \sigma_1; \tau = 0$.

In other words, for any values of $\sigma_1$ and $\sigma_3$, we can determine $\tau$ graphically in any plane perpendicular to the $\sigma_1 \sigma_3$ plane and oriented at any angle $\theta$ with $\sigma_3$. Or,

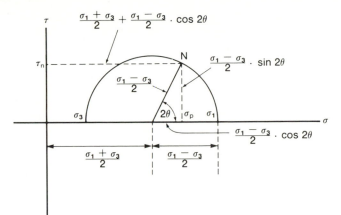

to find $\sigma_n$ and $\tau_n$, for any planar element that makes an angle $\theta$ with the direction of least normal stress, $\sigma_3$, we can draw a line on the Mohr diagram (figure 5.3) through the point on the circle corresponding to the least normal stress, $\sigma_3$, at an angle $\theta$ to the $\sigma$-axis. The other point at which this line intersects the circle corresponds to $\sigma_n$, $\tau_n$, which are the normal and shear stresses, respectively, on the element concerned.

The Mohr diagram also neatly shows the attitude of planes along which shear stress is greatest for a given state of stress; the point on the circle for which $\tau$ is a maximum ($\tau_{max}$) corresponds to a value of $\theta = 45°$. For the same point, $\tau$ is equal to the radius of the circle; thus $\tau_{max} = \frac{1}{2}(\sigma_1 - \sigma_3)$. By a common convention, shear stress in an anticlockwise sense ($\leftrightharpoons$) is positive; in the clockwise sense ($\rightleftharpoons$) it is negative.

If either principal stress is tension, it must be taken with a negative sign so that, in general, the center 0 of the Mohr circle may lie to either side of the origin in figure 5.3, but always on the $\sigma$-axis.

Note that it is the maximum *stress difference,* $\sigma_1 - \sigma_3$ (the diameter of the corresponding Mohr circle), which determines the value of $\tau$. Changes in the total normal stress, $\sigma_{1+}\sigma_3$, simply move the Mohr circle along the $\sigma$-axis. This reflects changes in the isotropic component of stress. In rocks, it corresponds to changes in confining pressure. The stress difference $\sigma_1 - \sigma_3$ is clearly unaffected by changes in confining pressure. We shall appreciate this relationship when we deal with pore pressures. The most important applications of the Mohr diagram are in fracture analysis, which we will discuss in chapter 13.

## Strain and Deformation

**Strain** is a measure of deformation in solids. A body has undergone strain when its constituent particles have been displaced relative to one another. This may result in change of shape or volume of the body, or both. Change of shape is called **distortion,** and change of volume, **dilatation** or **dilation.** Dilation is positive or negative, depending on whether the volume increases or decreases.

A strain is **homogeneous** when any two portions of the body that were similar in form and similarly oriented before the strain are still similar in form and similarly oriented after the strain. A good example of homogeneous strain is uniform extension of a rod. Any homogeneous strain may be the result of simple elongations or contractions of the body in three directions perpendicular to one another. A sphere will change into an ellipsoid representing the strain, the **strain ellipsoid.** The three lines in the body that pass through any designated point and are parallel to the three directions of simple elongation are the **principal axes** of the strain at that point. They coincide with the semiaxes ($X$, $Y$, and $Z$) of the strain ellipsoid, and they remain perpendicular to one another throughout the deformation.

The long semiaxis ($X$) of the strain ellipsoid normally represents maximum extension along that direction; the short ($Z$) semiaxis normally represents minimum elongation or maximum shortening. In normal or **dilatational strain,** particles in the strained body tend to either separate or crowd along straight lines; thus, dilatational strain can be either positive or

**Figure 5.4**
Some fundamental strains. (a) Simple extension. (b) Axial strain.
(c) Simple shear. (d) Pure shear.

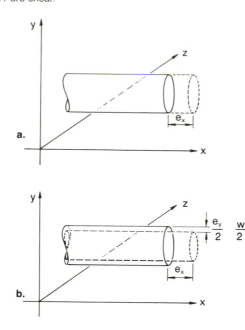

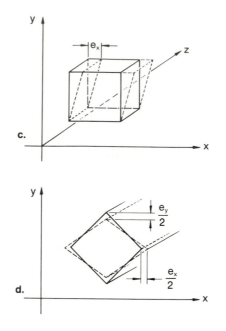

negative. In **shear strain,** particles move past one another. We shall use the conventional symbol $e$ for dilatational strain (extension); that for shear strain is $\gamma$ (following usage of Ramsay 1967). We shall label the largest, intermediate, and shortest principal strains $e_1$, $e_2$, and $e_3$ respectively.

Strain is a change in dimension divided by the original dimension, and as such is a pure ratio. For linear strain $e$ this is expressed $e = \Delta l / l$, where $\Delta l$ is the change in length, and $l$ the original length. For instance, in a stretched wire, strain may be 0.0001, or 0.01%. For volume (bulk) strain, the relationship is $\Delta V / V$, where $\Delta V$ is the change in volume, and $V$ is the original volume. Since strain is a pure ratio, it is dimensionless. Shear strain is measured by the **angle of shear** as illustrated in figure 5.4c. Note that an angle is also a ratio (the ratio between any arc subtended by the angle and the radius of the arc) and, hence, dimensionless.

The principal strain axes may or may not change their directions relative to an outside reference body. If their directions remain constant, the strain is called **pure, nonrotational,** or **irrotational.** If they rotate during deformation, the strain is **rotational.**

## Infinitesimal Strain

The mathematical theory of elastic strain assumes that all strains are so small (infinitesimal) that their squares and products may be neglected. Infinitesimal strain can be analyzed in much the same manner as stress. Extension (positive or negative) is analogous to normal stress, shear strain to tangential or shear stress. A Mohr diagram can represent the relationship between extensional strain, $e$, and shear strain, $\gamma$, on a plane in the same way that it represents the relationship between stresses $\sigma$ and $\tau$.

## Finite Homogeneous Strain

Most tectonic strains are finite. Mathematical treatment of finite strains is, unfortunately, very complicated. But for practical purposes it is often possible to assume that a given limited strain is homogeneous. In this chapter we consider only finite strains that are homogeneous.

Let us assume that, as a result of strain, a point $P_{(x, y, z)}$ has been displaced to position $P_{(x',y',z')}$, and let the displacement or strain components along the $x$, $y$, and $z$ axes be $(x' - x) = e_x$; $(y' - y) = e_y$; $(z' - z) = e_z$. In the most general case:

$$e_x > e_y > e_z$$

and a sphere deforms into a triaxial ellipsoid.

## Ideal Strains

It is convenient to refer strain to a number of ideal cases (figure 5.4). In **simple extension** (figure 5.4a), all points move parallel to a straight line. If we assume this straight line to be the $x$-axis, then,

$$e_x = cx \text{ and } e_y = e_z = 0$$

**Figure 5.5**
Simple shear and pure shear, as illustrated by strain ellipsoid.
(a) Simple shear, as in thrusting (chapter 15). (b) Pure shear, as in
crustal extension (chapter 14). (Courtesy Bradley Erskine.)

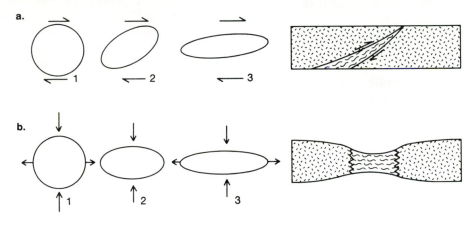

where $c$ is a constant. If $c > 0$, the net strain is an extension; if $c < 0$, the net strain is a contraction. In this type of strain, a sphere becomes an ellipsoid of rotation, prolate in extension and oblate in contraction. Since $e_y = e_z = 0$, simple extension or shortening involves a change in volume.

In **constant volume extension** (figure 5.4b), also called **axial strain,** the cross section must contract. In a cylindrical rod, $e_y = e_z$. In constant volume extension, an extension $e_x$ automatically involves shortening $e_y$ ($= e_z$), and the ratio of $e_x$ to $e_y$ is a constant, $-m$, for each material such that $-m = e_x/e_y$, where $m$ is known as **Poisson's ratio** or the cross-contraction ratio.

**Simple dilatation** amounts to pure volume change, and negative dilatation is contraction. In simple dilatation, a sphere remains a sphere.

In **simple shear** (figures 5.4c, 5.5a, and 5.6b), all points move parallel to a fixed direction, displacements being proportional to their perpendicular distances from a base plane. Let us assume displacement parallel to the $x$-axis, through distances proportional to $y$. Then

$$e_x = cy, \ e_y = 0, \ e_z = 0$$

This strain involves no volume change and no change along $z$. The body deforms by displacement in the $x$-direction along planes parallel to the $xz$ plane. The $xy$ plane is called the deformation plane, because all deformation takes place in it.

**Pure shear** consists of uniform extension in one direction and uniform contraction in a direction perpendicular to it (figures 5.4d, 5.5b, and 5.6a). Let us assume extension in $x$ and contraction in $y$. Then

$$e_x = cx, \quad e_y = -\frac{c}{1+c}y, \quad e_z = 0$$

**Figure 5.6**
Particle paths (a) in pure shear, and (b) in simple shear. (After
Hoeppener, 1964.)

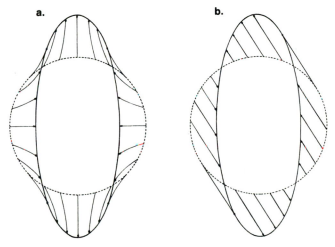

As in simple shear, there is no volume change and no change in z (figure 5.4d). A sphere becomes a triaxial ellipsoid.

Comparing figures 5.4c and 5.4d, we see that simple shear is equivalent to pure shear plus a rotation of the strain axes when considering initial and final shape. However, the path traveled by each particle is quite different in the two cases (figure 5.6).

In **plane strain** there is no change in length along the intermediate strain axis. This is sometimes referred to as *plane* or *biaxial deformation.* Here, too, a sphere becomes a triaxial ellipsoid, but the intermediate axis equals the radius of the original sphere. Plane strain can be conveniently described by the two-dimensional strain ellipse, rather than the three-dimensional strain ellipsoid. Unspecified strain along three mutually perpendicular axes is **triaxial strain.**

## Other Strain Parameters

So far, we have described strain by means of *extensions* or *elongations* (*e*), that is, changes in length, either positive or negative, in specified directions, such that

$$e = \frac{\Delta l}{l}$$

where *l* is the original length in a given direction, and $\Delta l$ is the change in length or difference between final length and original length (extension being positive, shortening negative).

We find it more convenient in some cases to use a quantity called quadratic elongation, $\lambda$, defined by the equation

$$\lambda = (l + e)^2 \qquad (5.2)$$

In other words, quadratic elongations or extensions are a measure of the length of the semiaxes *X, Y, Z* of a strain ellipsoid, where

$$X = \sqrt{\lambda_1}; \quad Y = \sqrt{\lambda_2}; \quad Z = \sqrt{\lambda_3}* \qquad (5.3)$$

We can use quadratic elongations to describe ideal strains (equation 5.4):

> Pure flattening: $\lambda_1 = \lambda_2 > 1 > \lambda_3$
> Plane strain: $\lambda_1 > \lambda_2 = 1 > \lambda_3$
> Pure constriction: $\lambda_1 > 1 > \lambda_2 = \lambda_3$ $\qquad (5.4)$

At the grain scale, strain may proceed by a number of different deformation mechanisms, which will be examined in chapter 6. These mechanisms may operate either simultaneously or in sequence. Thus, it may become necessary to determine the overall or **bulk strain** from partial strains due to separate deformation mechanisms. This is called **strain partitioning.**

## Measuring Strain

To determine strain at any point, or over any homogeneous domain, we must determine the orientation of the principal strain axes, and the ratios of the strains along each of the principal axes. To do so, we must find structures (discussed in more detail in chapters 9 and 10) that record the strain. Where they exist, such pervasive, repetitive structures form part of the internal structure or fabric of the rock; they are *fabric elements*. For instance, rock cleavage (chapter 10) is normally perpendicular to the *Z* direction of strain. That

means that *X* and *Y* lie within the plane of cleavage. Sometimes it is possible to discern a direction of elongation (stretching), usually within cleavage planes (e.g., figure 5.7); this gives us the direction of all three strain axes, for the direction of stretching would be along *X*, and the direction perpendicular to it within cleavage planes would be along *Y*. In order to determine the axial ratios of strain, however, we need additional clues. The best are three-dimensional markers, such as conglomerate pebbles (figure 5.7a), or fossils (figure 5.7b), whose original (prestrain) shape we can reconstruct. Many techniques for strain determination, using these and other fabric elements, are described in Ramsay and Huber (1983) and in other advanced works.

In chapter 10, we shall see how Ernst Cloos, in the first, now-classic paper on the subject (Cloos 1947), determined the strain in the South Mountain anticline in Maryland, by making use of strain markers in its fabric (figure 10.28).

## Natural Strain

So far, we have implicitly used elongations that represent infinitesimally small changes in length. Because the infinitesimal strain expressions neglect second and higher order products of the strains, for strains of the order of 0.1 (10%), the errors will be of the order of 0.01 (1%), which may be acceptable for some geological purposes. But for the larger strains we observe in much geological deformation, it leads to larger errors. It therefore is more accurate to use what is called **natural strain,** or **true strain,** $\epsilon$ (following Ramsay 1967), which integrates infinitesimally small changes over a finite interval and is defined by the equation

$$\epsilon = l_n (1 + e) \qquad (5.5)$$

For strains smaller than about 0.1, $\epsilon = e$, but for strains greater than 0.1, $\epsilon < e$.

## The Deformation Path

In finite strain we compare an initial state with a final state, without considering intervening states. But between the initial state, conventionally represented by a sphere, and the final state, represented by the corresponding (finite) strain ellipsoid, there is an infinite number of intermediate ellipsoids representing infinitesimal strains along the deformation path. These intermediate strains may have varying orientations in simple shear. In pure shear, a given fiber in the material may suffer, in turn, first extension and then contraction, because its orientation with respect to the principal strain axes will vary along the deformation path.

Homogeneous fabrics leave no trace of the deformation path, but in nonhomogeneous fabrics we may find clues to the strain history of the rock (see especially Ramsay 1967 and Ramsay and Huber 1983).

---

*For a full description of three-dimensional strain, we not only need elongation measurements along the three principal strain axes, we also need angular measurements to describe rotations that may have taken place during the strain. This can become quite complicated mathematically, and we will not deal with it here. For a full mathematical treatment of strain, see Ramsay (1967), Means (1976), or Ramsay and Huber (1983).

**Figure 5.7**
Three-dimensional strain indicators. (a) Pebble elongation parallel with minor fold axis. Shabu sedimentary series, Patricia district, Ontario. Schematic relations. $B$ = axial line of folds (see chapter 7); $S$ = planar elements (schistosity, cleavage etc.); $L$ = linear elements (elongated grains, stretched conglomerate boulders); $SS$ = stratification; $M_A$ = major dimensional axes of conglomerate boulders; $M_I$ = minor dimensional axes of conglomerate boulders; $F_R$ = shear fractures in conglomerate boulders. $X, Y, Z$ = principal strain direction; $a,b,c$, symmetry axes of fold, as defined in chapter 7 (p. 128). (b) Fossil deformation as strain indicator. Cambrian (Upper Tremadoc) slate with distorted *Angelina sedgwicki* (Slater); from Garth, Caernarvonshire, Wales. Note lineation in matrix, parallel to direction of elongation of the trilobite. (a. From Fairbairn, 1936, © 1936 by the University of Chicago; b. U.K. Crown copyright, British Geological Survey photograph.)

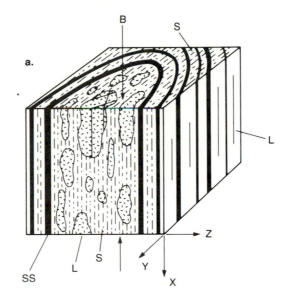

## Rheological Relationships

We have considered both stress and strain, and we are aware that strain is a function of stress. In nonrigid bodies, stress does not occur without strain. We must now consider the nature of the relationship between stress, strain, and time. The branch of materials science that examines this relationship is **rheology.** Rates of deformation are important in this field.

### Extension and Dilatation

The simplest rheological relationship is that of a normal stress, $\sigma$, producing simple extension, $e$ (positive or negative), where for all practical purposes $e$ is proportional to $\sigma$. This may be expressed algebraically as

$$\sigma_x = Ee_x \tag{5.6}$$

where $E$ is a constant of proportionality called **Young's modulus.** $E$ has the same dimensions as stress. Equation 5.6 is known as Hooke's law. Such deformation—proportional, instantaneous, and reversible—is **elastic deformation.** Reversibility implies that the work expended in extension is available as potential energy for recovery to the original dimension. The constant of proportionality in dilatation is called the **bulk modulus, $K$.** Thus, in dilatation,

$$\sigma = K \frac{\Delta v}{v} \tag{5.7}$$

### Distortion

Shear stress imposes a change in shape, a distortion. This is expressed by the relationship

$$\tau = f(\gamma) \tag{5.8}$$

where $\tau$ is shear stress, and $\gamma$ is shear strain. In rheology, one studies the different forms that this equation may take. The simplest case is elastic deformation, as in equations 5.6 and 5.7. Thus, for shear stress,

$$\tau = G\gamma \tag{5.9}$$

where $G$ is a constant proportionality, the **shear modulus,** sometimes called the modulus of rigidity. Like the modulus of elasticity, $E$, it has the dimensions of stress. In an ideal elastic body, also called a **Hooke or Hookean body,** strain is proportional to stress, reversible without loss of energy, and instantaneous. The analog of the Hooke body is an elastic spring (figure 5.8a).

In an ideal elastic body, all the energy expended in mechanical work is stored as potential energy and is called the **strain energy.** Originally, only elastic deformation was termed "strain," so this term has had a connotation of stored energy.

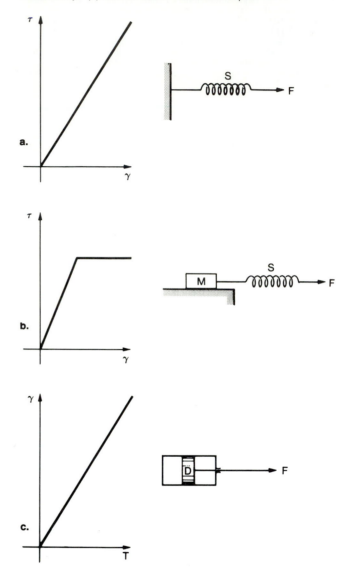

**Figure 5.8**
Some ideal stress-strain relationships. $\tau$ = shear stress; $\gamma$ = shear strain; $T$ = time elapsed; $F$ = applied force; $S$ = spring; $M$ = mass; $D$ = dashpot. The diagrams on the right show the respective models. (a) Elastic strain, according to Hooke's law: "Hooke body." (b) Ductile strain preceded by elastic strain up to a threshold value of the stress: the ductile part of the deformation is that of a "St. Venant body." (c) Viscous strain: "Newtonian liquid."

### Ideal Ductile Deformation

We must now become acquainted with the properties of materials that do not follow the simple Hooke relationship of equation 5.9 between stress and strain in shear. Consider a prism of modeling clay. Small, increasing shear stress will deform the prism elastically, according to equation 5.9. At a certain critical stress, proportionality will cease, and deformation will proceed indefinitely under constant stress (figure 5.8b).

**Figure 5.9**
Two derived models. $D$ = dashpot; $S$ = spring; $F$ = applied force.
(a) Elastico-viscous behavior; spring and dashpot in series. (b) Firmo-viscous behavior.

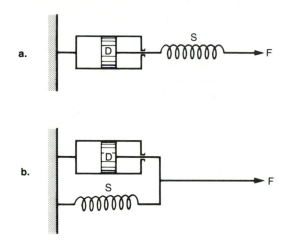

This critical stress is the **yield stress (Y)** of the material under given conditions, and the deformation that follows the stress-strain relationship beyond the yield stress is **ductile** deformation in the strict sense—irreversible distortion at constant stress. The analog of a ductile body in this sense is a mass moving over a flat surface. It needs a threshold force to begin moving against friction, and a constant force to keep moving.

### Viscous Deformation

In ideal viscous deformation, the amount of deformation is a function of time. Thus, shear stress is proportional to rate of shear:

$$\tau = \eta\dot{\gamma} \qquad (5.10)$$

where $\gamma$ is the rate of shear, and $\eta$ is a constant of proportionality called the **viscosity** of the liquid.

To obtain the dimensional expression for viscosity, we must remember that shear, a strain, is a ratio and hence, dimensionless. So rate of shear has the dimensions of $1/T$ or $[T^{-1}]$. We know that $\tau$ has the dimensions $[ML^{-1}T^{-2}]$, and therefore $\eta$ has the dimensions $[ML^{-1}T^{-1}]$, that is, the dimensions of stress multiplied by time. The unit of viscosity is the poise (after Jean-Louis Poiseuille, a nineteenth century French physician), which is 1 dyne/cm² × 1 second (sec). An ideally viscous body has been called a **Newtonian liquid**. Viscous deformation is irreversible, and a Newtonian liquid has no shear strength; it will deform indefinitely under any shear stress in such a way that strain is proportional to time elapsed (figure 5.8c). The analog of viscous deformation is a dashpot—a porous piston moving inside a hollow cylinder filled with a fluid. The resistance of the fluid to the piston moving through it represents viscous resistance to flow.

### Derived Rheological Models

So far, we have described ideal models. We may combine them in such a way that they more closely resemble real conditions, and we may visualize them by combining the standard analogs of figure 5.8.

Two combinations of ideal bodies are of particular interest in rock deformation: a viscous and an elastic body in series, and the same bodies in parallel. The spring-and-dashpot analogs of these two combinations are shown in figure 5.9. In figure 5.9a, spring $S$ will deform instantaneously as soon as force $F$ is applied and will extend to the length corresponding to $F$. Force $F$ is then transmitted to $D$, which will move through the fluid at a constant rate for as long as $F$ is applied. When $F$ is removed, $S$ reverts to its original length, but $D$ remains where it stopped. If the spring is extended and then held at a given extension, it stores a stress; this slowly relaxes as $D$ moves through the fluid, until the spring has returned to its unstressed length. The series arrangement simulates what is known as **elastico-viscous** deformation. In figure 5.9b, when $F$ is applied, both $S$ and $D$ move simultaneously; $D$ retards extension of $S$ and cannot move beyond an extension of $S$ corresponding to $F$. The ideal deformation corresponding to the parallel arrangement has been called **firmo-viscous** deformation.

### Elastico-viscous Deformation

Elastico-viscous deformation is elastic at the instant of stress application. It then continues as a function of time, exactly like a viscous liquid. When deformation stops, stress neither vanishes nor persists: it relaxes logarithmically, as shown in figure 5.10. The mathematical derivation of this relationship is beyond the scope of this treatment, but it can be understood empirically by studying the figure 5.9a analog. When total extension is held constant at a finite value, the viscosity of the liquid retards elastic recovery of piston displacement. The rate of recovery is rapid at first but decays as the tension in the spring decays. The time taken for the stress to decay to $1/e$ of its original value is known as the **Maxwell relaxation time**, where $e$ is the base of natural logarithms. It can be obtained by dividing viscosity by the shear modulus (rigidity), or $\eta/G$.

### Firmo-viscous Deformation

In firmo-viscous behavior, some strain energy dissipates during deformation. In the model of figure 5.9b, the dashpot impedes displacement of the spring. It is an impedance of this kind in firmo-viscous deformation that produces phenomena such as the damping of elastic oscillations. A water-soaked sponge is a good example. A load on the soaked sponge is distributed between the water and the sponge. During deformation the partial load on the liquid, which yields by flow, is transferred to the solid, which eventually will support the entire load elastically. Some unconsolidated geological materials approximate this behavior. When this happens, the partial stresses on the solid and on the liquid are additive, but the deformation is the same for both phases. In contrast, in elastico-viscous deformation, both components take up the same stress, but the deformations are additive.

### Some Definitions

Use of the terms *ductility, plasticity, brittleness,* and *creep* does not seem to be consistent. Since these terms come from materials science and engineering, we will, for a start, give definitions most often used in those sciences. Accordingly, ductile behavior or **ductility** is the ability of a material to undergo large strains without fracturing; it is the opposite of brittleness. **Brittle** behavior is characterized by failure in fracture following elastic strain. (We shall see in the next chapter that these simple definitions must be modified.) **Creep** is time-dependent flow of solid material under constant stress, below the short-term yield point. Creep can be by viscous or by plastic flow. **Plasticity** refers to the property of crystals to deform permanently by slip along lattice planes (see chapter 6). Whereas *viscous* flow is

strictly time dependent in a straight-line relationship and is irrecoverable, *plastic* flow is not necessarily proportional to time and may be partially recoverable.

When plotting strain resulting from creep against time, it is possible to divide the resulting curve into three regions representing three distinct types of creep: primary, secondary, and tertiary (figure 5.11). In *primary,* or transient creep, the slope of the curve (i.e., the strain rate) decreases with time. In *secondary,* or steady-state creep, the strain rate remains constant for a long time. In *tertiary* creep (not always observed), the strain rate increases rapidly, probably because of the loss of cohesion along microcracks; this eventually leads to fracture (see chapter 13). Figure 5.12 shows a spectrum of behaviors from brittle to ductile.

### Rheids

Carey (1954) gave the name **rheid** to "a substance whose temperature is below the melting point and whose deformation by viscous flow during the time of [observation] is at least three orders of magnitude greater than the elastic deformation under the given conditions." Thus, a material is a rheid by virtue of the time of observation. Rocks are not rheids in quarry operations, but they are during geological deformation. Carey also defined a threshold time for fluid behavior in deforming rheids: when the viscous term in a deformation becomes 1,000 times greater than the elastic term, the latter becomes negligible. The time necessary for this to happen is the **rheidity** of the material under specified conditions. The rheidity time threshold is analogous to the Maxwell relaxation time (figure 5.10) and is given by the ratio of viscosity ($\eta$) over shear modulus ($G$) multiplied by 1,000:

$$\text{Rheidity} = \frac{\eta}{G} \cdot 10^3 \text{ sec} \qquad \textbf{(5.11)}$$

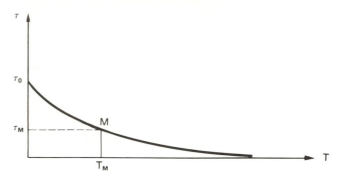

**Figure 5.11**
Generalized strain-time curve, showing primary (I), secondary (II), and tertiary (III) creep.

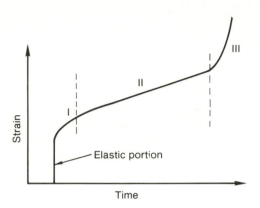

**Figure 5.12**
Transition from brittle to ductile behavior, expressed in stress–strain curves. A, Very brittle, strain essentially elastic before sudden rupture; B, brittle, small permanent strain before rupture; C, transitional, peak signals faulting without total loss of cohesion; D, moderately ductile, faulting accompanied by distributed deformation; E, ductile, elastic strain below well-defined yield stress, permanent uniform flow beyond; F, ductile, uniform flow with poorly defined yield stress and work hardening. (From Handin, 1966.)

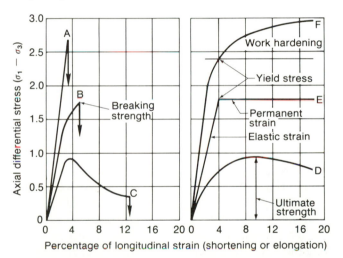

The factor 1,000 is arbitrary and is chosen for convenience. The rheidity of rock salt, for instance, is 10 years.

We must bear in mind that the "viscosity" involved in solid flow is not the same as that in a true liquid. True liquid viscosity is a physical constant for a given temperature and pressure, and it has no threshold. The *equivalent* or *apparent* viscosity of a flowing solid is a property that has the same dimensions as viscosity, but is not a physical constant in the same sense: it varies with the deforming stress. And, of course, it does not govern the whole range of stress-strain relations of the material. Table 5.1 gives some representative viscosities.

**Table 5.1.** Some Representative Viscosities and Equivalent Viscosities (in poises)

| Honey | 40 | Ice | $10^{13*}$ |
|---|---|---|---|
| Glycerine | 830 | Rock salt | $10^{18*}$ |
| Lava | $10^2$–$10^5$ | Sandstone slab | $10^{19*}$ |
| Asphalt | $10^6$ | Earth's crust | $10^{22*}$ |
| Pitch | $10^{10}$ | | |

* Averages and estimates.

**Figure 5.13**

Idealized stress–strain curves showing limiting stresses. A is a typical curve for a rock deformed under low confining pressure. B is a curve for an ideally elastico-plastic material. C is a typical curve for a rock deformed under high confining pressure, with a poorly defined yield stress and strain hardening. (From Handin, 1957.)

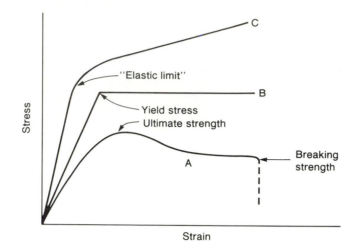

## Strength

One important variable in materials science is the limiting stress at which a material "fails." What is failure? In mechanics, failure is the inability to support a load (i.e., stress), by yielding or by rupture. Rupture is loss of cohesion. To reword our earlier definition, if a material ruptures below the yield point, it is said to be **brittle.** If a material ruptures above the yield point, it is **ductile.**

Strength is simply the limiting stress just before a specified failure (figure 5.13). Thus, we have **yield strength** and rupture or **breaking strength. Ultimate strength** is the greatest stress a material can support under any given condition. It is normally identical to rupture strength. **Fundamental strength** represents a (hypothetical) threshold below which there can be no inelastic deformation, however long the time over which the stress acts.

Solid materials fail under different stresses in compression, tension, and shear. Thus, we have **compressive** (crushing) **strength,** $S_K$, **shear strength,** $S_G$, and **tensile strength,** $S_T$. Their relative values are normally in the following order:

$$S_K > S_G > S_T \qquad (5.12)$$

## Conclusion

The fundamental quantities in the dynamics of rock deformation are stress, strain, and the rate of strain. Stress induces strain at a rate that depends on rock properties and on environmental factors, such as temperature, pressure, and chemical environment. In the next chapter we will see how all of these variables influence the flow of rocks.

## Review Questions

1. What are the fundamental dimensions of mechanics? Give step-by-step derivations of dimensional expressions for strength, energy, and viscosity.
2. Distinguish between the following terms: isotropic, anisotropic, homogeneous, and heterogeneous.
3. How are normal stress and shear stress on a plane related to the general (three-dimensional) state of stress?
4. What are the "principal axes of stress?"
5. Derive dimensional expressions for strain and for strain rate.

6. A test sample has stresses $\sigma_3 = 20$ MPa and $\sigma_1 = 45$ MPa applied to it. Construct a Mohr circle for this situation. Assume that a pore pressure of 15 MPa is applied. Construct a second Mohr circle on the same $\sigma$-axis for this new situation. Does it affect the shear stress on any plane? What would happen if the pore pressure were increased to more than 20 MPa?

7. In the situation in question 6, how must $\sigma_3$ be changed in order to increase the shear stress on any arbitrarily oriented plane?

8. Construct a Mohr circle for the stress state given by $\sigma_1 = 40$ MPa, $\sigma_3 = 10$ MPa, and determine the shear stress and normal stress on a plane oriented at 35° to $\sigma_3$ and parallel to $\sigma_2$.

9. Relate quadratic elongations to the axes of the strain ellipsoid and to simple elongations.

10. Explain the difference between infinitesimal strain and finite homogeneous strain.

11. Explain the difference between pure shear and simple shear, both in terms of external form and of internal displacement of particles.

12. By means of labeled, annotated stress-strain diagrams, explain (a) Hooke's law and (b) ductile strain.

13. By means of a labeled, annotated strain-time diagram, explain viscous deformation.

14. Combine initial elastic and subsequent viscous deformation in one strain-time diagram.

15. Explain Maxwell relaxation time and the concept of rheidity.

16. What is "strength" in mechanics? Explain the differences between several kinds of strength.

17. Differentiate between primary, steady-state, and tertiary creep.

18. Differentiate between strain coordinates $X, Y, Z$ and $e_1, e_2, e_3$ by comparing the manner in which they are defined.

## Additional Reading

De Paor, D. G. 1986. A graphical approach to quantitative structural geology. *Jour. Geol. Ed.* 34: 231–236.

Durney, D. W., and Ramsay, J. G. 1973. Incremental strains measured by syntectonic growths. in *Gravity and Tectonics*, K. DeJong and R. Scholten, eds. Wiley, New York: 67–96.

Feather, N. 1961. *Mass, Length and Time*. London and New York: Penguin. 358 pp.

Groshong, R. H., Jr. 1972. Strain calculated from twinning in calcite. *Geol. Soc. Am. Bull.* 83: 2025–2038.

Haimson, B. C., 1975. The state of stress in the earth's crust. *Rev. Geophys. Space Phys.* 13: 350–352.

Jaeger, J. C., and Cook, N. G. W. 1979. *Fundamentals of Rock Mechanics*. 3d ed. London: Methuen. 593 pp.

Means, W. D. 1983. Application of the Mohr circle construction to problems of inhomogeneous deformation. *J. Struct. Geol.* 5: 279–286.

Ramberg, H. 1975. Particle paths, displacement and progressive strain applicable to rocks. *Tectonophysics* 28: 1–37.

Ramsay, J. G. 1976. Displacement and strain. *R. Soc. London, Philos. Trans.* A283: 1–25.

Ramsay, J. G., and Huber, M. I. 1983. *The Techniques of Modern Structural Geology* (Vol. 1): *Strain Analysis*. London: Academic Press. 307 pp.

Reiner, M. 1960. *Deformation, Strain and Flow*. 2d ed. New York: Wiley-Interscience. 347 pp.

Swolfs, H. S. 1984. The triangular stress diagram; a graphical representation of crustal stress measurements. *U.S. Geol. Surv. Prof. Pap.* 1291.

Treagus, S. H. 1981. A simple-shear construction from Thompson and Tait. *J. Struct. Geol.* 3: 291–293.

Verhoogen, J., Turner, F. J., Weiss, L. E., Wahrhaftig, C., and Fyfe, W. S. 1970. *The Earth, an Introduction to Physical Geology*. New York: Holt, Rinehart and Winston. 455–520.

# 6

# Flow of Rocks

The ancient Greeks had a phrase for it: *Panta rhei,* "everything flows." This is certainly true of rocks. Even though we cannot actually watch solid rocks flow, we know it happens because we see rock structures that must result from flow (e.g., figures 8.3 and 8.17). How and under what conditions does it happen?

Experimentation may help us to understand the process. There are two experimental approaches. We may try to reproduce tectonic forms in scaled-down models, and assume that the processes that produced the forms in the model would closely resemble those occurring in nature. Or we may try to reproduce the actual deformation mechanism, and deform natural rocks and minerals. In this chapter we will examine each of these approaches in turn, and then go on to some natural tectonic models such as glaciers.

## Historical Introduction

Sir James Hall (1815) was the first to reproduce a reasonable facsimile of folded strata showing that lateral compression is one process that might fold rocks. To do this he used a box in which he laterally compressed a pile of cloths under a load. Later, he used stiff clay for the same purpose. The implication was that rocks do not behave like brittle materials while they are being folded.

The fabric of metamorphic rocks suggested to G. P. Scrope (1825) that these rocks had been deformed by flow in the solid state. A few years later, some observers noted fossil deformation in cleavage planes. In 1878 Albert Heim recognized the implication that most continuously deformed (folded) rocks must have been perfectly consolidated at the time of deformation. He pointed out that, under long, continued stresses, "brittle" rocks could flow. Heim thus introduced time as a controlling factor in rock deformation, which he compared to glacier flow, perceiving the main difference between the two to be rate of deformation. On a planet with greater gravitational attraction, rocks would flow like glaciers, and no mountains would subsist. This was an important advance in the understanding of rock deformation, and it should have pointed the way to better-informed modeling. Heim's insight, however, went unheeded for many years.

Daubrée (1879) wrote an important treatise on geological experimentation. He was aware of the importance of selecting appropriately weaker model substances to represent rock material, but we now know that his estimate of appropriate weakness was unrealistic. Daubrée used zinc, sheet iron, and laminated lead for his folding experiments (and, by implication, criticized Hall for using material that was too weak). But he also used wax of different degrees of ductility, particularly for fracturing experiments, and he was able to reproduce fracture systems in blocks of wax under compression (figure 6.1).

Nevertheless, Daubrée was aware of the ability of solids to flow, and he devised an experiment in which he demonstrated the formation of rock cleavage by extrusion flow of clay. Bailey Willis (1893) also recognized that geological model substances must be weaker than the rocks they represent, but his model materials, though more realistic than Daubrée's, were still too strong. Neither he nor any of his predecessors made any quantitative estimates of the necessary reduction factors.

### Scale Models

Otto Ampferer, in his classic monograph on deformation in orogenic belts (Ampferer 1906), strongly emphasized the necessity of scaling down all relevant dimensions to make meaningful models—real or conceptual—of geological processes. He pointed out that, on the human time scale, geological materials appear weak. He proposed calling the earth's crust its "skin," to emphasize its weakness in deformation over geological time; he thought that the term "crust" conveyed too much strength.

Koenigsberger and Morath (1913) applied model theory quantitatively to tectonic models. Morath deduced strength reduction factors by applying dimensional analysis to geological materials. He used a mixture of iron powder and oil, since relatively high density is an advantage.

## The Concept of Similitude

We construct scale models of objects whose natural dimensions are too large or too small for effective and convenient study. In mechanical models, the changes in dimensions include not only those of length, but also time, and physical properties involving mass and mass distribution. To scale mechanical models, therefore, we make use of each of the fundamental dimensions $[M]$, $[L]$, and $[T]$ (p. 66). A true mechanical scale model must be properly scaled in each of these dimensions and in proportions fixed according to circumstances. These proportions are *model ratios* $[M_m] : [M_n]$, $[L_m] : [L_n]$, $[T_m] : [T_n]$, where the subscript $n$ denotes the dimension of the prototype in nature, and the subscript $m$ denotes the corresponding dimension in the model. In practice, we may encounter difficulties in selecting a model substance with physical properties that will fit any particular chosen set of ratios. For instance, a basic property containing $[M]$ is density. Here our choice ranges only from about 1 to about 10, the range of most natural rocks and minerals.

Hans Cloos (1929) intuitively came to the conclusion that the material representing rock in geological models must be reduced in strength in accordance with the scale of reduction in length, and that it must also have low cohesive strength in order to fracture. Cloos therefore chose one of the most successful materials used in tectonic model experiments: weak, *wet* clay. Figure 14.12 shows some of the resulting structures. Cloos advised that the experiments not be too tidy; nature is not tidy, he said, and nature's "mistakes" may contain a lesson. These experiments, Cloos pointed out, benefit us only if we can watch them take shape or can carry them out ourselves, for that is the only way that we can clearly link movement and structure, bringing rocks to life. Cloos's intuitive understanding of tectonics was unique. It shows itself particularly well in his superb drawings of geological structures on all scales, some of which are reproduced in this book.

In 1934 Nettleton suggested that salt domes must become emplaced essentially by fluid mechanical processes. His reasons were that salt is lighter than the surrounding rocks, and that salt domes of the Gulf Coast have a rounded outline. These circumstances alone suggest yielding and flow. Nettleton supported his hypothesis by building models of salt domes in which the model substances for both the salt and the country rock were fluids: thick asphalt oil and syrup. The resulting shape was very close, indeed, to the actual form of a salt dome, scaled-down (see pp. 86 and 106).

Finally, in 1937 Hubbert gave the first complete theoretical treatment of geological models, using the methods of dimensional analysis.

**Figure 6.1**
Fracture system produced in wax by Daubrée. (From Daubrée, 1879.)

**Table 6.1.** Dimensional Expressions for Some Mechanical Quantities

| 1<br>Quantity | 2<br>Dimension | 3<br>Model Ratio | 4<br>Model Ratio, where $\lambda\tau^{-2} = 1$, and $\mu\lambda^{-3} = \delta$ | 5<br>Model Ratio, where $\delta = 1$ |
|---|---|---|---|---|
| Volume | $[L^3]$ | $\lambda^3$ | — | — |
| Density | $[ML^{-3}]$ | $\mu\lambda^{-3} = \delta$ | — | 1 |
| Acceleration | $[LT^{-2}]$ | $\lambda\tau^{-2}$ | 1 | — |
| Force | $[MLT^{-2}]$ | $\mu\lambda\tau^{-2}$ | $\delta\lambda^3$ | $\lambda^3$ |
| Stress<br>Pressure<br>Strength<br>Rigidity<br>Elastic modulus | $[ML^{-1}T^{-2}]$ | $\mu\lambda^{-1}\tau^{-2}$ | $\delta\lambda$ | $\lambda$ |
| Viscosity | $[ML^{-1}T^{-1}]$ | $\mu\lambda^{-1}\tau^{-1}$ | $\delta\lambda\tau$ | $\lambda\tau$ |
| Strain<br>Angle<br>Ratio | $[L^\circ] = 1$ | 1 | — | — |

NOTE: Expressions are derived from the fundamental dimensions mass [$M$], length [$L$], and time [$T$].

It is convenient (following Hubbert [1937]) to express model ratios for each of the dimensions by Greek letters; thus, for the fundamental dimensions,

$$\left[\frac{M_m}{M_n}\right] = \mu; \quad \left[\frac{L_m}{L_n}\right] = \lambda; \quad \left[\frac{T_m}{T_n}\right] = \tau \qquad (6.1)$$

Table 6.1 gives a list of common derived mechanical dimensions expressed in terms of the fundamental dimensions [$M$], [$L$], and [$T$], and of model ratios in columns 2 and 3.

## A Simple Model: Kinematic Similitude

In order to obtain a working model that will provide the answers to our questions, which dimensions must we change, and in what ratios? That depends on the questions we ask.

Let us consider a simple case. We intend to fill a certain irregularly shaped reservoir with water flowing at the rate of 1,000 cubic meters per minute (m³/min). This is an enormous inflow. How long will this process take, and how high will the water level be at the end of each day?

If we construct a model based on the map contours of the reservoir, we can fill the model at a proportional rate, measure the time taken to fill it, and the height at the end of each day. Let the following be our conditions: the length $l_n$ of the prototype is 10 km; the length $l_m$ of the model is 10 m. We want to fill the model in exactly one-tenth the time it would take to fill the prototype. So the model ratio of length is $l_m:l_n = 1:10^3$,

and the model ratio of volume is $l_m^3:l_n^3 = \lambda^3 = 1:10^9$. Our model will therefore hold $10^{-9}$ times as much water as the original. Evidently, for one-tenth of the elapsed time, the rate of inflow into the model must be reduced. By how much? Flow has the dimensions of volume per unit time, or $[L^3\,T]^{-1}$. Our model ratio of time is $\tau = 10^{-1}$. Hence, the model ratio of flow is $\phi = \lambda^3\tau^{-1} = 10^{-8}$, and the rate of feed into the model must be 1,000 multiplied by $10^{-8}$m³/min, or 10 cm³/min, a mere trickle. Evidently, height measurements made on the model each day must be multiplied by $\lambda$, or 1,000, to give the true rise in water level in the reservoir. Of course, the time taken to fill the model must be multiplied by 10 to give the time to fill the reservoir.

Thus, to obtain the model ratio for flow, we must take into account the dimension of time. In general, for similitude in motion, or **kinematic similitude,** the time taken for displacements in a model must be in a fixed ratio to the times taken for corresponding displacements in nature; a reduced-size kinematic model of a locomotive runs more slowly than the prototype.

## Dynamic Similitude

So far, we have neglected mass distribution in our models. For geometric and kinematic similitude, mass and its derived dimensions are irrelevant. When we take mass distribution into account, in addition to the conditions for geometric and kinematic similitude, we obtain **dynamic similitude.** Mass distribution in the model must be similar in a fixed ratio to mass distribution in

**Figure 6.2**
Structural elements of typical Gulf Coast salt dome. (From Dennis, 1967.)

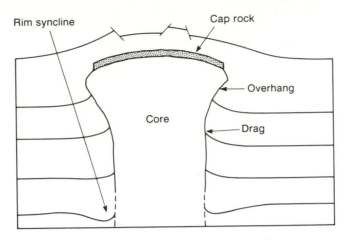

nature; in addition, of course, the conditions for geometric and kinematic similitude must be met. This is the most general case of mechanical similitude: all the derived dimensions of mechanics are in fixed ratios, and these ratios are directly related to the corresponding dimensional expression. A true dynamic model reproduces a geometrically similar sequence of movements; its final form and fabric must be geometrically similar to those of the prototype. This can be achieved only in models with true dynamic similitude.

From equation 6.1, for any dimensional expression of the form $[M]^a[L]^b[T]^c$, the model ratio has the form:

$$\mu^a \cdot \lambda^b \cdot \tau^c$$

Hubbert (1937) pointed out that, in dimensional expressions and model ratios applied to geology, we may substitute density $[D]$ for $[M]/[L]^3$ (or $\delta$ for $\mu/\lambda^3$). Furthermore, most model substances have densities of the same order of magnitude as common rocks and minerals (between 1 and 10). Thus, $\delta = 1$ for many geological examples. Also, the forces due to gravity are identical in most models and in nature; acceleration due to inertial forces (including gravity) is negligible. Thus, $\lambda\tau^{-2} = 1$. Substituting accordingly, we obtain the list of geological model ratios in column 5 of table 6.1. Morath recognized that, since the model ratio for time in geological experiments is exceedingly small, the actual time taken in model experiments is not important, provided the material is allowed to relax adequately throughout the experiment.

## Representative Models

The application of the concept of similitude will become clearer through the study of several representative models. By reducing phenomena of geological dimensions in space and time to a human scale, these models should allow us a more intuitive grasp of the nature of tectonic deformation. Let us begin with salt domes.

### Salt Dome Models

Original deposition of rock salt is in stratiform bodies. In many areas of the world, however, salt is now emplaced as dome-shaped, diapiric bodies in otherwise relatively undisturbed country rock. A salt bed—from which the domes are presumably derived—exists or is assumed to exist beneath all salt domes.

Since the discovery in 1901 of a rich accumulation of oil in traps associated with the famous spindletop dome in Texas, the primary economic incentive for the exploration of salt domes has been in connection with petroleum. Deep drilling, seismic exploration, and gravity surveys have provided unusually accurate details of many salt structures, and the question of the origin of salt domes has assumed considerable practical importance.

Figure 6.2 is a cross section of a typical Gulf Coast salt dome, indicating the standard structural features: core or stem, overhang, drag, rim syncline, and caprock. In most Gulf Coast salt domes, the height of the

**Figure 6.3**

Cross section of Bethel dome, Texas.

(Source: Halbouty, M. T. and G. C. Handin, Jr., "A Geological Appraisal of Present and Future Exploration Techniques on Salt Domes of the Gulf Region of the United States." 5th World Petroleum Congress, Section L, paper 5, p. 13, 1959.)

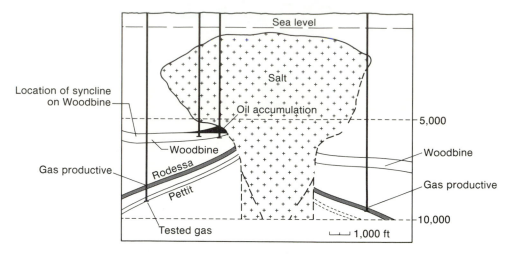

core above the mother salt varies from little about 5 km to 10 km and more. Diameters average 3 km, varying between limits of about 1½ to 10 km. Figure 6.3 shows the Bethel dome in Texas. In the explored section, the rim syncline is developed only on the Woodbine. Overhang may or may not be present, and exploration has revealed the presence of subsidiary overhangs at depth. The peripheral sink in the mother salt around the core induces a circular syncline, the *rim syncline*. Upturning of sedimentary beds around the core strongly suggests upward movement of the salt during emplacement. A system of normal faults commonly dislocates the rocks immediately overlying the salt dome (figure 14.10).

The history of concepts of salt dome formation is well summarized by Nettleton (1955). In the hundred years or so that salt domes have been known, two main formation processes have received serious consideration: flow due to density contrast between salt and country rock, and flow due to tectonic forces. These processes have not been considered to be mutually exclusive. In 1934, in a striking scale-model experiment,

Nettleton showed that viscous flow induced by density contrast could explain all the known features of Gulf Coast salt domes:

[He] constructed several models to illustrate the fluid mechanical concept, with movement depending only on internal forces due to differences in density. These were the first models to produce artificial "salt domes" by purely gravitational forces and without external pressure or constraints. The essential components of these models were two fluids of different density with the lighter fluid initially beneath the heavier one. The most successful models were made with a layer of very heavy crude oil or thin asphalt with a density of about 1.0 as the lighter medium to simulate the salt, and a thick, viscous, boiled-down corn syrup with a density of about 1.4 to simulate the overburden. The syrup was clear and the model was made in a glass box so the form of the "salt" could be observed throughout its development. In the normal or upright position, the fluids were in a stable condition with the lighter fluid on top. The normally top surface of the box was closed with a synthetic rubber diaphragm in contact with the asphalt. To cause a "dome" to grow, the box was simply inverted. A small cork or other protuberance or a mechanical lever arrangement caused

**Table 6.2.** Measurements of Prototype and Model Salt Domes

| | Prototype | Model Ratios | Model |
|---|---|---|---|
| Radius | $10^5$ cm | $\lambda = 5 \times 10^{-5}$ | 5 cm |
| Height | $6 \times 10^5$ cm | $\lambda = 5 \times 10^{-5}$ | 30 cm |
| Thickness of mother salt | To be determined | $\lambda = 5 \times 10^{-5}$ | Assume different values |
| Time of formation | $6 \times 10^7$ yr or $2 \times 10^{15}$ sec | $\tau = 10^{-11}$ | $2 \times 10^4$ sec (5.5 hr) |
| Viscosity | Salt: $10^{18}$ CGS units | $\psi = \delta \lambda \tau = 5 \times 10^{-16}$ | $5 \times 10^2$ CGS units |
| | Sediments: To be determined | $\psi = \delta \lambda \tau$ $\tau$ to be measured experimentally | Viscosity of model to be chosen arbitrarily |
| Shear strength | Sediments: $10^8$ CGS units | $\sigma = \delta \lambda = 5 \times 10^{-5}$ | $5 \times 10^3$ CGS units |
| Density | Salt: 2.2 Sediments: 1.9–2.5 | $\delta = 1$ | Same as prototype |

an initial "uplift" of the diaphragm (now on the bottom of the inverted box) and the overlying asphalt or "salt" layer. Thereafter, the growth and form taken by the dome resulted entirely from the gravitational forces due to the difference in density between the two fluids (Nettleton 1955).

The triggering arrangement is necessary to disturb the metastable equilibrium between the two fluids; it initiates, in effect, a slight contrast in lateral density.

In 1937 Hubbert applied dimensional analysis to a hypothetical model salt dome. His data and computations are given in table 6.2.

The length dimensions of the prototype are from field observations. Time of formation is from geological observation and reasoning. The viscosity of solid rock salt is from Gutenberg (1931). The thickness of the mother salt is indeterminate and must be found by a series of trial runs. Since liquids have no shear strength, Nettleton had not considered it in his experiments. Solids that flow *do* have shear strength, however, and Hubbert included it in his analysis. Computations show that the viscosity of the model under the given conditions must be $5 \times 10^2$ poises, and the intruded sediments must have a shear strength of $5 \times 10^3$ dynes per square centimeter (cm²). In practice it is usually difficult, if not impossible, to provide model materials of both correct viscosity and correct shear strength. The experimenter must adjust the conditions of the experiment to suit available materials and must be satisfied with correct orders of magnitude. Great precision is not obtainable, but is not needed for this type of experiment.

## Other Model Materials

### Clay
Hubbert computed the dimensions and properties of a model earth. For $\lambda = 10^{-7}$, and both $\delta$ and $\lambda \tau^{-2} = 1$, as previously justified, he found that a model earth having a radius of 63.7 cm would have mechanical properties very similar to those of weak clay. We have already seen that Hans Cloos constructed many weak clay models. Clay has the advantage of low cohesiveness, in contrast to asphalt and syrup, and can thus sustain fracturing. Its physical properties can be varied by the addition of water. Ernst Cloos (1955) undertook an analysis of fracture patterns in clay models, some of which are shown in figures 13.26 and 13.27.

### Stitching Wax
In 1956, Bucher showed the advantages of using stitching wax in certain models of rock deformation. On sudden impact, this strikingly elastico-viscous substance rings and fractures just like an elastic solid. But if a block of stitching wax is left alone for about a day, it flows like a viscous fluid. It thus simulates, at a human time scale, the properties of rocks that deform elastically and fracture under sufficiently high instantaneous stresses, but flow under very small stresses if allowed to act over a long time.

Bucher (1956) used the Jura mountains as his prototype and performed a model experiment, using alternating layers of stitching wax and grease to find the equivalent viscosity of the rocks during deformation.

**Table 6.3.** Computations for Bucher's (1956) Model of Jura Folding

$$\frac{\text{Density of model}}{\text{Density of prototype}} = \frac{1}{2.5} = 2 \cdot 5^{-1} = \delta$$

$$\frac{\text{Length of model}}{\text{Length of prototype}} = \frac{30 \text{cm}}{3 \times 10^6 \text{ cm}} = 10^{-5} = \lambda$$

$$\frac{\text{Time (duration) in model}}{\text{Time (duration) in prototype}} = \frac{2.7 \times 10^{-3} \text{ yr}}{10^6 \text{ yr}} \simeq 2 \cdot 5 \times 10^{-9} = \tau$$

From table 6.1, column 4 $\quad \dfrac{\text{Viscosity of model}}{\text{Viscosity of prototype}} = \dfrac{\eta_m}{\eta_p} = \delta \lambda \tau$

Hence $\eta_p = \dfrac{\eta_m}{\delta \lambda \tau} = \dfrac{10^6}{10^{-14}} = 10^{20}$ poises

This is the overall viscosity for Jura deformation according to Bucher's experiment.

---

**Figure 6.4**

(a) Top view and (b) cross section of initially horizontally layered, subsequently centrifuged model. Note areas of compression and of extension. All deformation occurred because of adjustments between layers of contrasting densities. (From Ramberg, 1982.)

a.

b.

His data and computations are given in table 6.3. Thus, $10^{20}$ poises is the overall viscosity for Jura deformation according to Bucher's experiment.

The rock sequence in the Jura mountains includes some marls and shales, and these certainly have a much lower viscosity than the dominant limestones of that mountain range. This is why Bucher added low-viscosity grease layers to his models. One would need to supplement this experiment with laboratory determination of strain rates of the rocks involved, as in Müller et al. (1981), for instance.

## Scaling Gravity

All of the above experiments suffer from a severe restriction: gravity is the same in the experiment as in nature. This limits the choice of materials, and the duration of experiments may become inconveniently long. We may obviate these difficulties, however, by subjecting the experimental setup to centrifugal force.

Ramberg (1982) has constructed centrifuged scale models for geological processes in which deformation occurs because of an uneven distribution of densities. Such processes are not only important in themselves; they are peculiarly difficult to analyze theoretically. They are responsible for structures such as diapirs, domes, isostatic adjustments, and migration of magma. Ramberg has been able to reach accelerations of up to 4,000 *g*, achieving very useful model ratios for acceleration. Figure 6.4 shows one model obtained in a centrifuge, illustrating a possible mechanism of doming. Density contrasts are likely to play an important role in tectonic deformation: on all scales, centrifuged models help to clarify the effects that may be expected.

**Figure 6.5**
Schematic diagram of triaxial compression apparatus and states of stress in homogeneous cylindrical jacketed specimens in compression and in extension tests. The experimenter may vary $p_c$, $p_p$, or $\Delta\sigma$ to obtain the desired relationships. (After Handin, 1966.)

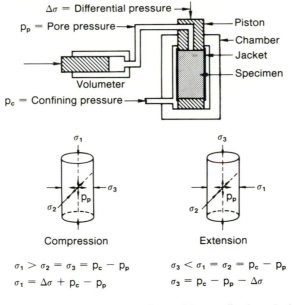

$\Delta\sigma$ = Differential pressure

$p_p$ = Pore pressure

Piston

Chamber

Jacket

Specimen

Volumeter

$p_c$ = Confining pressure

Compression

Extension

$$\sigma_1 > \sigma_2 = \sigma_3 = p_c - p_p$$
$$\sigma_1 = \Delta\sigma + p_c - p_p$$

$$\sigma_3 < \sigma_1 = \sigma_2 = p_c - p_p$$
$$\sigma_3 = p_c - p_p - \Delta\sigma$$

$\sigma_1, \sigma_2, \sigma_3$ = Maximum, intermediate, minimum effective principal stresses

$p_c - p_p$ = Effective confining pressure

## Experiments on Natural Rocks and Minerals

Scale models do not enable us to investigate deforming rocks as *materials*. But we need to know as much as possible about the material behavior of rocks to relate strain rates with stress, and we need to do this under realistic environmental conditions of temperature, pressure, effects of pore fluids, and time.

The first systematic experiments designed to study flow in rocks were undertaken by Adams and Nicholson (1901). To obtain flow rather than fracture, they simulated confining pressure by jacketing a cylindrical specimen of marble in a tightly fitting iron sleeve, and then applied compression along the axis of the cylinder. Deformation of the marble caused the iron sleeve to bulge, and the marble failed both in pervasive flow and by crushing along discrete zones. The experiments were repeated at higher temperatures, for longer durations, and in the presence of pore water under pressure. Each of these factors increased the ductility of the material. This method of introducing confining pressure is rather crude, however, and the quantitative results are open to question.

Von Kármán (1911) introduced the principle on which all current experiments are based: confining pressure is exerted by a fluid under pressure. It has now become standard practice to stress small cylindrical specimens, with diameters one-half or one-third their length, allowing a predetermined stress to be applied along each of the three principal axes, as illustrated in figure 6.5. This is known as a *triaxial test*. For practical reasons, two of the principal stresses must remain equal. In such a test, the specimen is held in a pressure chamber, surrounded by a fluid that transmits uniform confining pressure, $p_c$, to the specimen through an impermeable, flexible sleeve called a jacket usually made of copper (or sometimes rubber). In many experiments, a second fluid is admitted directly to the specimen to provide internal (pore) pressure, $p_p$ (see figure 6.5). The difference $p_c - p_p$ is known as the **effective pressure.** Sometimes the jacket is omitted, in which case the confining pressure is also the pore pressure. An axial force applied by pistons to the ends of the test cylinder results in either maximum or minimum stress along the axis of the cylinder, depending on the relative magnitude of

## Figure 6.6

Change of temperature and pressure in the crust and upper mantle with depth. *ABC* is the temperature–depth curve in the crust and upper mantle; curve *BD* is an adiabatic temperature gradient.

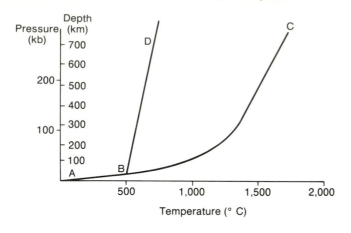

## Figure 6.7

Composite diagram showing the range of strain rates for various geological processes. (From Price, 1975.)

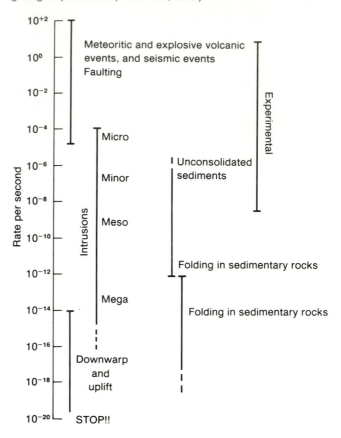

$p_p$, and on the axial force. Both of the remaining principal stresses will be equal to the confining pressure. By varying any or all of the pressures (axial load, confining pressure, or pore pressure), one can obtain different stress configurations.

### Scope of Experiments on Rocks

A triaxial test apparatus enables us to vary *stress* and hence, *strain* as well as *strain rate* in small rock specimens, under the influence of controlled environmental variables, essentially *temperature, confining pressure, pore fluid permeation,* and *time* (duration). Let us look at the role of these variables in tests and at some of the results and their significance.

Variations in temperature and confining pressure are introduced to simulate temperatures and pressures deep in the crust and mantle. The temperature-depth and confining pressure-depth curves of figure 6.6 will help us to gain some perspective on the actual values used in the different experiments. Confining pressure is measured in kilobars, or megapascals (see Appendix C). To gain perspective on temperature, we may note that the temperature range of granitization is estimated to be from about 400 to 550° C, and the melting temperature of basaltic lava varies between 1,100 and 1,250° C.

Price (1975) showed the likely ranges for various geological strain rates (figure 6.7). Strain rate is usually expressed as a fraction (change in dimension) per unit time, in the form $10^n$/sec. Realistic average strain rates in sedimentary rocks, for instance, would seem to

range from $10^{-10}$ to $10^{-15}$/sec. Carter (1976) suggests $10^{-13}$ to $10^{-15}$/sec. Clearly, a rate of $10^{-21}$/sec is negligible, even over geologic time; hence the STOP!! at the end of the scale in figure 6.7.

Determinations of the apparent viscosity of rocks involve timing a given deformation. Griggs (1939) determined the equivalent viscosity of the very fine grained, homogeneous Solnhofen limestone at 1 atmosphere (atm) confining pressure and a stress difference of 1,400 kg/cm². The result was $2 \times 10^{22}$ poises. By changing certain conditions, such as increasing the confining pressure, or keeping the limestone saturated with moisture, this figure can be reduced to as low as $10^{14}$. Table 5.1 gives some common viscosities for comparison.

Many experiments have shown that increasing the confining pressure increases strength (figure 6.8). Increasing the temperature (figure 6.9) decreases yield

**Figure 6.8**
Compression stress–strain curves of Solnhofen limestone at various confining pressures (indicated in atmospheres) at (a) 25° C and (b) 400° C. (From Heard, 1960.)

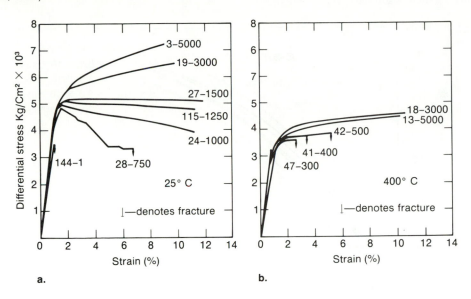

a.

b.

**Figure 6.9**
Compression stress–strain curves of Solnhofen limestone at various temperatures (indicated in ° C) at (a) 1 atmosphere confining pressure and at (b) 400 atmospheres confining pressure. (From Heard, 1960.)

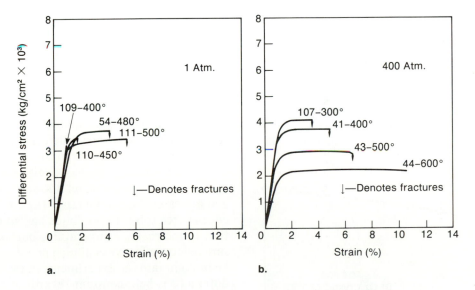

a.

b.

**Figure 6.10**

Stress–strain curves for similarly oriented specimens in weakly foliated Yule marble at various constant strain rates: (a) at 25° C. (b) at 500° C. Note, particularly, in (b) the effect of lowering the strain rate by a factor of $10^7$. (From Heard, 1963.)

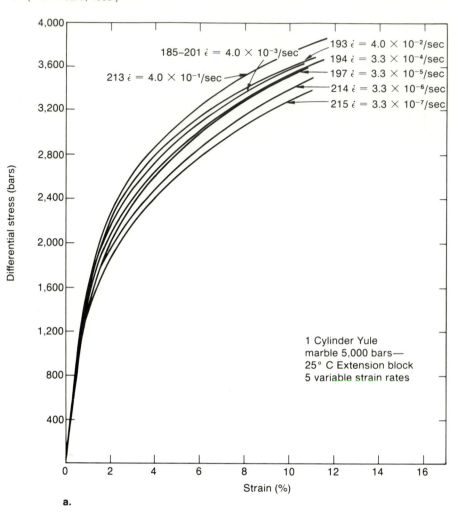

185–201 $\dot{\varepsilon} = 4.0 \times 10^{-3}$/sec
213 $\dot{\varepsilon} = 4.0 \times 10^{-1}$/sec
193 $\dot{\varepsilon} = 4.0 \times 10^{-2}$/sec
194 $\dot{\varepsilon} = 3.3 \times 10^{-4}$/sec
197 $\dot{\varepsilon} = 3.3 \times 10^{-5}$/sec
214 $\dot{\varepsilon} = 3.3 \times 10^{-6}$/sec
215 $\dot{\varepsilon} = 3.3 \times 10^{-7}$/sec

1 Cylinder Yule
marble 5,000 bars—
25° C Extension block
5 variable strain rates

Differential stress (bars)

Strain (%)

a.

**Figure 6.11**

Strain–time curve of hot creep experiment. Synthetic single crystals of quartz. The strain rate became constant ($5 \times 10^{-8}$ per second) toward the end of the experiment. (Griggs and Blacic.)

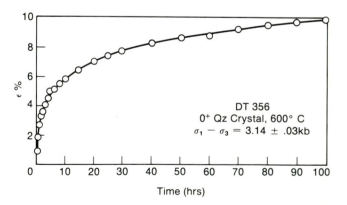

DT 356
0+ Qz Crystal, 600° C
$\sigma_1 - \sigma_3 = 3.14 \pm .03$kb

$\varepsilon$ %

Time (hrs)

strength and increases the ductile range, and also the rate at which the material deforms under a given applied stress, the rate of strain ($\dot{\varepsilon}$). On the other hand, a decrease in strain rate has an effect similar to that of increasing the temperature: it lowers the yield strength and increases the ductile range (figure 6.10). The presence of free solvent pore fluids has an effect similar to that of increasing temperature, but excess pore pressure reduces effective confining pressure. Strain rate can be brought into the experiments in two ways: (1) stress difference is held constant regardless of rate of strain (figure 6.11); and (2) strain rate is held constant while allowing differential stress to vary (figure 6.12).

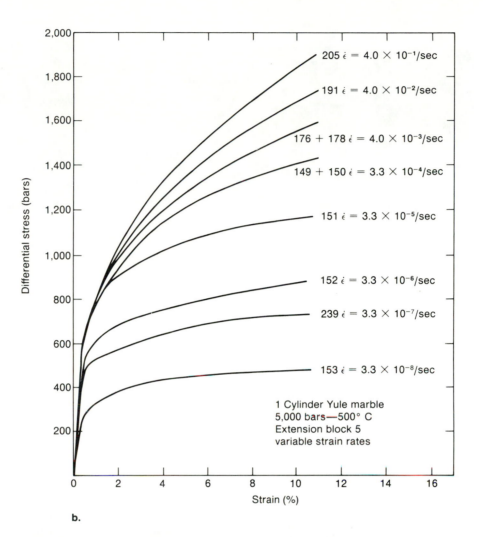

205 $\dot{\epsilon} = 4.0 \times 10^{-1}$/sec

191 $\dot{\epsilon} = 4.0 \times 10^{-2}$/sec

176 + 178 $\dot{\epsilon} = 4.0 \times 10^{-3}$/sec

149 + 150 $\dot{\epsilon} = 3.3 \times 10^{-4}$/sec

151 $\dot{\epsilon} = 3.3 \times 10^{-5}$/sec

152 $\dot{\epsilon} = 3.3 \times 10^{-6}$/sec

239 $\dot{\epsilon} = 3.3 \times 10^{-7}$/sec

153 $\dot{\epsilon} = 3.3 \times 10^{-8}$/sec

1 Cylinder Yule marble
5,000 bars—500° C
Extension block 5
variable strain rates

b.

### Figure 6.12

Schematic relations between constant
strain–rate and creep tests. (a) Stress–strain
curves for constant strain–rate tests.
(b) Constant strain–rate tests plotted on
strain–time coordinates. (c) Creep tests
plotted on stress–strain coordinates.
(d) Creep curves plotted on usual
strain–time coordinates. Curves 1–5 are in
order of decreasing strain rate. Differences
between creep tests and constant
strain–rate tests are brought out by
comparing (a) with (c), and (b) with (d).
(From Heard, 1963.)

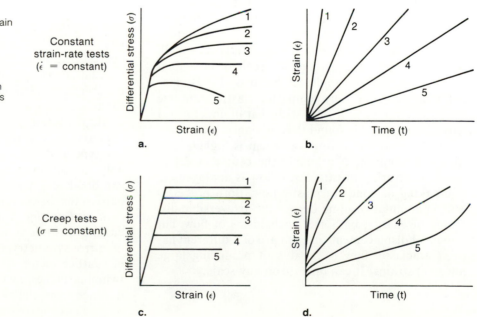

Constant
strain-rate tests
($\dot{\epsilon}$ = constant)

Creep tests
($\sigma$ = constant)

**Figure 6.13**
Extension experiment showing cataclastic flow in inner core of Luning dolomite surrounded by more ductile Italian marble. (Griggs and Handin, 1960.)

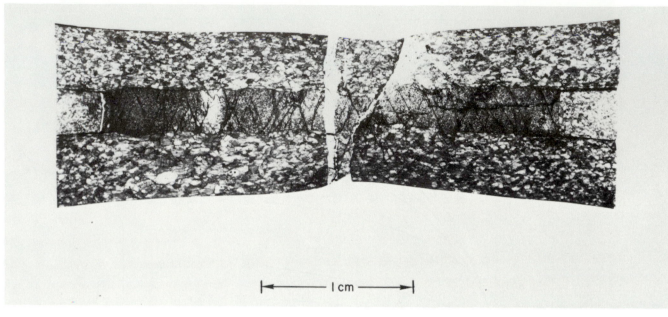

|← —— 1 cm —— →|

## The Mechanism of Flow in Rocks

Flow in solids, or creep, proceeds by several mechanisms, which can be grouped into three broad categories: (1) cataclastic flow, (2) plastic flow, and (3) diffusive mass transfer.

### Cataclastic Flow

In consolidated rock, grains are locked in place by irregularity of shape, by cohesive grain boundaries, and by cement. Mutual displacement of rigid elements, therefore, is usually accompanied by fracture. The overall or bulk deformation in such cases is ductile (e.g., figure 6.13); this process is known as **cataclastic flow.** The individual components in cataclastic flow may be of any size, from fine clastic grains to large fracture-bounded blocks. The essential characteristic is bulk ductile behavior. Fracturing on the scale of individual grains may be difficult to detect. Stearns (1969) managed to do so, however, by comparing clastic grain size distribution in hinges of folds with that in undeformed limbs of one layer. He found that in hinges, grain sizes are smaller, and the sorting coefficient is higher, indicating comminution of grains in the course of deformation. Fracturing in cataclastic flow is analogous to the breaking of bonds in plastic flow. Both processes result in cumulative, time dependent, irreversible dissipation of energy; both are mechanisms of flow, but component displacements in each are of a different nature. Cataclastic flow is capable of producing large geological strains. It can operate on any scale.

Thus, brittleness and ductility are not mutually exclusive, and Rutter (1986) urges that these terms should not be used in a mutually exclusive way. Rather, the *scale* of observation should be indicated, for example, brittle on the grain scale, macroscopically ductile.

### Plastic Flow

Plastic flow takes place by crystal lattice deformation. In rocks, plastic flow proceeds mainly in two modes: *dislocation glide* and *dislocation creep.*

#### Dislocation Glide (Crystal Plasticity)

Gliding along crystal lattice planes can proceed by translation and by twinning (figure 6.14). In *translation gliding,* atoms nearest the glide plane break nearest-neighbor bonds and form new bonds with next-nearest neighbors (figure 6.14b). It is clear that, for a translation glide plane of any extent, a very large number of bonds must be broken simultaneously, and for this a very high shear stress would be necessary. Natural crystals, however, are not perfect; they contain several defects in their lattices. Figure 6.15 shows how one type of defect, an *edge dislocation,* can considerably reduce the energy requirement for gliding: bonds can break singly and in sequence, atom pair by atom pair, as the dislocation travels along the glide plane. In *twin gliding* (figure 6.14c), a segment of the crystal changes attitude across the twin plane, which is a plane of mirror symmetry. Atom displacements are a fraction of a lattice vector. Atom planes that are parallel to the twinning plane move a distance proportional to their distance from the twin boundary. This results in homogeneous shear strain.

## Figure 6.14

Modes of lattice deformation: (a) perfect crystal; (b) translation gliding; (c) twin gliding. (From Edelglass, 1966.)

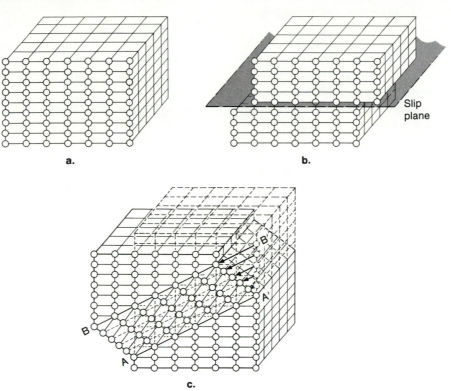

## Figure 6.15

Geometry of lattice deformation by an edge dislocation in a simple cubic crystal. (a) The linear defect at the line of intersection of the extra half plane (plane 4) is an edge dislocation. The atoms above the slip plane to the left of the extra half plane tend to push the dislocated atoms of the half plane to the right, while the atoms to the right push to the left. As a result, relatively little applied shear stress is needed to (b) push the extra half plane to regular lattice sites by forcing the occupying atoms into a new set of dislocated positions. (c) A unit of slip results when the extra half plane emerges at the surface, and the dislocation annihilates itself. (From Edelglass, 1966.)

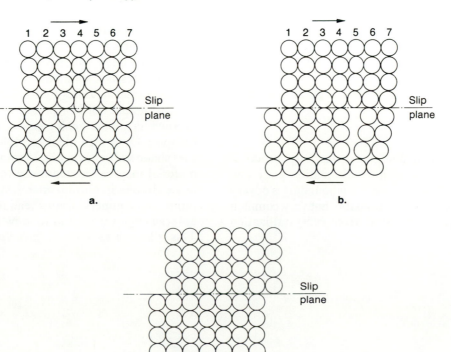

Crystals contain *preferred glide planes*—the planes of closest packing—within their structures. Glide planes also have a preferred glide *direction,* in the closest-packed direction. Thus, the orientation of crystals in an aggregate has an important effect on ductility.

During plastic deformation in crystals, dislocations migrate through individual grains and tend to pile up at grain boundaries. This effect reduces the ductility of the material, a phenomenon known as **strain hardening.** It corresponds to the primary creep of figure 5.12. Strain rate in dislocation glide is a nonlinear function of stress.

### Dislocation Creep
A dislocation may climb out of its original glide plane by a process called *dislocation climb.* This contributes to an effective creep mechanism in higher temperature ranges; in conjunction with dislocation glide and some diffusion, it results in dislocation creep. Dislocation creep is also known as *power law creep,* because in this mode, strain rate varies as a power of differential stress. The relationship between strain rate, $\dot{\epsilon}$, differential stress or flow stress $\sigma$ ($= \sigma_1 - \sigma_3$), and absolute temperature, T, is the **flow law** of a material under specified conditions; it must be determined experimentally in each case. A representative relationship has the form

$$\dot{\epsilon} = A\sigma^n \exp \frac{-Q}{RT} \qquad (6.2)$$

where $\dot{\epsilon}$ is the strain rate, $Q$ is the thermal activation energy, which must be determined experimentally; $R$ is the gas constant; $T$ is the absolute temperature, usually in the range 30 to 1,100 °K; $\sigma$ is the differential flow stress ($\sigma_1 - \sigma_3$); and $A$ is an experimentally determined material constant. The exponent $n$ must also be determined experimentally; it commonly varies between 3 and 5.

Equation 6.2, serves to illustrate an important feature of plastic flow: it is *thermally activated.* That is, where there is no heat, there is no flow. This agrees well with what we know of the structure of matter: in addition to differential stress, heat energy is needed to break bonds.

### Superplasticity
A third mode of plastic flow, *superplasticity,* proceeds at higher temperatures than do dislocation glide and dislocation creep, by slip along grain boundaries of very fine-grained aggregates. It is probably not very common in rocks. The fine grain results from recrystallization during deformation, called *dynamic recrystallization.* It permits large, near-steady-state strains, and in effect is closely related to diffusion creep.

### Diffusive Mass Transfer
In rocks, flow may occur by internal mass transfer. There are two distinct flow mechanisms: diffusion at high temperatures, and pressure solution.

### Flow by Diffusion at High Temperatures
This group of mechanisms involves (1) diffusion of ions (or vacancies) to and from dislocations and grain boundaries, and around grain boundaries (known as *Coble creep,* solid-state flow by grain-boundary diffusion); or (2) volume diffusion through grains (known as *Nabarro-Herring creep*). All of these processes may produce stress-induced changes in grain shape, but they are effective only in very fine grained aggregates at very high temperatures. Somewhat related models with melt or other fluid on solid grain boundaries involve diffusion in the fluid phase, or viscous flow. All of these diffusion mechanisms suggest that the resulting flow should be Newtonian (i.e., $\dot{\epsilon} \propto \sigma$).

High-temperature diffusional flow requires much less energy than do either cataclastic or plastic flow; strain rates can be higher, and flow stress very much lower. This situation is strikingly reflected in the deformation style of rocks that deform by that mechanism: the pattern is extremely fluid (figure 8.24) and can best be simulated by low-viscosity scale models. Furthermore, recrystallization generally results in strong preferred orientation of crystals, reflecting the state of stress at the time of crystallization (see chapter 9). For instance, platy minerals tend to recrystallize perpendicular to $\sigma_1$.

### Pressure Solution
At low temperatures and in the presence of fluids, a geologically important mechanism of mass transfer may operate: **pressure solution.** Pore fluids selectively dissolve material in zones under high pressure and redeposit it in zones of low pressure. This manifests itself in such common geological structures as stylolites (chapter 4), spaced cleavage (chapter 10), and pressure shadows (figure 6.16). In some instances, dissolved material may be carried away for a considerable distance. In one important manifestation of pressure solution, some mineral grains tend to dissolve along faces at large angles to $\sigma_1$, and to grow in the direction of $\sigma_3$. This is known as *Riecke's principle.*

**Figure 6.16**
Photomicrographs of fibrous pressure shadows in Martinsburg Slate composed of white mica and lesser chlorite on a subspherical framboidal pyrite. (a) View in *XZ* plane of the strain ellipsoid. (Width of view = 250$\mu$.) (b) View in *XZ* plane of the strain ellipsoid. (Width of view = 100$\mu$.) (Both views are from different specimens of the same rock.) (From Beutner and Diegel, 1985.)

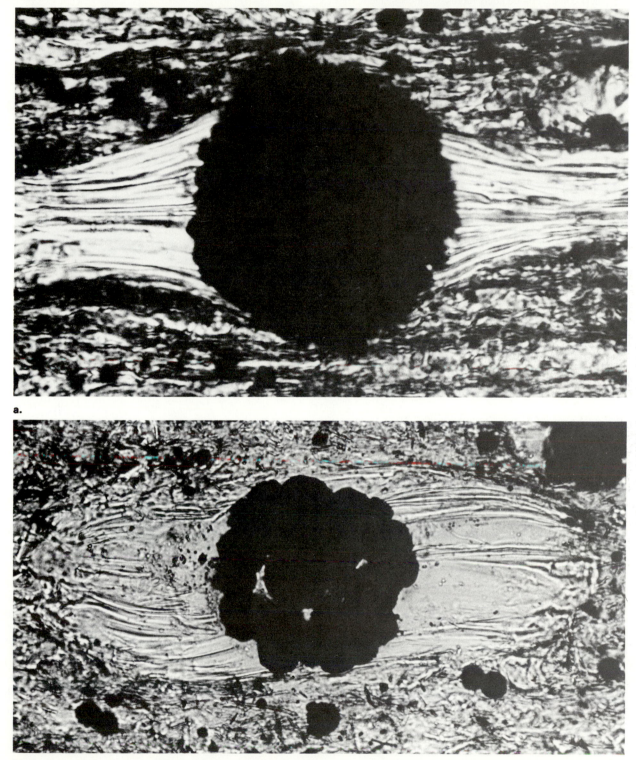

a.

b.

**Figure 6.17**
Deformation mechanism maps for quartz. (a) without a pressure solution field. Contours of −log strain rate are shown. σ is differential stress. (b) modified by the addition of a pressure solution field. The region of dashed strain-rate contours represents the inhibition of pressure solution through decrease in pore water concentration. (From Rutter, 1976.)

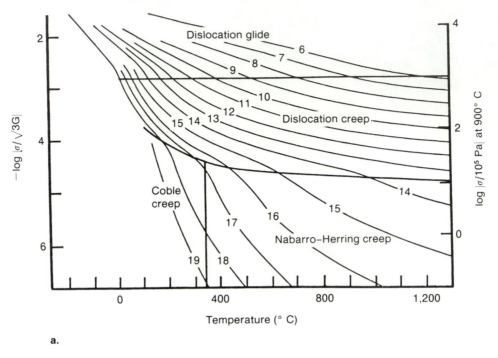

a.

## Representation of Flow Regimes

The different flow or creep regimes can be represented on a diagram that clearly shows over what ranges of flow stress and temperature they apply for a given mineral, at a stated confining pressure. This diagram is called a *deformation map* or *Ashby diagram* (figure 6.17). The coordinates are not simple flow stress and temperature, but *normalized* stress and temperature. A normalized physical quantity is a ratio between a variable and a material constant (with the same physical dimensions) measured in the same units. In this case, flow stress σ is normalized to the rigidity $G$ of the material ($\sigma/G$), and temperature $T$ (absolute, in °K) is normalized to the melting temperature $T_m$ of the material ($T/T_m$); the ratio $T/T_m$ is generally called "homologous temperature." This formulation allows more meaningful comparisons between different materials. For instance, in flow, the relationship of temperature to melting point is much more significant than is absolute temperature.

On the deformation map, we plot lines of constant strain rate, as shown in figure 6.17 for quartz. Part of the diagram can be constructed from experimental results, but some regions are outside the range of laboratory experiments, and so we must extrapolate. This is comparatively easy where a linear (Newtonian) relationship exists between ε and σ, as in diffusional flow. For other regions, mainly in the plastic flow range, we must find the applicable flow laws through experimentation.

The boundaries between the different fields of the Ashby diagram are not as clear-cut as they seem. Two adjacent mechanisms operate simultaneously along the boundary region; the one that generates the faster flow rate is the one that predominates. The diagram shows domains of what we may expect in nature. We must bear in mind geologically realistic strain rates (p. 90). We must also remember that the presence of fluids has an important influence. So has grain size: under similar conditions, fine-grained aggregates deform much faster than coarse-grained ones.

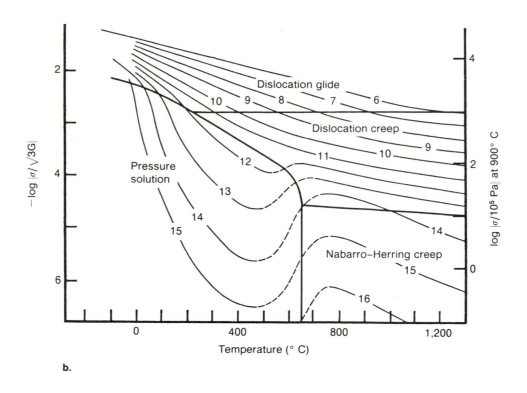

b.

We now see that flow in rocks is not as simple as the ideal mechanisms presented earlier might suggest. Still, large geological strains can generally only be achieved by steady state flow, and that proceeds as viscous flow or, more correctly, as *pseudoviscous flow.*

True viscous flow rate (in Newtonian liquids) is directly proportional to stress. In general, the flow of solids is not Newtonian, and the steady-state flow rate is proportional to stress raised to some power $n$ (commonly $n \sim 3-5$). Only in steady-state flow does a unique relationship exist between stress and strain rate. Thus, the extrapolations of flow laws, discussed above, assume steady-state (secondary) creep behavior (p. 76). Large geological strains could be achieved either by primary (transient) creep followed by extensive secondary (steady-state) creep, or by successive episodes of transient creep, or even by a more complex history of primary, secondary, and even tertiary creep (which commonly ends in fracture).

## Microstructures

The transmission electron microscope (TEM), commonly used for imaging submicroscopic defects such as dislocations in crystals, permits imaging of microstructures at magnifications of up to $\times 500,000$, with a resolution of a few Ångstrom units ($1\text{Å} = 10^{-10}m$). However, this high resolution is not usually necessary for examination of dislocation structures, and more conventional TEM examination of dislocations is done at magnifications of $\times 10,000-\times 50,000$. For examination by TEM (which generally reveals the defect structures by diffraction contrast of lattice distortions associated with the defects), samples are prepared by special processes so that they are sufficiently thin (a few $\mu$m) to transmit the electron beam.

The microstructure of mineral grains may give valuable clues to the deformation mechanism. Microcracking is characteristic of some cataclastic flow. Dislocation creep is revealed by dislocation traces, brought

**Figure 6.18**
Transmission electron micrographs, showing microstructures resulting from plastic deformation. (a) Dislocation lines, loops, and arrays in experimentally deformed olivine crystal. (b) Grain boundary triple junctions with fringes and dislocations in a quartzite experimentally shortened about 50% and recrystallized at 800° C; strain rate = $10^{-7}$ per second. (See Ardell et al., 1973.) (a. From Phakey et al., 1972. Copyright by the American Geophysical Union; Photomicrographs courtesy J. M. Christie.)

a.

b.

out by etching (figure 6.18a). In fabrics resulting from high-temperature diffusion flow, grain boundaries tend to be straight and to form characteristic triple junctions (figure 6.18b). Microstructural studies may also help in extrapolating experimental results. If microfabrics produced in triaxial experiments are similar to microstructures found in naturally deformed rocks, then there is a high probability that the experimental samples and the field samples were deformed by similar mechanisms.

## Some Applications

So far, we have discussed experiments that have enabled us to gain a better understanding of general deformation characteristics of rocks. Most rocks and mineral deformation experiments address specific problems. For instance, laboratory deformation of peridotite, dunite, and olivine crystals has the aim of increasing our understanding of the rheological behavior of mantle material. This is needed to constrain models of mantle convection and flow. It is not realistic to perform creep experiments on peridotite at mantle strain

**Figure 6.19**
Stress–strain curves of anhydrite specimens deformed at (a) 300° C, and (b) 450° C, at 1.5 kbar confining pressure. AV C, AW P, and AR denote different specimens. (From Müller et al., 1981.)

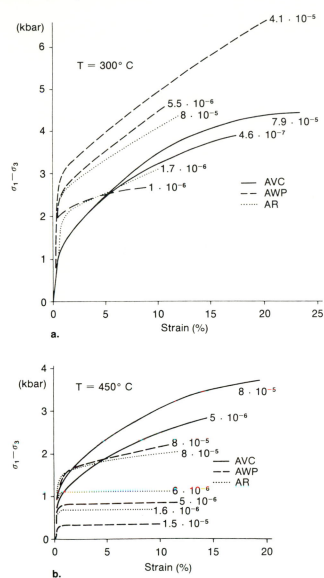

rates ($10^{-14}$–$10^{-15}$/sec). To obtain measurable strain rates, $10^{-7}$/sec or faster, temperature and differential stress would need to be unrealistically high. However, single crystals of olivine have a melting point appreciably above the solidus for peridotite, so that higher temperatures (allowing lower stresses for acceptable strain rates) are attainable. The test characteristics of olivine appear to represent mantle peridotite quite well. Confirmation comes from examination of the microstructure of experimentally deformed olivine. Transmission electron micrographs show that microstructures that form under experimental conditions are not unlike deformation structures in natural mantle peridotite crystals.

A rock that has been submitted to experimental deformation, with a specific problem in mind, is anhydrite. Layers of anhydrite form the base of several well-known detachments and thrust sheets. Investigation of the flow behavior of anhydrite will help in developing a model for thrust and detachment mechanics in which anhydrite flow was involved (see chapter 15). Figure 6.19 contrasts stress–strain curves for different strain

**Figure 6.20**
Synoptic diagram of log stress vs. temperature illustrates the relative strength at 10% strain of halite, anhydrite, and limestone expected at geological strain rates. (From Müller et al., 1981.)

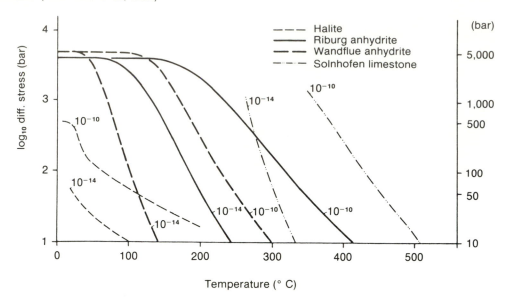

rates at 300 and at 450° C. The sensitivity of anhydrite flow to temperature is quite striking. Figure 6.20 compares the effect of temperature on stresses needed for different strain rates, in a number of sedimentary rocks. This shows which rock in an alternating sedimentary sequence would yield at the lowest stress at any given temperature. In other words, it shows which rock would tend to form a yielding or detachment horizon. We will discuss the implications of this in chapter 15.

## Grain Fabrics

In 1953, F. Turner initiated experimentation on stressed test specimens, in order to study the fabric that results from deformation. Turner's basic thesis was that, for small strains, twin or translation gliding occurs only along crystal lattice planes that are oriented to carry a high resolved shear stress. To deform plastically, therefore, grains must be favorably oriented within an aggregate. This thesis has been amply tested and confirmed for a number of minerals, beginning with calcite and dolomite, and later extending to quartz and several rock-forming silicates (Carter and Raleigh 1969). The experimental procedure is to deform a small test cylinder—usually of monomineralic rock—but in some cases single crystals—under controlled conditions, and then to measure the orientation of deformation structures so formed on a universal stage. The

results are assembled on an equal-area stereographic plot, which gives the statistical orientation of the features that have been measured (chapter 9, p. 156). By knowing the relation of the maximum compressive stress direction relative to the specimen from the experimental setup, one can readily obtain an angular relation between the orientation of measured fabric elements and that of the stress axes (Turner 1953). For instance, Carter, Christie, and Griggs (1964) experimentally deformed quartz crystals. They obtained deformation lamellae parallel to basal planes of quartz and found that these were parallel to directions of high resolved shear stress in the experimental sample. Other structures, such as oriented undular extinction and kink bands, formed in grains so oriented that (0001) was a plane of high resolved shear stress. These and similar results can be applied to give a dynamic interpretation to naturally deformed rock fabrics (see chapter 9).

## Natural Scale Models

Somewhere between the controlled, man-made experiments with rocks and artificial model substances, and actual rock deformation resulting from tectonic causes, we may encounter nature's own models of rock deformation in nontectonic processes. In these natural scale models, we can observe weak natural materials deform at measurable rates.

**Figure 6.21**
Malaspina glacier, Alaska. The dark bands that are folded are
morainal septa. Note axial plane attitude of foliation and cross joints
perpendicular to flow direction. (Photograph by U.S. Coast Guard;
from Sharp, 1958.)

## Glaciers

Glaciers provide a good example of a crystalline solid
(actually, a true rock) that flows at humanly detectable
rates. We can relate their fractures, foliation, grain
fabric, and folded morainal septa to observable movement.

A good example, the Malaspina glacier in Alaska,
has been thoroughly investigated (Sharp 1958). A typical piedmont glacier, the Malaspina flows from a main
feeder valley and spreads out at the foot of a mountain
range. Its fabric broadly reflects this mode of flow
(figure 6.21). Fractures include radial crevasses and
transverse crevasses at right angles to them. Both are
essentially vertical. Tight fractures consisting of two sets
at approximately 75–90° to one another form variable
angles with the crevasses. Since they do not make a
constant angle with the crevasses, they appear to have
been formed under a different stress system.

A regular alternation of thin whitish and bluish
layers forms the glacier foliation. The foliation is def-
initely not primary, because it is symmetrically related
to folds formed in morainal septa and cannot be traced
back into primary névé layers. Past investigations leave
little doubt that glacier foliation is an expression of
pseudoviscous flow, but the exact relation of the foliation pattern to observed modes of glacier flow is
not always easy to determine. Axel Hamberg (1932)
described interesting examples of "ice conglomerates":
spherical ice "pebbles" from an ice fall become strained
in a matrix of glacier ice, flattened in foliation planes,
and elongated in the direction of glacial motion.

Glacier ice is an excellent example of a material
that reacts both as a fluid and as a solid undergoing
fracture, under the same force (gravity). The foliation
is an expression of "fluid" properties, whereas the fractures are indications of brittle ("solid") behavior.

Looking back at the different ideal bodies cited in
chapter 5, we now see even more clearly that no substance *is* a certain rheological body but, rather, *reacts*
as a particular rheological body under given conditions
in space and time. Thus, ice in a glacier supports brittle
fracture as it flows: it exhibits structural features characteristic of more than one rheological state, depending
on duration, magnitude, and direction of stress application.

## Salt Domes

A little earlier, we demonstrated how the movement of
salt domes can be studied by means of scale model experiments. In itself, rock-salt deformation is a natural
model of solid flow in rocks (figure 6.22).

Salt diapirism normally proceeds as a result of the
density contrast between the salt and the overlying
rocks. Figures 6.23 and 6.24, show how individual salt
structures may develop from a mother salt. Diapirism

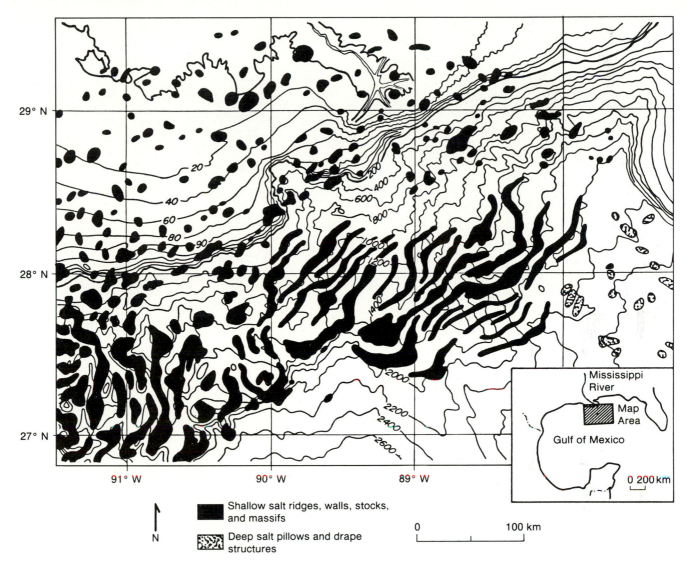

**Figure 6.23**
Map showing pattern of salt structures near the Mississippi delta.
(Courtesy R. G. Martin and M. P. A. Jackson.)

Shallow salt ridges, walls, stocks, and massifs

Deep salt pillows and drape structures

Mississippi River

Map Area

Gulf of Mexico

0 200 km

N

0         100 km

**Figure 6.24**
The main types of large salt structures. Structure contours are in arbitrary units. (From Jackson and Talbot, 1985.)

Extrusive salt dome

Piedmont namakier

Namakier

Detached diapir

Diapiric salt stock

Salt ridge

Salt roller

Cap

Salt ridge

STEM

Concordant low amplitude

Structural evolution

Diapiric salt wall

Root

Pre-salt floor

Source layer

Salt pillow

Intrusive high amplitude

**Figure 6.25**

Salt structures in cross section, from seismic reflection. Offshore Aquitaine basin, France. (From Curnelle and Marco, 1983.)

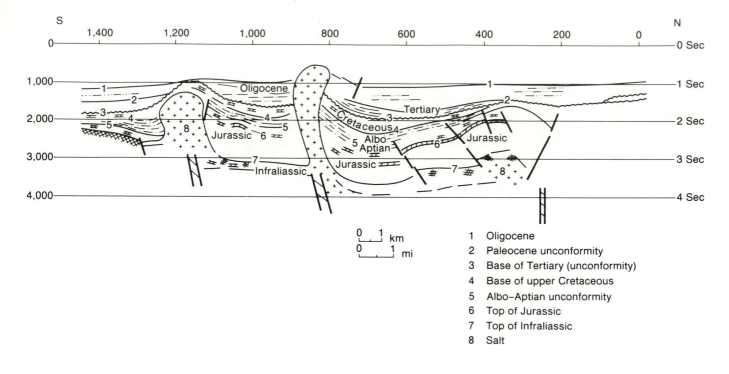

1  Oligocene
2  Paleocene unconformity
3  Base of Tertiary (unconformity)
4  Base of upper Cretaceous
5  Albo–Aptian unconformity
6  Top of Jurassic
7  Top of Infraliassic
8  Salt

seems to be a function of depth of overburden: only where the mother salt is deep enough will salt domes individualize. Trusheim (1957) called such deformation, resulting entirely from solid flow along pressure gradients, **halokinesis.**

What starts the movement in the first place? Many evaporite beds form part of an undisturbed stratigraphic sequence, with little or no signs of diapiric behavior. Two conditions seem essential for diapir initiation. First, density contrast must become an effective driving force, and this requires a minimum depth of burial. Depths of mother salt layers under true salt domes range from 6 to 10 km. In addition to depth, however, there must be a reason why a salt dome rises in a particular location: local instability. Highly mobile (weak) material such as evaporites will tend to flow along pressure gradients, from high pressure to low pressure. Salt "pillows" and "waves" (figures 6.23 and 6.24) tend to build up where salt flowage is guided along pressure gradients toward the highest level of some

tilted zone (no matter how gentle the tilt). Good examples of guiding structures include flexure zones and faults (which may be associated with flexure zones), as in figure 6.25. Salt diapirs may individualize from salt waves. Where salt diapirs are found aligned in zones, this explanation seems satisfactory.

The relatively rapid flowage of evaporite along pressure gradients is strikingly recorded in the internal structure of salt diapirs (figure 6.26) and in the internal structure of many evaporite or other rock bodies that have undergone flowage.

## Shale Bulges and Related Structures

Natural models of rock flowage may be found in some mine openings. Where a shale layer is overlain by a less-yielding rock, and an opening is cut to expose the shale (figure 6.27), the shale tends to squeeze out through the opening and produce minor local folding. In a mine opening, the exposure is man-made. But this behavior is much like that of *"valley bulges,"* some of which are found near Northampton, England; here the opening is

**Figure 6.26**
Internal structure of Jefferson Island salt dome, Louisiana. Room is nearly 100 ft. high. Note closures, fold traces, and implied steep fold axes. (From Balk, 1953.)

**Figure 6.27**
Section of carboniferous strata at Wallsend, Newcastle, showing "creeps" (a–d) in progressive stages of development and effect on lower seams (e–h). (Drawings by J. Buddle) (From Lyell, C. *Principals of Geology*, 5th edition (2 Volumes). Philadelphia, 1837.)

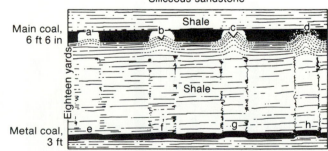

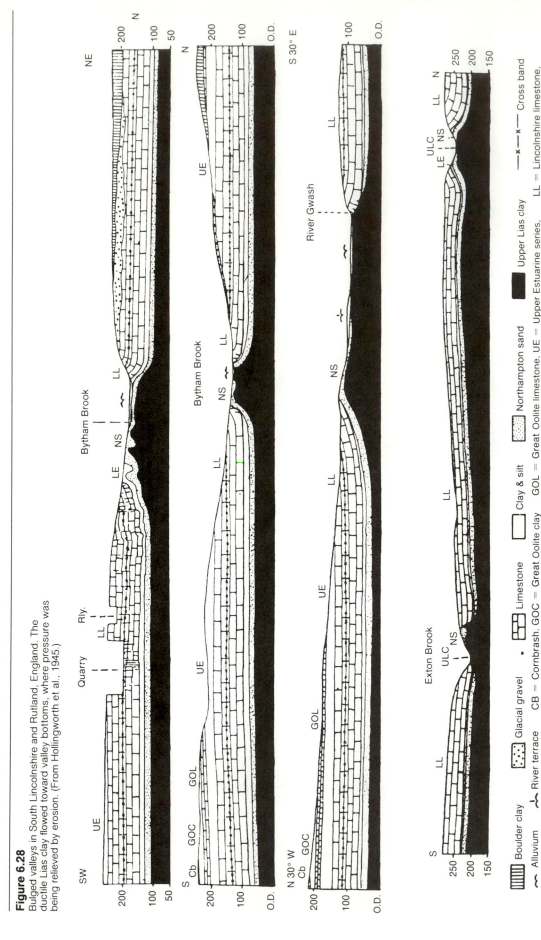

**Figure 6.28**
Bulged valleys in South Lincolnshire and Rutland, England. The ductile Lias clay flowed toward valley bottoms, where pressure was being relieved by erosion. (From Hollingworth et al., 1945.)

Boulder clay

Alluvium

River terrace

Glacial gravel

Limestone

Clay & silt

Northampton sand

Upper Lias clay

Cross band

CB = Cornbrash.  GOC = Great Oolite clay  GOL = Great Oolite limestone,  LL = Lincolnshire limestone,
LE = Lower Estuarine series, NS = Northampton sands, ULC = Upper Lias clay.  UE = Upper Estuarine series,

NS = Northampton sands, ULC = Upper Lias clay.  O.D. = Ordinance datum = Sea level

Horizontal scale  0                    500 Ft    Vertical scale 2× horizontal

**Figure 6.29**
Intrusive relationship of coal seam. Nappe de Morcles sequence, near border between France and Switzerland. A high degree of differential ductility is present, resulting in relationships very similar to those of magmatic intrusive rocks. Elsewhere, the coal is missing, and its place is taken by a detachment surface. (From Feugueur, 1953.)

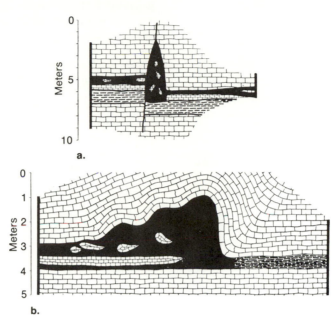

the result of erosion (figure 6.28). Rock flowage in this area is not the result of density contrast, as in salt domes, but of unequal distribution of load on a ductile layer.

Extreme variants of the above phenomenon have been observed in the Alps, where coal has been squeezed into fractures to produce true "coal dikes" (figure 6.29).

## Viscosity of the Lithosphere

A well-known means of determining the equivalent viscosity of the lithosphere uses the isostatic rebound of recently glaciated areas of the northern hemisphere, assuming this rebound to be linear with time (viscous flow). Because of the many factors involved, the procedure is somewhat complex and possibly of doubtful reliability; yet there is widespread agreement on the figure of about $10^{22}$ poises for the equivalent viscosity of the continental lithosphere so determined. Equivalent viscosity may be determined at a specific temperature and pressure, and differential stress, for non-Newtonian materials ($\eta = \sigma/3\dot{\epsilon}$). True viscosity of Newtonian materials is independent of $\sigma$. One complicating factor is the probability that tectonic movements occur in pulses, rather than uniformly, so that the episodic rate of deformation attains a greater value than that found in scale-model experiments.

## Conclusion

Flow of rocks has resulted in many of the geological structures we see today. It is possible to unravel, at least partially, the processes involved and to document the manner in which flow takes place. To do this, we needed to know ranges of temperature, confining pressure, stress, and strain rate that are realistic under given conditions. What is realistic in the mantle is not possible close to earth's surface, and vice-versa. But if we understand the necessary limitations of rock deformation, we may be able, as Hans Cloos put it, to "see mountains grow."

In the next few chapters we shall study some of the structures that result from the flow of rocks.

## Review Questions

1. Before 1920, what important and still valid insights were gained by means of geological experimentation?
2. (a) Derive, from first principles, model ratios for strain rate and for viscosity (assuming that the effects of gravity and density can be neglected).
   (b) Explain why it is reasonable to neglect the effects of gravity and of density.

3. What advantages are gained by accelerating tectonic models in a centrifuge?

4. In a manner similar to the procedure used in table 6.2, work out the viscosity of a model earth as described on page 87.

5. Differentiate clearly between confining pressure, pore pressure, and effective pressure in triaxial experiments.

6. What is the difference between true viscosity and apparent (or equivalent) viscosity?

7. What parameters and environmental variables can be controlled in triaxial tests? What ranges of values for each of these might be reasonable (1) in nature, and (2) in experiments?

8. Contrast and compare geological and experimental strain rates. To what extent are we justified in extrapolating from experimental to geological strain rates?

9. Discuss the different internal flow mechanisms of rocks and minerals. What textural clues could reveal which of these mechanisms has operated?

10. What information can be gained from a deformation map? How is such a "map" constructed?

11. Discuss the relationships illustrated in figures 6.7–6.10.

12. Explain the essential difference between creep tests and constant strain-rate tests.

13. What relationship did Turner find between crystal lattice deformation and stress?

14. What driving force formed the "valley bulges" in England? By what means did it act?

15. How is it that some substances can react to stress both as a fluid and as a solid at the same time?

## Additional Reading

Ashby, M. F., and Verrall, R. A. 1978. Micromechanisms of flow and fracture, and their relevance to the rheology of the upper mantle. *R. Soc. London, Philos. Trans.* A288: 59–93.

Benghazi, J. H., and Rutter, E. H. 1983. The low-temperature brittle-ductile transition in a quartzite and the occurrence of cataclastic flow in nature. *Geol. Rundsch.* 72: 493–509.

Berner, H., Ramberg, H., and Stephansson, O. 1972. Diapirism in theory and experiment. *Tectonophysics* 15: 197–218.

Christie, J. M., and Ardell, A. J. 1976. Deformation structures in minerals. Pages 373–403 in: H. R. Wenk et al. eds. *Electron microscopy in mineralogy.* Berlin: Springer-Verlag.

Dixon, J. M., and Summers, J. M. 1985. Recent developments in centrifuge modelling of tectonic processes: Equipment, model construction techniques and rheology of model materials. J. Struct. Geol. 7: 83–102.

Donath, F. A. 1970. Rock deformation apparatus and experiments for dynamic structural geology. *J. Geol. Educ.* 18: 3–12.

Gretener, P. E. 1981. Reflections on the value of laboratory tests on rocks. Pages 323–326 in: N. L. Carter, M. Friedman, J. M. Logan, and D. W. Stearns, eds. *Mechanical Behavior of Crustal Rocks.* Am. Geophys. Union, Geophys. Monogr. 24.

Hubbert, M. K. 1981. Mechanics of deformation of crustal rocks: Historical development. Pages 1–10 in: N. L. Carter, M. Friedman, J. M. Logan, and D. W. Stearns, eds. *Mechanical Behavior of Crustal Rocks.* Am. Geophys. Union, Geophys. Monogr. 24.

Kelly, A., Cook, A. H., and Greenwood, G. W. eds. 1978. Creep of engineering materials and of the earth. *Symp. R. Soc. London, Philos. Trans.* A288: 1–236.

Kirby, S. H. 1983. Rheology of the lithosphere. *Rev. Geophys. Space Phys.* 21(6): 1458–1487.

Kohlstedt, D. L., and Goetze, C. 1974. Low-stress high-temperature creep in olivine single crystals. *J. Geophys.Res.* 79: 2045–2052.

Lister, G. S., and Williams, P. F. 1981. The partitioning of deformation in flowing rock masses. *Tectonophysics* 992; 1–34.

Means, W. D. 1986. Three microstructural exercises for students. *Jour. Geol. Ed.* 34: 224–230.

Molnar, P., and Tapponnier, P. 1981. A possible dependence of tectonic strength on the age of the crust in Asia: Earth and planet. *Sci. Lett.* 52: 107–114.

Nicolas, A., and Poirier, J. P. 1976. *Crystalline Plasticity and Solid-State Flow in Metamorphic Rocks.* New York: Wiley, 444 pp.

Ord, A., and Christie, J. M., 1984. Flow stresses from microstructures in mylonitic quartzites of the Moine thrust zone, Assynt area, Scotland. *J. Struct. Geol.* 6: 1–15.

Pfiffner, O. A. 1982. Deformation mechanisms and flow regimes in limestones from the Helvetic zone of the Swiss Alps. *J. Struct. Geol.* 4: 429–442.

Rutter, E. H. 1983. Pressure solution in nature, theory and experiment. *J. Geol. Soc. (London)* 140: 725–740.

Schmid, S. M. 1983. Microfabric studies as indicators of deformation mechanisms and flow laws operative in mountain building. Pages 95–110 in: K. Hsu et al. *Mountain Building Symposium* Zurich: Academic Press.

Talbot, C. J., and Rogers, E. A. 1980. Seasonal movements in a salt glacier in Iran. *Science* 208: 395–397.

Tullis, J. A. 1979. High temperature deformation of rocks and minerals. *Rev. Geophys. Space Phys.* 17(6): 1137–1154.

# Part Two

# Continuous Structures

# 7

# Geometry of Folds

One of the most common manifestations of flow in rocks is folding of layered rocks. But before we can understand folding as a process, we must understand and be able to describe fold forms. Few folds are exposed in one complete, coherent outcrop; so to reconstruct reasonable folded structures, we must know the range of possible forms.

## Anatomy of Folds

A straight line that moves through space so that it always remains parallel to itself sweeps out what is known as a **cylindrical surface** (figure 7.1). We may treat most geological folds as cylindrical surfaces; or we may consider them in segments, each of which is very nearly cylindrical.

A cylindrical surface is made up of an infinite number of lines parallel to a "generating line"; that is the line which, moved parallel to itself through space, sweeps out a cylindrically folded surface. The generating line is the **axis** of the fold. The axis of a cylindrical fold is a direction valid over the entire fold and not a discrete line that can be singled out and identified.

The locus of greatest curvature in a folded surface, usually a line, is the fold **hinge** or hinge line (figure 7.2). Where the greatest curvature is not along a line, but is distributed over a surface area with a circular arc cross section (as in figure 7.1), the hinge is the line joining midpoints of circular cross-section arcs. The whole area of maximum curvature may then be called the *hinge area.*

### Anticlines and Synclines

An elementary classification of folds distinguishes between *anticlines* and *synclines*. In some introductory texts, anticlines are said to be convex upward ("upfolds"), and synclines are said to be concave upward ("downfolds"). The scientifically correct definitions, first introduced by Bailey and McCallien (1937), are stricter: they take into account the sequence of folded beds. Thus, in an **anticline** the rocks get progressively older toward the core of the fold. In a **syncline,** the rocks get progressively younger toward the core of the fold (see figure 7.3). The more weakly curved or uncurved portions of a fold that join anticlinal and synclinal hinges are the **limbs** of the fold, sometimes called its *flanks.* Where the sense of the stratigraphic succession is not relevant or remains doubtful, a fold that is convex upward is properly called an **antiform,** and a fold that is concave upward is called a **synform** (figure 7.3).

**Figure 7.1**
Concept of a cylindrically folded surface. (From Wilson, 1982.)

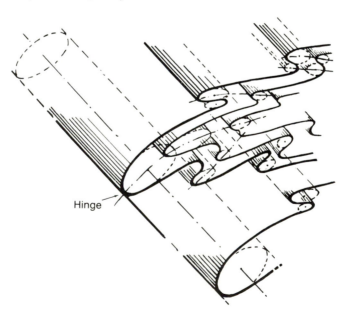

Hinge

## More Definitions

Let us return to figure 7.2. In an antiform, the axial line that is topographically higher than any other is the **crest** of the fold. In a synform, the axial line that is topographically lower than any other is the **trough.** Crest and trough may or may not coincide with hinge lines. The surface that is the locus of all the hinges of stacked, folded surfaces in a fold is the **axial surface** or **hinge surface.** Penetrative planar elements that are parallel to the hinge surface are **axial planes.** The **axial trace** is the trace of the axial surface on the topographic surface or on the map. It does not necessarily have the same trend as the fold axis (see figure 7.15).

Fold axes are rarely straight lines. They vary in attitude and thus change in trend and plunge. A curvature in the axis that is convex upward is called an **axial culmination.** A curvature that is concave upward is an **axial depression.**

In folds with asymmetric cross sections, the shorter limb is known as the **fore limb** and the longer limb as the **back limb.** A fore limb that has been rotated through the vertical, reversing the stratigraphic succession, is known as the **reversed, inverted,** or **overturned limb,** and in that case the back limb is known as the **normal limb.** The angle between the limbs is the **interlimb angle.** Folds with overturned limbs are **overturned folds** (e.g., figure 7.2). A fold with parallel limbs is an **isoclinal fold.**

**Figure 7.2**
Geometrical elements of a fold. (From Dennis, 1967.)

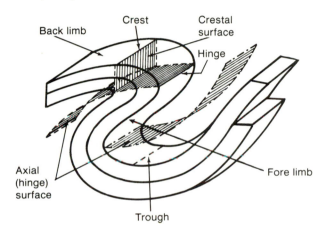

**Figure 7.3**
Terminology for antiforms and synforms. (From Dennis, 1967.)

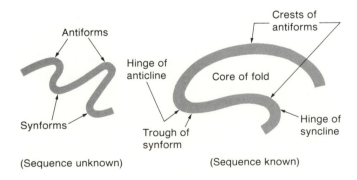

## Figure 7.4
Fold classification by attitudes of hinge and axial surface. (From Rickard, 1971.)

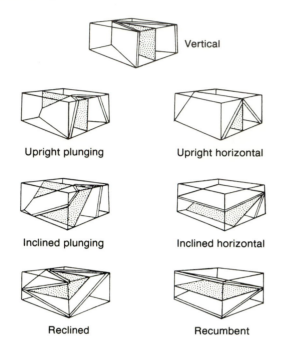

## Figure 7.5
Classification of fold shapes by interlimb angle and hinge flexure. (a) diagrammatic representation of fold shapes; (b) descriptive terminology. (From Williams and Chapman, 1979.)

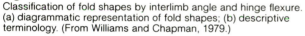

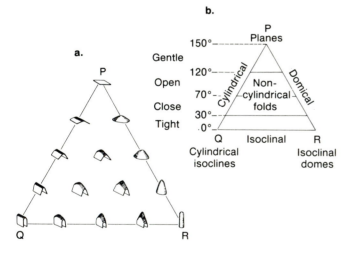

## Classification by Attitude of Folds

Folds may be classified according to hinge and axial plane attitudes. A convenient classification is illustrated in figure 7.4. According to axial plane dips, we have **upright folds** (dips of 80–90°), **inclined folds** (dips of 10–80°), and **recumbent folds** (dips of 0–10°). An inclined fold with an overturned limb is also an overturned fold. According to plunge of fold axis, we have **horizontal** (plunge of 0–10°), **plunging** (plunge of 10–80°) and **vertical folds** (plunge of 80–90°). A further, special category consists of inclined folds in which the pitch of the fold hinge in the axial plane is between 80 and 90°; this type of fold is a **reclined fold.**

## Classification by Interlimb Angle and Hinge Flexure

Fold shapes may be classified by interlimb angle and curvature of the hinge line. Figure 7.5 shows a very useful classification diagram. Most common, and many not-so-common, fold shapes have a place on it. It is convenient to characterize folds by their location on this *PQR* diagram, where *P, Q,* and *R* are arbitrary designations. Some descriptive names are shown in the inset. Minor folds near the *R* corner are sometimes called **sheath folds.**

## Classification by Mode of Stacking of Folded Surfaces

A fundamental classification of folds, first proposed by Van Hise (1896), is by the mutual relationship of successive folded surfaces. There are two ideal types: **parallel folds,** in which successive folded surfaces remain mutually parallel throughout (figure 7.6); and **similar folds,** in which successive folded surfaces are ideally congruent (figure 7.6b). Figure 7.6 shows that for parallel folds, the thickness, *t,* of each folded layer is constant at right angles to the layer; but the distance, *T,* (across any layer measured along axial planes), varies. In a profile of similar folds the distance *T,* measured parallel to axial planes, remains constant between any two folded surfaces; but thickness of layers, measured at right angles to layers, varies. In parallel folds, distances measured along layers, at right angles to fold **axes,** remain the same before and after folding. In similar folds, these distances increase as a result of folding.

## Figure 7.6

(a) Parallel folds. Note that parallelism must eventually break down in the cores of folds. The manner in which this happens depends largely on the mechanical properties of the folded rock layers.
(b) Similar folds. $t$ = layer thickness, and $T$ = layer width measured along axial planes. In ideal cases, $t$ remains constant for (a), and $T$ remains constant for (b). (From Ramsay, 1962.)

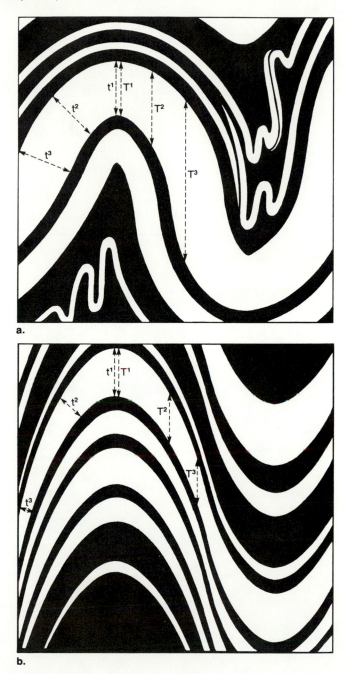

a.

b.

## Figure 7.7
Fundamental types of fold classes. Dip isogons at 10° intervals, from the lower to the upper surfaces X and Y. (From Ramsay, 1967.)

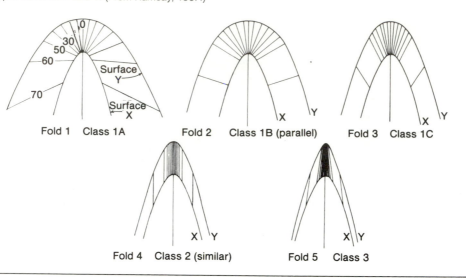

Fold 1   Class 1A      Fold 2   Class 1B (parallel)      Fold 3   Class 1C

Fold 4   Class 2 (similar)      Fold 5   Class 3

## Figure 7.8
Chevron folds in Franciscan chert beds, Marin County, California. (Photograph by Mary Hill.)

**Figure 7.9**
Kink folds in Gile Mountain formation, near St. Johnsbury, Vermont.
(Photography by J. F. de Chadenedes; from Eric and Dennis, 1958.)

Parallel folds and similar folds are special cases in a more general classification first considered by Ramsay (1967), who recognized three main classes (figure 7.7). In *class 1* the curvature of successive folded surfaces increases toward the inner arc of the fold. *Class 1* includes three subclasses. In *class 1A,* thicknesses of folded layers in hinges are thinner than those on limbs (supratenuous folds). These folds tend to be comparatively open, with gently curved crestal regions and no readily identifiable hinge. Here the crest dominates the geometry. *Class 1B,* parallel folds, is a special case. In *class 1C,* thicknesses in hinges are greater than on limbs. In *class 2* the curvature of successive folded surfaces remains the same. This is a special case of similar folds. In *class 3,* the curvature of successive folded surfaces decreases toward the inner arc of the fold.

Changes in curvature between folded surfaces in fold profiles can be brought out most effectively by joining points of equal dip; the lines so constructed are **dip isogons** or *isogonal lines.* In *class 1* folds they diverge, in *class 2* they are parallel, and in *class 3* they converge (figure 7.7).

### Terminology of Some Other Fold Shapes

**Chevron folds** have straight limbs and angular hinges (figure 7.8). **Kink folds** are chevron folds in schistose rocks (figure 7.9).

Where the wavelength of folds varies between successive layers, the folding is said to be **disharmonic** (e.g., figure 8.11). The core of an anticline that pierces overlying rocks is a **diapir fold,** or **diapir,** many of which are dome- or plug-shaped. The most common examples are salt domes (figures 6.2 and 6.3).

## Figure 7.10
Diagrammatic cross sections of (a) homocline and (b) monocline.

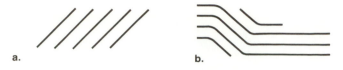

a.                    b.

## Figure 7.11
Sharp anticline in Coal Measures (Pennsylvanian), Saundersfoot, Wales. The fold axis is almost perfectly horizontal, so that both limbs have parallel outcrops on the horizontal wave-cut terrace. (U.K. Crown copyright, British Geological Survey photograph by T. Clifford Hall.)

**Figure 7.12**
Sheep Mountain anticline, Wyoming, a young anticlinal mountain.
Note "flatiron" facets on the steep limb (contrast with triangular
facets of figure 2.8), and antecedent stream in foreground.
(Photography by John S. Shelton.)

We may also distinguish folds according to their axial extension. Anticlines of relatively short axial extension are **brachyanticlines. A dome,** strictly speaking, is an upwarp of any size that is approximately equidimensional in plan. Many elongated anticlines, however, have been called domes.

On a regional scale, a steeply dipping zone in an otherwise gently dipping sequence constitutes a **monocline.** A local, shelflike flattening of the dip is a **structural terrace.** A succession of beds with uniform parallel attitudes over a large area forms a **homocline.** The terms monocline, structural terrace, and homocline (figure 7.10) are used only for structures on a macroscopic scale.

## Map Pattern of Folds

Folds with zero plunge form parallel outcrop bands (figure 7.11); the stratigraphic succession reverses across crests and troughs. Plunging folds are normally revealed by sharp bends in the outcrop pattern (figure 7.12), and these bends are called **noses.**

**Figure 7.13**

**Figure 7.13**
A folded aluminum sheet dipped in tinted water shows that map pattern does not necessarily correspond directly with fold shape. Water surface represents level topographic surface, intersected by plunging overturned fold. Hinge of fold, brought out by reflection contrast at left, barely affects map trace of fold. Nose of fold is slightly to the rear of the synclinal trough (whose location the photograph cannot show). (Photograph by Audio-Visual Department, California State University, Long Beach.)

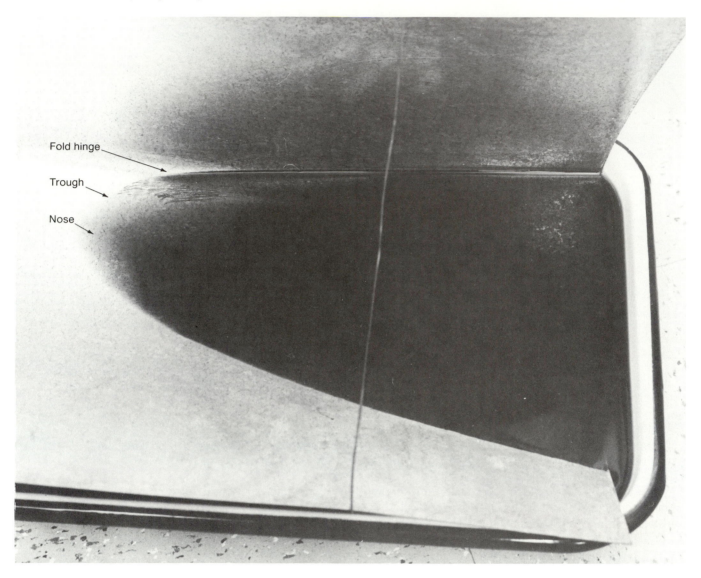

The shape of folds in map pattern and in most cross sections is a function of two geometrical variables: the true curvature (in normal cross section) and its distortion due to projection. Figure 7.13 illustrates the principle involved and shows that the nose of a fold may not coincide with the trace of the hinge line. The cross sections through a log (figure 7.14) further illustrate the relationship. Curvatures evidently become distorted in all but normal cross sections: the circular cross section of the log projects onto the map as an ellipse whose "nose" has its greatest curvature in the direction of the plunge. Thus, a plunging fold with a perfectly

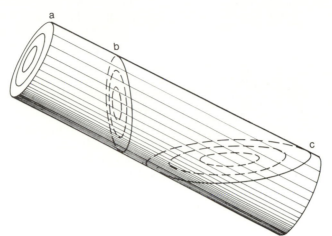

circular cross section acquires a nose in map pattern in the direction of the plunge, but this nose is not related to any hinge line on the fold. We may call this the **nose effect** (figure 7.13).

In natural folds with defined hinges, the nose coincides with the hinge trace only where hinge and crest (or trough) coincide, or where no crest or trough exists, as in reclined folds. In folds in which crests and troughs do not coincide with hinges, the nose occupies a position intermediate between the traces of hinge and crest (or trough). For a given separation between hinge and crest, the exact location of the nose depends on the angle of plunge. In folds with perfectly circular hinge-area cross sections, the nose, of course, coincides with the trace of the crest (or trough). Clearly, where hinge and crest (or trough) do not coincide, the line joining the noses of folded contacts on a map has no geometric significance; it is certainly not the "axis," nor is it the outcrop of the axial surface. Thus, the hinge of a fold cannot always be located from the map pattern alone. However, along crests and troughs, the dip of folded surfaces is a minimum, and their strike is perpendicular to the fold axis. Also, along troughs and crest lines, the dip of the folded surfaces is identical to the plunge of the fold. Of course, on maps the crest line is within the oldest bed, and the trough line is within the youngest bed along any cross section.

## Cross Sections

When constructing cross sections from maps, one must be sure to use all the information included on the map. In the rather common case of plunging cylindrical folds, the map pattern can be projected along parallel axial lines into any convenient plane, such as a vertical cross-section plane. In figure 7.14, cross sections along vertical and horizontal planes are random sections that give variously distorted views, according to the attitude of the log. But there is one cross-section orientation that yields the true curvature of the tree rings: that which is at right angles to the axis of the log. This is called the **normal cross section.** All other cross sections are ellipsoidal distortions. Normal cross sections give the only undistorted representation of the internal structure. In geology, a region with parallel, straight fold axes can be treated in exactly the same way. The normal cross section, also known as the **tectonic profile,** gives the best view of the structure: angles and thicknesses are undistorted. The map and vertical cross sections are random sections. Each cross section can be projected along the plunge to yield any other desired cross section. The most common aim is the construction of a normal cross section or a vertical cross section from the map (e.g., figures 7.15 and 7.16). The axial line along which projection is made need not be straight, since there are geometric methods for projecting along curved

## Figure 7.15

Block diagram illustrating plunging overturned cylindrical folds. *AT* = axial trend; *RS* = "regional strike"; *AP* = axial plunge. Note divergence between axial trend and regional strike, as well as the varying attitudes of bedding with constant axial plunge. Divergent strikes are plainly part of the total geometry. Note that the strike of vertical beds is parallel to the axial trend of folding. (From Wegmann, 1929.)

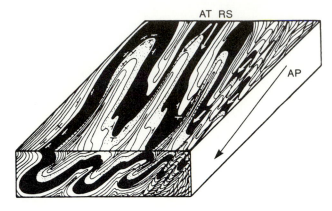

paths. Actual deviations from the theoretically predicted profile occur because of departures from true parallel cylindrical curvature.

The style of a structure constructed in cross section should reflect the style of its parts that can be observed in outcrop. This results in what Elliot (1983) calls an *admissible* cross section. If a cross section of deformed rocks can be restored to the predeformation state, it is a *viable* cross section. A cross section that is both *admissible* and *viable* is a **balanced cross section** (e.g., figure 15.9).

All cross sections are interpretations; and several interpretations are usually possible. A balanced cross section narrows the scope for alternative interpretations; it has a high degree of validity. We will learn more about balanced cross sections in chapter 15.

## Figure 7.16

Plan and normal cross section of Sheila Lake fold, Sherritt Gordon mine area, Manitoba. The cross section below was constructed by projecting the map along axial lines onto a vertical plane. Note relationship of local bedding attitudes to axial attitude. (From Stockwell, 1950.)

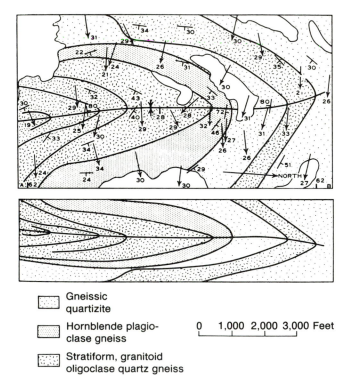

☐ Gneissic quartizite

☐ Hornblende plagioclase gneiss

☐ Stratiform, granitoid oligoclase quartz gneiss

0   1,000  2,000  3,000 Feet

## Symmetry of Folds

We can describe fold symmetry in much the same way as crystal symmetry. This helps to describe the fold, and to understand how associated structures, such as fractures, relate to it. We assume a cylindrical fold or portion of a fold.

In cylindrical folds, the plane normal to the fold axis is always a plane of symmetry. If the profile is also symmetrical about the hinge surface, we have two mutually perpendicular planes of bilateral symmetry, and hence **orthorhombic** symmetry. Such folds are sometimes called "symmetrical folds." Cylindrical folds that are asymmetrical about the hinge surface, retain only one plane of bilateral symmetry, and hence are **monoclinic** (figure 7.17). This is by far the most common

fold symmetry. Most folds with triclinic symmetry can be subdivided into domains of approximate monoclinic symmetry.

The form of the normal profile represents "frozen movement," a "still" of the movement that generated the fold form. Sander (1948) called this the *movement picture* of the fold. Stille (1930) gave a special name to the one-sidedness of such folds—*vergence*—which is the direction toward which the fold is turned. The term actually includes all structures that clearly express one-sided regional movement. In English we may also say that the folds **face** or **verge** in a given direction. Strictly defined, a fold faces in the direction toward which its constituent beds become progressively younger along the hinge surface. Figure 7.18 shows how the term applies in different situations.

**Figure 7.17**
Fold symmetry. (a) Orthorhombic, two mutually perpendicular symmetry planes; (b) monoclinic, one plane of symmetry, (c) Triclinic, no plane of symmetry. (From Wilson, 1982.)

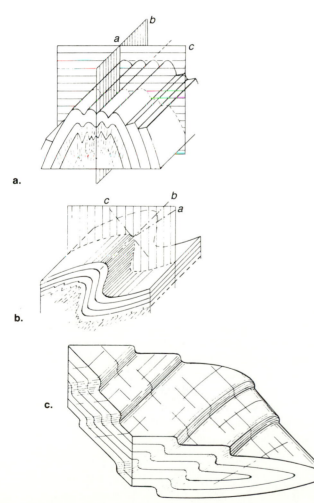

**Figure 7.18**
The concept of facing (shown in cross section). (a) Stratigraphic sequence facing west. (b) Stratigraphic sequence facing east. (c) Fold facing (verging) west. (d) Fold facing (verging) east. (e) Minor (subsidiary) fold facing downdip. (f) Minor (subsidiary) fold facing updip.

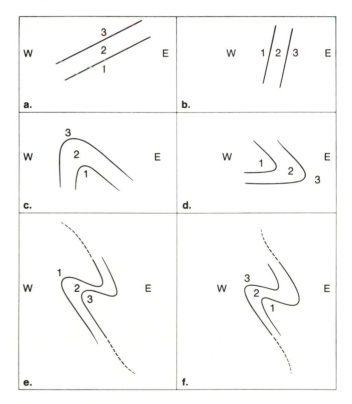

**Figure 7.19**
Symmetry axes of folds. (From Dennis, 1967.)

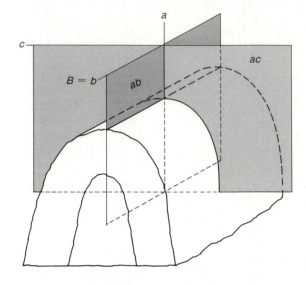

## Minor Folds

Many major folds have superimposed subsidiary folds called **minor folds** (figure 7.1) which may or may not be related genetically to the major folds. To avoid controversy as to origin, we shall first consider them strictly as geometrical features.

*Longitudinal minor folds* are subparallel to the main fold axis. They may share axial planes with the main folds, and some minor folds can be used very effectively for extrapolation in field work. Pumpelly formulated a rule of thumb: "The degree and direction of pitch of a fold are often indicated by those of the axes of the minor plications on its sides" (Pumpelly et al. 1894). This is known as **Pumpelly's rule.** Instead of "degree and direction of pitch," today we should say "plunge and trend." The orientation of the common fold axis can be measured wherever these minor folds are exposed. Furthermore, the orientation of the short limb of the minor fold indicates whether an adjacent hinge of the main fold is antiformal or synformal (anticlinal or synclinal, if the stratigraphy is known). The rule in such cases (see figure 7.1) is that the anticline of the minor fold is directed toward (faces) the anticline of the main fold; and the syncline in the minor fold is similarly directed toward the syncline in the main fold. Such folds we call **concordant minor** (or subsidiary) **folds. Discordant minor folds** have axes at an appreciable angle to the major fold axis.

## Reference Axes

Crystallographers use a system of reference axes, *a, b,* and *c,* to describe the form and symmetry of crystals. In the same way, it is convenient to fix reference directions for the symmetry of folds. As in crystallography, we will use a coordinate system or *axial cross,* with three mutually perpendicular axes, *a, b,* and *c,* (figure 7.19) following Sander's nomenclature (Sander 1930, 1948). The *b* direction is the direction of the fold axis. The plane containing *a* and *c,* the *ac* plane, is defined as normal to *b.* It is a plane of bilateral symmetry of the fold form, provided all fold axes are parallel. Where this is so, the folding is *homoaxial.* Axial planes, as defined earlier, are parallel to the plane containing *a* and *b,* the *ab* plane. Thus, *a* is defined as that direction in the axial plane which is perpendicular to *b.* It follows that *a* can only be uniquely defined for a fold with perfectly straight and parallel axial planes. Where these planes are curved, we can define the *a* direction only for specified points or for straight axes. The *c* direction is, of course, normal to *ab* (figure 7.19). Plane *ab* may be a plane of bilateral symmetry, as in the so-called symmetrical folds.

We have defined reference axes for fold forms only. In practice, the axial cross of fold symmetry is useful mainly for relating fracture patterns to the fold form (chapter 11). In chapter 9, we will encounter similar sets of reference axes to describe internal fabric symmetry. Following the practice of Sander (1948), we may use the same symbols, *a, b,* and *c,* but they will not be derived or defined in the same way.

## Conclusion

We are now in a position to describe and analyze fold forms geometrically. In the next chapter, we will see how fold form reflects the movements that give rise to it.

## Review Questions

1. How is a cylindrically folded surface generated?
2. Distinguish between anticline and antiform.
3. Distinguish between crest of a fold, hinge of a fold, and nose of a fold.
4. Distinguish between inclined, reclined, and recumbent fold.
5. How do $T$ and $t$ vary in similar and parallel folds?
6. Illustrate Pumpelly's rule.
7. Show symmetry elements in monoclinic folds, and in orthorhombic folds.
8. What is fold *vergence?*
9. Demonstrate the geometric relationship between map pattern and cross section of plunging folds.
10. What are the ingredients of a balanced cross section?

## Additional Reading

Busk, H. G. 1929. *Earth Flexures*. Cambridge Univ. Press. 106 pp.

Carey, S. W. 1962. Folding. *J. Alberta Soc. Pet. Geol.* 10: 95–144.

Higgins, C. G. 1962. Reconstruction of flexure folds by concentric arc method. *AAPG Bull.* 46: 1737–1739.

Keppie, J. D. 1976. Structural model for the Saddle Reef and associated gold veins in the Meguma Group, Nova Scotia. *Can. Mining Metallurg. Bull.* October, 1976: 1–14.

Langstaff, C. S., and Morrill, D. 1981. *Geologic Cross Sections*. Boston: IHRDC. 108 pp.

Lisle, R. J. 1980. A simplified work scheme for using block diagrams with the orthographic net. *Jour. Geol. Ed.* 28: 81–83.

Shaw, C. E. 1976. Large-scale recumbent folding in the valley and ridge province of Alabama. *Geol. Soc. Am. Bull.* 87: 407–418.

Weiss, L. E. 1959. Geometry of superposed folding. *Geol. Soc. Am. Bull.* 70: 91–106.

Whitten, E. H. T. 1966. *Structural Geology of Folded Rocks*. Chicago: Rand-McNally. 663 pp.

# 8

# Folding of Rocks

Since folds in layered rocks are very common geological structures, they must reflect a very common mode of ductile deformation. How do stacked layers of rock come to be folded? We certainly cannot watch the process in nature, so our approach must be theoretical and experimental. In this chapter, we will discuss mechanisms of folding and some of the tectonic settings in which it occurs.

## Basic Premises

Flat layers can become curved in two ways: by **buckling** and by **bending.** In buckling, relative compression is parallel to the layer surfaces; in bending it is at right angles to the surfaces (figure 8.1). In all cases of practical importance, rock layers are part of a succession; they are embedded in a mass, and their deformation must be considered as part of the overall deformation of the mass.

We shall assume that deformation occurs at constant volume. This is not always true, because both intergranular adjustment and deformation during metamorphism may, and almost certainly do, involve volume changes. The assumption of constant volume, however, does not seriously diminish the validity of an elementary discussion. We shall also initially consider only portions of a rock body that were, during deformation, sufficiently far removed from its outer limits that special boundary effects may be ignored. Finally, for a first discussion, we shall disregard strain components *along* fold axes; that is, we shall look principally at deformation in the plane perpendicular to the fold axis.

## Folding Mechanisms

Rock layer boundaries may have one of two principal roles during folding. They may give the rock mechanical anisotropy, which guides the process called **flexural folding.** Or they may be completely passive reference horizons known as **passive folding.** These processes may operate together, but we shall first consider each ideal mechanism separately.

### Flexural Folding

Flexural folding normally results in parallel folds, and layers ideally remain parallel throughout the deformation process. The only mechanism that will accomplish this is displacement parallel to layer boundaries, as when a sheaf of papers is bent. This is called **flexural slip folding.** Flexural slip may occur along discrete surfaces, usually bedding surfaces; but it also occurs *within* layers, in ductile flow. The basic strain is simple shear. Ideally, it would be greatest along limbs and zero along hinges. But deformation in hinges is rarely ideally flexural (figure 8.2a and b). On the convex side of a "neutral surface," extension occurs parallel to the flexed layer; on the concave side, compression occurs. In some flexural folding, material flows from the limbs into hinges, and layer boundaries are no longer parallel (figure 8.3). Donath and Parker (1964) have named this process **flexural flow folding,** to distinguish it from flexural slip folding, which, ideally, conserves parallel layers.

Some of the evidence for slip along bedding surfaces is similar to that for faulting (chapter 11). Slickensides, for instance, are common.

---

**Figure 8.1**
Folding of a thin plate. (a) Bending. (b) Buckling. (From Dennis, 1967.)

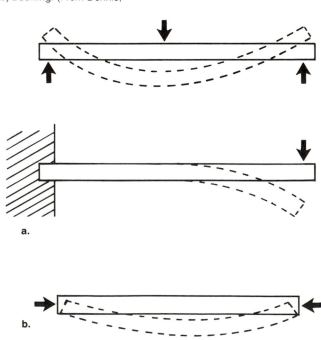

## Figure 8.2

(a) Distribution of strain within a thick folded layer, indicated by strain ellipses. (b) Distribution of strain in hinge region of folded multilayer consisting of a thick ductile layer sandwiched between the two thinner competent layers. (From Ramberg and Ghosh, 1968.)

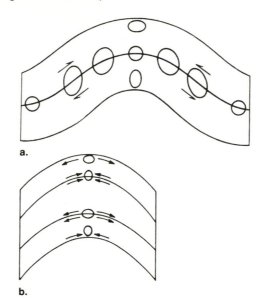

a.

b.

## Figure 8.3

Flexural flow folding in Mesozoic limestone, Diablerets nappe, Swiss Alps.

**Figure 8.4**
Cross section through the Mont Terri and Clos du Doubs chains,
Swiss Jura. This is a characteristic style of folding in the Jura
mountains. Note décollement, and broad synclinal troughs. (From
Laubscher, 1961.)

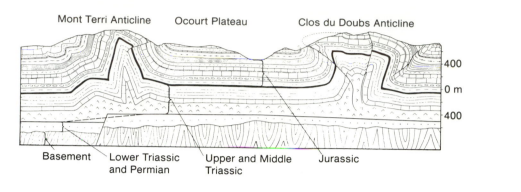

**Figure 8.5**
Folding by heterogeneous simple shear.

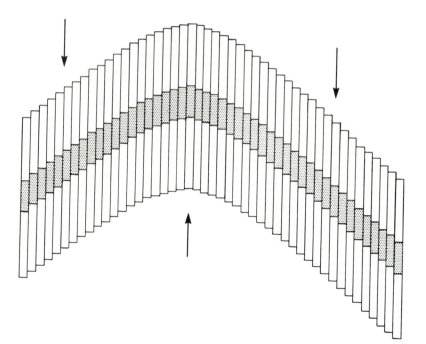

If the slip on all beds is uniform, or nearly uniform, the concave side of sequences folded by flexural slip must adjust to decreasing space. Figure 7.6a illustrates some effects of crowding, which vary depending on the relative thickness and ductility of the layers involved. In sequences with contrasting ductilities, disharmony results. All these features are evidence of bedding slip along some of the folded surfaces.

Figure 7.6a also indicates that flexural slip folds cannot be extrapolated to depth. Their characteristic habitat is in shallow levels of the crust, and they must terminate downward at some surface of **detachment.** The folds of the Jura mountains of Switzerland and France are typically cited as the best example of such detachment, or *décollement.* The oldest rocks in any of the Jura anticlines are from the middle Triassic period;

they consist mainly of highly ductile gypsiferous beds and marls, which allow the overlying units to deform independently and to become detached as an independently moving sheet (figure 8.4).

Flexural slip folding is most common in the upper levels of the crust in rocks that have undergone little or no metamorphism. It requires a detachment horizon at the base of the folded sequence. It thus seems reasonable to assume that this type of folding is a superficial phenomenon.

## Passive Folding

Passive folding occurs in relatively ductile and mechanically isotropic rock units. Usually, folds of similar type (chapter 7) result. Theoretically, such folds can form in two ways: by heterogeneous simple shear (figures 8.5 and 8.6) and by pure shear (figure 8.7). Carey

**Figure 8.6**
Rheid folding. Flow is guided by a pattern of flow planes, which
need not be entirely straight and parallel. (After Carey, 1954.)

Flow
lines

Original beds

**Figure 8.7**
An originally folded layer undergoing pure shear (flattening); (a) initial
state; (b) 50% shortening perpendicular to axial plans; and (c) 80%
shortening. (The change in shape of the black layer cross section
results solely from pure shear, there being no simple shear along
axial planes.) (d) Shows 90% shortening (only half a wavelength is
shown). (From Ramberg, 1964.)

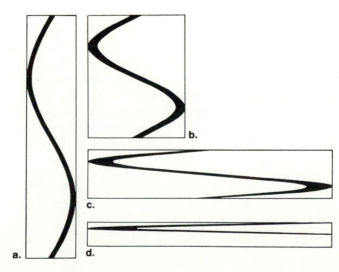

a.    b.    c.    d.

**Figure 8.8**
State of strain and cleavage orientation in folds in Helvetic limestones, Swiss Alps. Decimal fractions are extensions ($\sqrt{\lambda}$); 1 = massive limestone, well-bedded on decimeter scale; 2 = ferruginous oolitic limestone, well-stratified on decimeter scale, but no bedding separation planes; 3 = limestone, bedded on 30-cm scale. (a) Open fold in autochthonous cover of Aar Massif. (b) Tight minor fold in same general area, on normal limb of a larger fold. (From Pfiffner, 1980.)

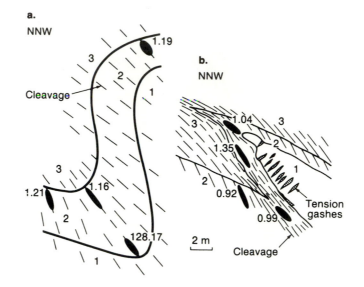

(1954) shows how similar folds may result from heterogeneous laminar flow. He calls such folds **rheid folds** (figure 8.6). Accordingly, where flow lines remain parallel during deformation, the thickness of a bed measured in the direction of flow remains constant. Where flow lines diverge or converge, the thickness of a bed measured in the direction of flow decreases or increases in inverse proportion to the normal distance between flow lines. The cross section would not contract laterally, as it would have to in pure shear.

How can we distinguish similar folds formed by simple shear from those formed by pure shear? And, where both mechanisms may have operated, is it possible to ascertain the relative role of each? The only clues we have are deformed structures in the fold that may serve as indicators of strain geometry (Ramsay and Huber 1983).

In simple shear folding, strain axes would have to be at a finite angle to axial planes of the folds. In pure shear (flattening), strain directions $X$ and $Y$ remain essentially parallel to fold axial planes. Although simple shear certainly contributes to folding deformation, it is

very difficult to conceive of a geological mechanism that alone might produce large-scale folding of the pack of cards type, as in figure 8.5. Passive folding occurs predominantly in highly ductile rocks, especially in metamorphic terrains. Certainly, passive folding may well depart from the ideal similar form of figure 7.6b. Only strain analysis gives a true picture of the displacements involved as in figure 8.8.

Ideally passive folding is, of course, a paradox; even passive folding must be initiated by some ductility contrast or detachment somewhere in the succession or at its boundaries, and this evidently limits passivity in folding to given domains.

## Ductility Contrast

Many stratigraphic successions are made up of layers of contrasting ductilities that react differentially in the course of deformation (e.g., figures 8.9–8.11). The behavior of the less ductile beds has been called **competent,** that of the more ductile beds, **incompetent.** This nomenclature is not apt mechanically, as originally proposed by Willis (1893), but its use is widespread.

**Figure 8.9**

Sandy slate (*below*) in a ductile matrix of fine-grained, shaly slate, near Ilfracombe, Devonshire, England. The sandy layer has buckled in compression, while the shaly slate has yielded in almost homogeneous strain. Note the departure from ideal homogeneous strain in the vicinity of the dominant sandy layer.

(Source: Sorby, H. C., "On the Origin of Slaty Cleavage" in *Edinburgh New Philosophical Journal, 55,* pp. 137–148, 1853.)

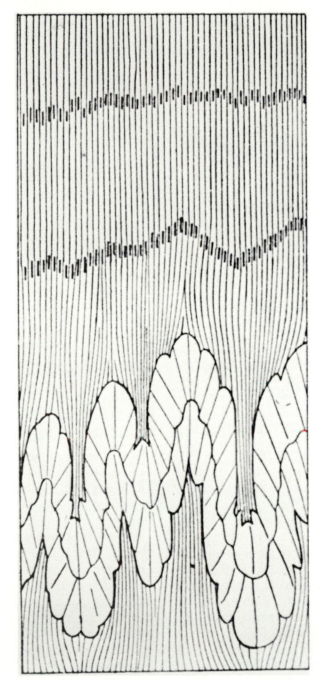

**Figure 8.10**
Computed small amplitude folding of a layer 2 ft. thick, viscosity $10^{21}$ poises in a matrix of $10^{18}$ poises, under compressive stress of 1,450 psi. Amplitude shown after 70,000 years. (From Biot, 1961.)

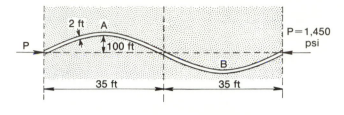

**Figure 8.11**
Idealized sketches of structural lithic units. (From Currie et al., 1962.)

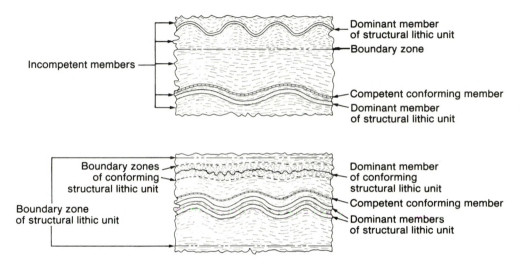

## Buckling of layers of "competent" rock

How and why does folding originate when a layer of material is compressed longitudinally? It has long been known that any "thin" material (that is, material that is very long compared with its thickness) will buckle under compression and develop a set of waves with characteristic wavelength for a given set of conditions. This is true whether the layer deforms elastically or viscously. For a viscous layer confined in a viscous medium, ductility contrast is an essential factor in buckle folding. As long ago as 1911, Sander established his "rule of buckle fold size" which states that for a given ductility contrast of a competent layer embedded in a less viscous homogeneous medium, *the wavelength of folds formed by buckling is proportional to the thickness of the folded layer.* In geology, we assume viscous buckling, although the deformation may have been elastic initially. We must remember that "viscosity" here is equivalent viscosity. It varies with strain rate and with stress.

Let us examine the case of a thin viscous layer embedded in a less viscous matrix. Biot (1961) and Ramberg (1963b) showed analytically that, under longitudinal compression and for a given thickness and viscosity ratio between layer and matrix, a characteristic *dominant wavelength* of buckle folding always develops (figures 8.10 and 8.11). The fold wavelength is determined at the initiation of folding and is independent of surrounding competent layers, provided these are more than a wavelength apart when folding begins. Where competent layers are closer together, one dominant layer will impose its wavelength on the others, within what has been called a *structural lithic unit* (Currie et al. 1962). A structural lithic unit contains a dominant member or members whose thickness and physical properties, relative to surrounding material, determine the wavelength of major folds within the unit (figures 8.10 and 8.11). Where layers in a stacked sequence are able to fold independently and acquire different wavelengths (as in figures 8.11 and 8.12), the folding is said to be **disharmonic.** High-amplitude disharmonic folding may result in detachment between adjacent layers.

**Figure 8.12**
Small-scale disharmony in anhydrite, Permian Castile formation, Delaware Basin, Texas. Light layers are anhydrite; dark layers are calcite rich in organic material. Note the effect of relative thickness of anhydrite layers, and the role of spacing between them (about 3 times normal size). (Photograph by R. Y. Anderson.)

**Figure 8.13**
Computed fold model, wavelength = 42.1, after 100% shortening.
(a) Stress field. Short lines are drawn perpendicular to $\sigma_1$. (b) Strains.
Short lines are drawn perpendicular to axes of maximum total
shortening (i.e., parallel to directions of maximum total elongation).
(From Dieterich, 1969.)

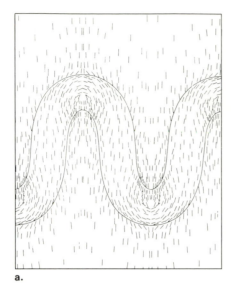

a.

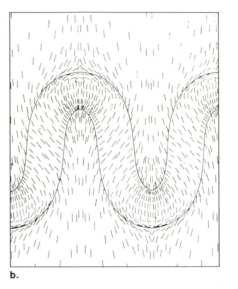

b.

## Computer Experiments with Folds

It is possible to simulate model experiments entirely
with numerical formulations. Thus, instead of using actual model materials and measuring effects of deformation, we can translate the necessary boundary
conditions, properties, and selected effects into a set of
mathematical equations. These can be solved to obtain,
for any set of numerical values for the parameters, relationships between stress, strain rate, and finite strain.
The process consists of two steps: first we set up mathematical relationships between all the parameters; and
then we solve for chosen sets of values with the aid of
a suitable computer program.

Two applications of this method are instructive. In
one, Dieterich and Carter (1969) and Dieterich (1969,
1970) mapped the stress distribution in an embedded
single competent layer during viscous folding. This
problem is practically insoluble by analytical methods
except for small amplitudes of folding. Dieterich, using
numerical methods, obtained stress and strain distributions for folds with steep limb dips (figure 8.13).
These are in remarkable agreement with stress orientations obtained from the dynamic analysis of grain deformation features in natural folds.

In another numerical experiment, Chapple (1970)
showed that, with increasing limb dips of folds, overall
viscosity of the competent layer and the matrix must
change. This seems clear empirically, for at low amplitudes the competent layer controls the strain rate,

owing to its higher viscosity, but at finite limb dips the
resistance of the competent layer diminishes rapidly.
Most of the stress within it then is concentrated at the
hinges (figure 8.14). When the fold becomes closely appressed, resistance to deformation increases once more,
as the limbs of the competent layer approach one another. In his computer experiments, Chapple was able
to recognize four phases in the deformation sequence.

1. Homogeneous contraction parallel to
   compression.
2. Establishment and growth of buckle folds (figure
   8.10).
3. Rapid weakening in compression, which he calls
   high-amplitude instability (figure 8.14).
4. Almost homogeneous bulk strain with tightly
   appressed fold limbs.

Biaxial strain and strain rates for successive stages in
this process (*not* corresponding to phases 1 through 4)
are shown in figure 8.15. Since stress and strain rate
are interdependent in viscous deformation, "weakening" may reflect increasing strain rate at constant
applied stress, or decreasing stress at constant strain
rate, or a combined effect of both.

In the above discussion we considered only single-
layer models. However, stacked-layer models (to represent sequences of strata) tend to confirm the same
principles.

## Figure 8.14

2 feet thick competent ($10^{21}$ poises) layer embedded in a ductile ($10^{18}$ poises) matrix, yielding at hinges as limb dips steepen (compare figure 8.10) (From Biot, 1961.)

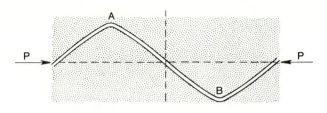

## Figure 8.15

Axes of finite strain ellipses and principal extension rates in the matrix for the folding of a competent viscous layer embedded in a less viscous matrix. Strain (crosses) and strain-rate (lines) of equivalent uniform compression are shown to the right of each diagram. Limb dips are: (a) 23°; (b) 46°; (c) 66°; and (d) 89°. (From Chapple, 1970.)

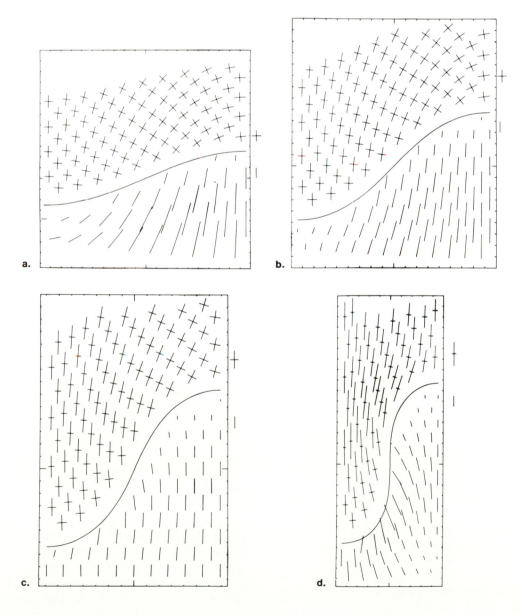

**Figure 8.16**
"Compression boudinage" (mullions) in steep limb of fold near
Dedenborn, West Germany. Competent sandstone layer in phyllite
(not visible), Siegenian (Lower Devonian).

## Boudinage

Deformation of stacked layers with contrasting ductilities may induce structures other than folds. Perhaps the best known such structures are *mullions* and *boudinage*. We shall discuss mullions later. **Boudinage** is a term derived from the French *boudin,* a kind of sausage popular in Europe. The term was first used informally in 1909, in the course of a field trip to the Belgian Ardennes, to describe a structure that looked like sausages lying side by side. The original outcrop no longer exists, but figure 8.16 shows a structure of the same type. It consists of a competent layer embedded in a less competent matrix, with constrictions in the competent layer at regular intervals.

## Origin of Boudinage

Where a low-ductility (competent) layer makes a large angle with the direction of shortening, the result is thinning and extension of that layer. The ductility contrast results in different rates of strain in the two lithologies involved, which in turn causes viscous instability in the competent layer. Where ductility contrast is low, we find "pinch-and-swell" structures (figure 8.17). This reflects a relatively lower strain rate in the competent material. Where ductility contrast is greater, rupture occurs along "necks" (figure 8.18). This is true boudinage. It indicates flattening at a large angle to layers and extension perpendicular to necking lines. Necks (constrictions) are parallel to the intermediate strain direction, and necking lines are sites of concentrated extension. Hence, they are favorable loci of mineralization by selective migration from within the deformed

## Figure 8.17
Flow of shale layer in recrystallized limestone resulting in pinch-and-swell structure. 1, shale; 2, laminated limestone; 3, calcite. (From Korn and Martin, 1959.)

## Figure 8.18
Boudinage structure; quartzite in Wissahickon Schist. (From E. Cloos, 1947b.)

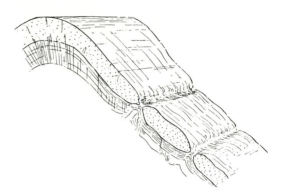

## Figure 8.19
"Boudinage within boudinage"; Ellison quartzite, Homestake Mine. (Photograph by W. Chinn.)

**Figure 8.21**
(a) Diagrammatic cross section of a folded layer in a ductile matrix. Cleavage traces indicate axial planes. Extension structures such as boudinage tend to develop along parts of limbs that make a small angle with cleavage; minor buckling folds develop in the hinges, where layers make a small angle with the contraction direction. Typical of some folding in Devonian schists of the Rheinische Schiefergebirge, West Germany. (b) Deformation of competent veins by boudinage and folding and the relationships of these structural styles to the directions of no finite longitudinal strain (n.f.l.s.) in the strain ellipse $l$ = finite longitudinal strain. (a. After Hoeppener, 1955; b. from Ramsay, 1967.)

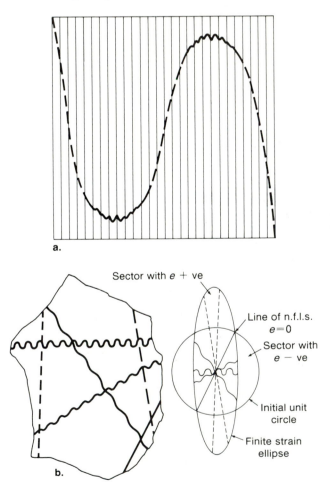

rocks (figure 8.19). In limestones, calcite nests are common; in quartzites secondary quartz enrichment is often found. Minerals such as sulphides also tend to migrate into the necks. This form of boudinage has become widely known through the classical works by Wegmann (1932) and E. Cloos (1947b).

Boudins are good strain indicators. Limbs of folds, for instance, may have boudinage neck lines oriented parallel to the fold axis, indicating extension of the less-ductile layers perpendicular to fold hinges. On the other hand, doming may result in neck lines oriented down the dip of dome flanks. This can easily be understood as stretching of the outer envelope of the dome in the course of its formation, with local extension greatest in a nearly horizontal direction. Boudinage occurs on all scales. Figure 8.19 shows an example of boudinage within boudinage.

## Boudinage in the Type Area

At the type locality and surrounding outcrops in the Ardennes, cleavage in the matrix around the boudins is at a large angle to the layering (figure 8.16). But, as we shall see in chapter 10, cleavage indicates shortening perpendicular, or nearly perpendicular, to cleavage planes at the time of formation. So, at the type locality, the competent layer has been shortened and thickened, and the shortening is at right angles to the axes (given by the lines of constriction) of the boudins. Here we have a rather curious situation: "boudinage" in the type area is not what is now commonly known as boudinage (e.g., figure 8.18). The structure in figure 8.16, which is what was first described as boudinage in 1909, is now known as *mullion structure*. We will discuss mullion structure in chapter 10.

## Minor Folds

Geologists once believed that almost all subsidiary or minor folds resulted from drag between beds in flexural slip folding. From this, the misleading genetic term "drag fold" came into use. In a great many cases, minor folds face toward the hinge of the associated major anticline, and this relationship has been explained by the assumption of "drag" along bedding toward the hinge. However, bedding slip does not normally produce minor folds that share the axial surfaces of the major structure. The most reasonable explanation of Pumpelly's rule (p. 128) and the reason that minor folds face toward the anticlinal hinge is simply that such concordant minor folds were produced in the course of the same strain as the major fold, and hence, share axial surfaces with it. Many concordant minor folds develop as buckle folds in a more ductile matrix (figure 8.20). Such folds may develop in the hinge area of the main fold, while the limbs, if at a large angle to the compression direction, may become boudinaged (figure 8.21). Some concordant minor folds may be an expression of heterogeneous laminar flow (see figure 8.6).

**Figure 8.23**
Minor folds and cleavage developed as a result of flowage of
mantling layers (2,3,4) down the flanks of a rising arch in the
basement (1). Such structures, common in the Appalachians of
eastern Vermont, are one kind of supratenuous fold.

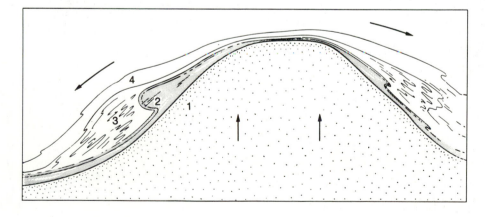

In migmatite terrains and other highly mobile environments, layers and veins that make a low angle with
a compression direction become buckled in characteristically tight, somewhat irregular, folds called **ptygmatic folds** (figure 8.22).

Some discordant minor folds may also be directly
related to the development of the major structure associated with them. Such is the case of subsidiary folds
developed in the course of flowage of mantling layers
down the flanks of a rising dome or arch (figure 8.23).

**Figure 8.24**
Repeated folding in Moine gneiss, Loch Monar, Scotland.

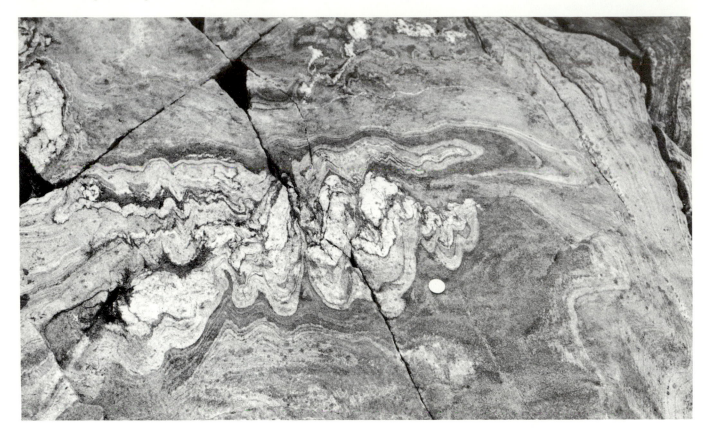

**Figure 8.25**
Structural map (with faults removed) of the Fiskefjord region, western Greenland, showing the interference pattern of domes, basins, luniform folds, and refolded isoclinal folds formed by triple to quadruple folding. (From Windley, 1969; in part from Berthelsen, 1960 and Lauerma, 1964; reproduced by permission of the Geological Survey of Greenland.)

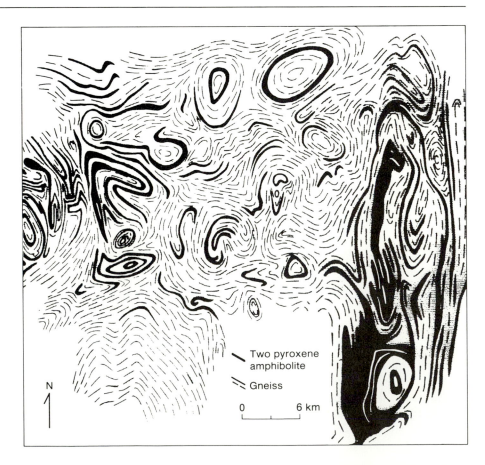

**Figure 8.26**
Comparison of orthogonal and oblique section of superposed folding
with variations of fold patterns (map views). (From Carey, 1962.)

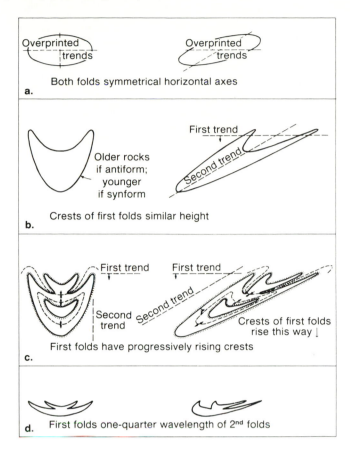

We may thus set up a separate genetic classification of minor folds: **dependent** minor folds, which are not necessarily concordant but form as part of the same deformation episode as the major structure; and **independent** minor folds, which are in no known way related to the dominant major structure. The latter are usually related to a different deformation phase, such as subsequent faulting.

## Successive Folding

Some folds show striking effects of successive phases of folding (e.g., figure 8.24). Folding of fold hinges along new axes results in complicated patterns where fold hinges cross. This may happen at any scale, from outcrop to map dimensions (figures 8.25 and 8.26). Ramsay (1967) gives an extensive account of repeated folding.

## Driving Forces

Most folds in layered rocks result from lateral shortening. The chief driving forces involved are (1) *lateral compression*; and (2) *gravity instabilities,* which are of two kinds: *density instabilities* and *relief instabilities*.

*Lateral compression* is normally caused (directly or indirectly) by plate convergence or igneous intrusion.

The most common geological examples of *density instabilities* arise where dense rock overlies less dense rock. To establish gravitational equilibrium (provided strain rates are adequate under the environmental conditions concerned), the denser material will sink and the less dense material will rise (see p. 190). In the course of this process, layered material may "crowd" to form folds (figure 6.25).

**Figure 8.27**
Cross section through part of a small trough filled with folded
Casper sandstone (Pennsylvanian). The trough filling plainly
"sagged" toward the southeast. (From Terra, 1931.)

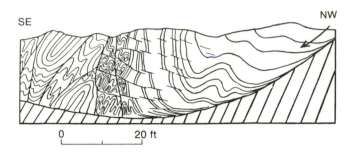

**Figure 8.28**
Recumbent fold in Nägelihorn, Switzerland, illustrating nappe style of
folding. Note antithetic faults in massive limestone at top, indicating
extension along axial planes. (Aerial photograph of the
Topographical Survey of Switzerland.)

*Relief instability* arises where rates of erosion are slow compared with strain rates of the rocks involved, so that high tectonic relief tends to be leveled by lateral flow toward lower ground. Such leveling may proceed by *sliding* along detachment surfaces, as in landslides, or by lateral *spreading* (see chapter 15), either along a detachment surface or without basal detachment (*rooted*). These processes are not mutually exclusive. The resulting lateral movement can and often does result in "crowding" of layered rocks and, hence, in folding (figure 8.27).

**Figure 8.29**
Tectonic cross section of the Helvetic nappes along the Reuss valley, Switzerland. (From S. Schmid in Funk et al., 1982.)

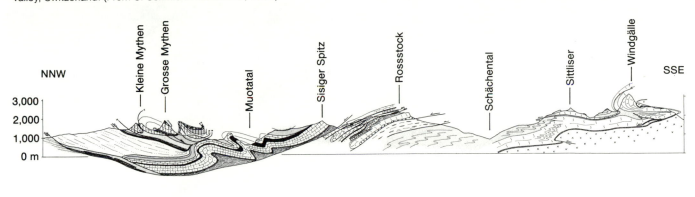

**Figure 8.30**
Cross section of Rattlesnake Mountain west of Cody, Wyoming. Paleozoic sedimentary rocks are draped over the faulted Precambrian basement, forming **drape folds** (descriptive term) or **forced folds** (genetic term). (From Stearns, 1978.)

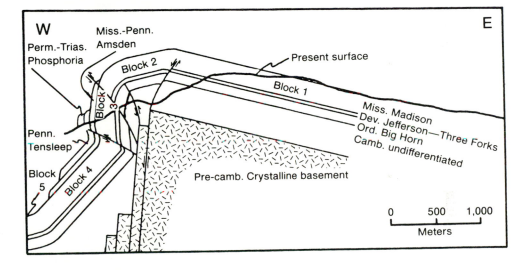

Many of the recumbent folds in the Helvetic nappes of the central Alps were formed by flexural slip and flow (figure 8.28). They present a movement picture of descent from the south (figure 8.29). Sliding under gravity is, at best, a partial explanation; conservative stratigraphic reconstruction of the strata involved in the folding seems to require at least, but probably much more than, 200 km of shortening by both folding and thrusting, to account for the pre–folding extent of the beds. Crustal shortening seems evident.

We shall discuss driving forces more thoroughly in chapter 15, because they assume considerable importance in the dynamics of thrusting.

Less commonly, folding can be caused by *vertical forces*. Diapirs are a good example (figure 6.25). But veneers of layered rocks may also become draped as folds over fault blocks in what are called *drape folds*, which are shaped by faulting of their core (figure 8.30). In these cases, the absence of lateral shortening should be apparent from the geological setting, though in some instances it may be controversial.

Before Nettleton's experiment (p. 86), many geologists believed that diapirs were the result of lateral compression. Some granite diapirs and gneiss domes (figure 16.14) may, indeed, be the result of both buoyancy and lateral compression.

## Folding Under Shallow Overburden

We have established some laws and rules governing folding in layered rocks. They are based on the assumptions that, in the course of folding, rocks behave pseudoviscously, and that flow occurs by grain-scale mechanisms that can be studied in experiments. However, much near-surface folding—the most commonly accessible style—has not occurred in this manner. At high crustal levels, strain rates of grain-scale plastic deformation mechanisms in carbonate rocks and sandstones are too slow to cope with the stresses induced by the driving forces we have considered. Many such rocks, where folded under shallow overburden, show little evidence of plastic deformation at the grain scale. In some shallow fold belts, such as those in the Jura mountains of France and Switzerland, and those in the Umbrian fold belt (figure 8.31) near Perugia, Italy, fracturing as well pressure solution (p. 96) in competent carbonate rocks are the mechanisms enabling these rocks to adjust to folding stresses. In these and other locations, less competent, interbedded marls tend to adapt passively to the folding. This alternation of competent and less competent beds facilitates fold formation because the competent layers are separated into thin, more easily folded stacks (figure 8.32).

Under shallow overburden, textbook-style folds with sinusoidal cross section are rare. In the Jura, Laubscher (1977) has found that many folds can be approximated by kink bands with rounded hinges. A characteristic cross-sectional shape is the *box fold* such as the Clos du Doubs anticline of figure 8.4.

Some rocks adjust by cataclastic flow. This involves microcracking and fragmentation ranging from the millimeter scale and less, to the meter scale and more; it allows the rock to adapt by bulk ductile flow and to form well-rounded folds. For such folds, no law exists that would allow us to extrapolate deformation rates, as we did for true plastic deformation mechanisms (chapter 6, p. 96). Furthermore, experiments show that at high crustal levels, folds do not form simultaneously in fold trains but, rather, in sequence, one fold piling up against the next (**serial folding**). Initiation of the first fold probably occurs over some topographic irregularity in the detachment surface underneath the folds.

**Figure 8.31**
Chevron fold in Cretaceous Scaglia Rossa, Umbrian Hills, Italy.

## Conclusion

Folding reflects strain in an inhomogeneously deforming mass. But what we see now is a "still". It is frozen movement and includes cases of "multiple exposure," in which the final states of several deformations are superimposed. The evolution of folding at all tectonic levels has to be reconstructed from such instantaneous pictures and from theoretical extrapolations.

**Figure 8.32**
Slightly overturned syncline in alternating sandstone and shale,
Flysch sequence in Allgäu Alps, West Germany.

With rare exceptions, folded layers occupy a shorter cross section than they did before folding. Understanding the strain history of folding can lead us to its causes. Shortening of cross section certainly implies lateral compression. But we must not extrapolate too far: compression may be on a crustal scale, as in some plate convergence, but it may also only involve a thin, detached upper layer in the crust, as in detachment folding (figure 8.4). In such cases, it may be more difficult to reconstruct the primary causes of folding, though it seems clear that gravity must have been involved at times (as in figures 8.27 and 8.29).

In our study of continuous structures, we now turn to the internal fabric of continuously deformed rocks and consider what it can tell us about the strain and strain history of folding.

## Review Questions

1. Distinguish between flexural folding and passive folding, emphasizing differences in mechanics rather than geometry.
2. Discuss the roles of simple shear and of pure shear in folding.
3. How does ductility contrast between rock layers lead to folding?
4. What is meant by the "rule of buckle fold size"?
5. Explain boudinage as a result of ductility contrast in rocks; apply the strain ellipsoid to an example.
6. Discuss the mechanics of formation of minor folds and of ptygmatic folds.
7. Under what geological conditions would you expect to find (a) flexural folding and (b) passive folding?
8. Discuss folding under shallow overburden.
9. What are some of the driving forces that might be involved in folding rocks?

Be sure you can *illustrate* your answers, where appropriate.

## Additional Reading

Davis, G. H. 1975. Gravity-induced folding off a gneiss dome complex, Rincon Mountains, Arizona. *Geol. Soc. Am. Bull.* 86: 979–990.

Elliott, D. 1972. Deformation paths in structural geology. *Geol. Soc. Am. Bull.* 83: 2621–2638.

Ghosh, S. K. 1974. Strain distribution in superposed buckling folds and the problem of reorientation of early lineations. *Tectonophysics* 21: 249–272.

Gwinn, V. E. 1964. Thin-skinned tectonics in the plateau and northwestern valley and ridge provinces of the central Appalachians. *Geol. Soc. Am. Bull.* 69: 863–900.

Hudleston, P. J. 1986. Extracting information from folds in rocks. *Jour. Geol. Ed.* 34: 237–245.

Hudleston, P. J. 1973. Fold morphology and some geometrical implications of theories of fold development. *Tectonophysics* 16: 1–46.

Johnson, A. M. 1977. *Styles of Folding: Developments in Geotectonics, 11.* Amsterdam: Elsevier. 406 pp.

Klein, J. 1981. Sequential development of fold orders in ptygmatic structures (Damara orogenic belt, Namibia). *Geol. Rundsch.* 70(3): 925–940.

Norris, D. K. 1971. Comparative study of the Castle River and other folds in the eastern Cordillera of Canada. *Geol. Surv. Can. Bull.* 205: 1–58.

Parrish, D. K., Krivz, A. L., and Carter, N. L. 1976. Finite-element folds of similar geometry. *Tectonophysics* 32: 183–207.

Prucha, J. J. 1968. Salt deformation and décollement in the Firtree Point anticline of central New York. *Tectonophysics* 6: 273–299.

Ramberg, H. 1964. Selective buckling of composite layers with contrasted rheological properties. *Tectonophysics* 1: 307–341.

Ramsay, J. G. 1967. *Folding and Fracturing of Rocks.* New York: McGraw-Hill. 568 pp.

Smith, R. B. 1975. Unified theory of the onset of folding, boudinage, and mullion structure. *Geol. Soc. Am. Bull.* 86: 1601–1609.

# 9

# Rock Fabrics— Fundamentals

In many rocks individual components form a systematic, repetitive pattern. For example, micaceous minerals are statistically parallel in slate; feldspars are aligned in some lava flows and other igneous rocks; and fractures are regularly oriented in most rocks. These are examples of repetitive ordering of rock components and structures, which we call the *fabric* of rocks.

## Definitions

The **fabric** of any object is the geometric relationship between repetitive constituent parts. These parts, or **fabric elements,** may be on any scale, the main characteristic being that they repeat themselves regularly in some definable pattern.

## Historical Notes

In 1824 and 1825, C. F. Naumann in Germany and G. P. Scrope in Great Britain drew attention to the ordering of grains in many igneous and metamorphic rocks. Scrope concluded that mineral alignment in metamorphic rocks indicates that they must have flowed in the solid state during deformation, in analogy with igneous rocks, in which crystal alignment reflects flow. Not surprisingly, both Scrope and Naumann saw rock deformation as solid flow; they were probably the first to suggest that folding might be the result of flowage and gliding under the influence of gravity. In his textbook, Naumann (1849) recognized two chief categories of fabric elements: planar structures (*Plattung*) and linear structures (*Streckung*). Their patterns were determined by a *vis directrix* (Latin meaning "directing force") or a *stress field*, in present terminology. Naumann was evidently well aware of the difference between force and stress, a difference that has been ignored not infrequently since then. He appears to have had a decidedly dynamic, almost modern, approach to tectonics. He saw rock structure as a kind of architecture, but an architecture which expresses movement.

In a classic work on rock deformation, Heim (1878), illustrated many fine examples of fabric resulting from strain in rocks (e.g., figure 9.1). He showed that these rocks must have flowed while undergoing strain and that ductile deformation imparts a fabric to the rock.

At the turn of the century, the Wisconsin school of structural geology under van Hise and Leith made an intensive, mainly qualitative, analysis of fabrics on all scales. Van Hise's *Principles of North American Precambrian Geology* (1896) and Leith's *Rock Cleavage* (1905) are classic works on the subject.

In 1911, Bruno Sander of Vienna demonstrated that metamorphic rocks are a particularly fruitful object of fabric study because their fabric expresses the pattern of particle movement that has taken place within them; this pattern is a more direct expression of tectonic strain than external form. The germ of this idea was implicit in Scrope's and Naumann's work on rock fabric, in Heim's (1878) beautiful studies of small-scale deformation, and in Becker's (1893) analysis of the strain of rocks. Sander named the study of rock fabric on any scale *Gefügekunde der Gesteine* (science of rock fabrics). Later, Walter Schmidt (1925, 1932) developed methods for the statistical evaluation of fabric data. Although the *Gefügekunde* of Sander and Schmidt encompassed rock fabrics in general, they were especially concerned with grain fabrics revealed by the microscope.

**Figure 9.1**
Crenulation cleavage, natural size, in dolomitic schist, part of the Röti Dolomite south of Piz Urlaun, Switzerland. Note that fold wave-length is controlled by layer thickness, according to Sander's rule of buckling fold size (p. 138). (From Heim, 1878.)

Rock fabric reflects a number of processes that rocks have undergone from the time of their deposition or emplacement. **Primary fabric** reflects genesis of the rock, either sedimentary or igneous (see figures 4.6 and 16.28). **Secondary fabric** reflects deformation, both in primary fabric and in the creation of new fabric elements.

**Fabric elements** of rocks are of two kinds: planar and linear. Planar elements include such structures as bedding, layering, schistosity, cleavage, and jointing; linear elements include flow lines, intersections of planes, fold axes, and directions of elongation of certain indicators such as conglomerate pebbles or mineral grains. A fabric made up of linear elements is known as a **lineation.** A fabric that consists of several sets of different fabric elements may be subdivided into **subfabrics** of like elements.

Measured orientations of fabric elements, or **fabric data,** are valid only within certain limits of confidence, called **domains.** A domain may extend over a given volume of rock or over a given order of magnitude. Thus, we may speak of fabric in the domain of recumbent folding, or in the domain of a fold limb, or in the domain of a few grains.

It is convenient to classify structural data and observations according to scale of observation. We shall adopt the classification proposed by Weiss (1959). The **microscopic scale** covers structures we can conveniently observe under the petrographic microscope, such as grain fabrics. The **mesoscopic scale** covers structures we can observe continuously, either directly or by means of a low-powered hand lens; this scale includes the range from hand specimens to large but essentially continuous local exposures. Some examples of mesoscopic structures are cleavage, jointing, and small-scale folds. The **macroscopic scale** includes structures too large or too poorly exposed to be seen directly in their entirety. The last two scales collectively are the **megascopic scale** of petrography.

## Homogeneity

Fabric data are usually measured at discrete points, or within limited domains. We can only extrapolate over a larger rock body if the fabric data are known to be repeated identically throughout, in other words, if the domain is homogeneous with respect to the fabric data concerned. Thus, homogeneity is relative, with respect to certain properties or data. In rocks, it can only be statistical, since no rock is perfectly homogeneous. Small-scale heterogeneities always form part of a rock body that is statistically homogeneous on a larger scale. Indeed, homogeneity depends on scale and domain. Figure 9.2 illustrates this point. Fracturing may introduce heterogeneity on the mesoscopic scale, but it may form a homogeneous pattern on a macroscopic scale (chapter 11).

**Figure 9.2**

Homogeneous domains with respect to both bedding *S* and fold axis *B*. Domain I is homogeneous with respect to both *S* and *B*. Domain II is inhomogeneous with respect to *S*, homogeneous with respect to *B*. Domain III is inhomogeneous with respect to both *S* and *B*. (From Weiss, L. E., "Structural Analysis of the Basement System at Turoka, Kenya" in *Overseas Geology and Mineral Resources, 7,* pp. 3–35 and 123–153, 1959. Reproduction by permission of the Director, British Geological Survey. UK Crown Copyright Reserved.)

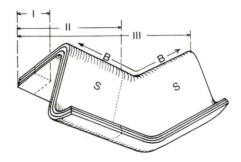

## Grain Fabrics

Mineral grains in rocks may become preferentially oriented as a result of emplacement and as a result of deformation. In the emplacement of sedimentary rocks, the orienting process is clear; it is governed by settling and compaction of platy and elongated grains, and, in some cases, by current flow. Interpretation of grain orientation in deformed rocks is based on the principles and mechanisms discussed in chapter 6. In grain-orientation analysis, three parameters are available for convenient measurement by optical methods: (1) actual *dimensional* grain orientation, usually obtained by measuring the attitude of grain boundaries; (2) orientation of *optic axes,* most commonly in uniaxial minerals such as calcite and quartz, where the optic axis coincides with the crystallographic *c*-axis; and (3) orientation of *cleavage planes* and *twin planes.*

### Indicator Minerals

Grain fabrics are defined by the orientation of certain **indicator minerals.** The most commonly used indicator minerals are quartz, calcite, and mica. *Quartz* is an abundant mineral in deformed rocks. Its dimensional orientation can be measured in a rather tedious procedure. But quartz is optically uniaxial, and the orientation of its optic axis (the crystallographic *c*-axis) can be much more easily measured. The relationship between the orientation of the *c*-axis and possible glide planes in quartz used in plastic deformation has been difficult to establish. Experimental work seems to indicate that (0001) is an active glide plane. However, quartz orientation may be used empirically to help in establishing strain symmetry.

Dimensional orientation of *calcite* usually results in a mesoscopically visible foliation, which can be readily measured in outcrop or on oriented specimens.

**Figure 9.3**
Mica girdle yields unique direction of lineation parallel to oriented
mica flakes of this fabric. (From Oertel, 1962.)

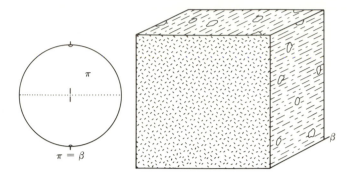

Calcite is far less abundant than quartz, but an impressive amount of experimental work has revealed a great deal of the glide mechanism in calcite, as well as in dolomite. The crystallographic parameters that can be measured in calcite include the optic axis ($c_v$) and twinning planes ($e$). The $e$ planes are easy to measure on twin lamellae (where present); they form a characteristic angle of 26° with the $c_v$ axis. The measurement of the calcite optic axis is rather tedious, but the orientation pattern of $c_v$ axes can be used empirically as a guide to subfabric symmetry.

The measurement of *mica* orientation is one of the easiest procedures in grain fabric analysis. Dimensional orientation and crystallographic orientation tend to coincide, and we can measure the orientation of mica flakes by measuring the attitude of their cleavage planes, which are crystallographic (001) planes. Where micas are part of a foliation, this foliation can usually be measured by mesoscopic methods. On the other hand, micas that are not part of a foliation may be parallel to one fabric axis and oriented randomly with respect to the other two axes. Microscopic analysis of mica orientation reveals the preferred direction (figure 9.3).

## Fabric Diagrams

Orientation of fabric elements is usually plotted in stereographic projection, the most convenient way of showing orientation of lines and planes on the surface of a sphere (see chapter 3). The traces of lines are points; the traces of planes are great circles. In fabric work it is usually convenient to plot planes as *poles* (i.e., the points of intersection of their normals with the surface of the reference sphere). Since data are statistical, we use the equal-area (Schmidt) net. The resulting diagrams are **fabric diagrams** or **pole figures.**

There are four significant patterns of point distribution for any homogeneous domain (figure 9.4).

1. The points may be randomly distributed, in which case the subfabric concerned is **isotropic** (figure 9.4a).
2. The points may be concentrated in one or more small areas of the projection, called **point maxima,** which represent statistical parallelism of the fabric elements concerned (figure 9.4b).
3. The points may be concentrated along a great circle zone, a distribution known as a **girdle.** If the points represent lines, then these lines are statistically coplanar; if they are poles of planes, then the planes concerned are in a statistically cylindrical zone as in figure 9.4c.
4. Points may be concentrated along a small circle of the projection. This distribution represents a conical zone around an axis; in other words, the fabric elements concerned make a fixed angle with a line (figure 9.4d). Girdles and small circles may have internal maxima, which express a preferred attitude, as in folds with straight limbs (figure 9.4e).

It is useful to emphasize point concentrations by contouring them. Contours represent concentrations per unit area. The unit area is usually 1% of the projection net area. Point concentrations are given either as the actual number of points counted per unit area, or as a percentage of total points plotted in the diagram. Contour steps need not be equal.

**Figure 9.4**
Point stereograms illustrating common distributions of lines and poles on the reference sphere. (a) Isotropic distribution. (b) Point maximum (axial symmetry). (c) Girdle (axial symmetry, if uniform). (d) Small circle (conical distribution of lines). (e) Girdle with internal maximum, reflecting preferred orientation within the girdle. This pattern is common in folds; maxima reflect concentration of attitudes along the limbs.

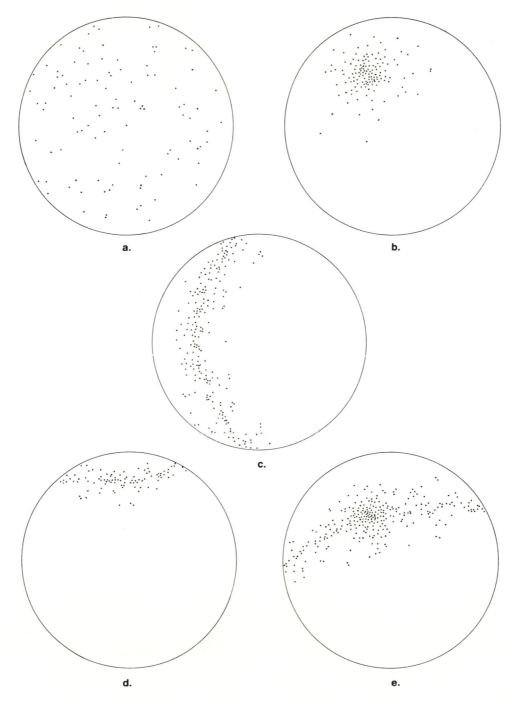

a.

b.

c.

d.

e.

**Figure 9.5**
Contoured equal-area projections illustrating different kinds of
symmetry observed in subfabrics (m = plane of symmetry). (a), (b)
Axial fabrics indicating parallel linear structures. (c), (d) orthorhombic
fabrics indicating three mutually perpendicular planes of symmetry.
(e), (f) monoclinic fabrics, one plane of symmetry. (g), (h) triclinic
fabrics. (From Paterson and Weiss, 1961.)

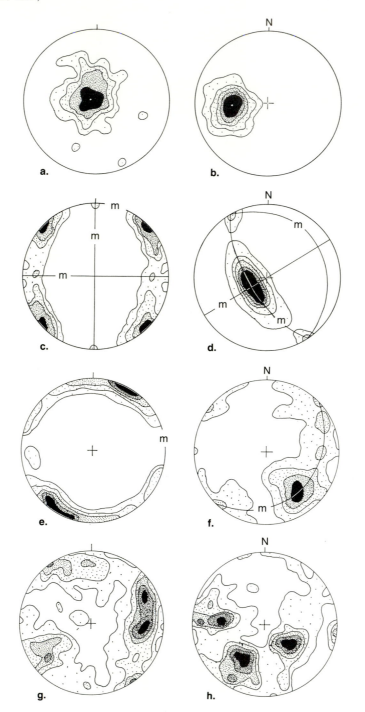

**Figure 9.6**
X-ray pole figures of chlorite (*top diagrams*) and mica (*bottom diagrams*) in Lower Carboniferous phyllites, Rheinische Schiefergebirge, West Germany. Note tight axial symmetry of (a), from a straight fold limb, and dispersion into a girdle in (b), from a hinge zone. (From Weber, 1980.)

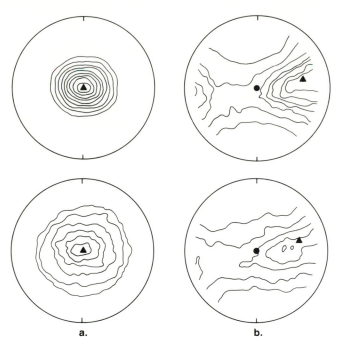

a.　　　　　　　　　　b.

## Statistical Significance

In normal tectonic analysis, tests of statistical validity are usually not necessary. A preferred orientation is reliable, if it is reproducible.

## The Role of Symmetry

The different kinds of fabric elements may be present singly or in any combination in a given rock body. Some groups of elements have a meaningful relationship to one another, and the most meaningful grouping is that of elements that are closely related to the same strain.

Fabric (other than depositional) reflects strain in rocks. Its symmetry reflects the cumulative strain symmetry. Just as fabric in rocks is defined statistically, so is fabric symmetry. Furthermore, since statistical homogeneity of fabrics varies with scale, symmetry must always be considered in conjunction with scale. In rock fabrics, four kinds of symmetry are encountered: orthorhombic, monoclinic, triclinic, and axial. These are best revealed in contoured pole figures (as in figure 9.5).

*Orthorhombic* symmetry has three reflection planes, mutually perpendicular. This symmetry is characteristic of homogeneous triaxial strain, and reflects a triaxial stress ellipsoid (figures 9.3, and 9.5c). Orthorhombic symmetry may be polar along one axis. *Monoclinic* symmetry has only one reflection plane. This is elementary bilateral symmetry, which is the symmetry of cylindrical folds, and by far the most common symmetry in tectonites (figure 9.5e). *Triclinic* symmetry has no symmetry planes, only a center of symmetry. Two or more superimposed monoclinic fabrics result in triclinic symmetry (figure 9.5g). *Axial* symmetry has one infinite-fold axis of symmetry, and one reflection plane normal to it (e.g., figure 9.6a). This is also the symmetry of homogeneous sedimentation. Axial symmetry is **polar** when properties change along the symmetry axis, as in graded bedding.

Preferred orientation in very fine-grained rocks, such as shales and slates, can be determined by X-ray diffraction. In this method, an X-ray beam slowly scans a rock specimen as it rotates in such a way that rays diffracted from every planar orientation can be recorded. The total intensity of diffracted rays in any direction over a given time span is a measure of the degree of preferred orientation in that direction. The results are recorded in contoured pole figures (e.g., figure 9.6). They represent preferred orientation and fabric symmetry in much the same way as optically determined pole figures (figure 9.5).

# Fabric Coordinates

In chapter 7 we introduced reference axes to describe fold form. Reference axes are also convenient to describe the symmetry and orientation of fabric. This has been a common practice in the past, and fabric axes have been used extensively in the mid-twentieth century literature on tectonic analysis. It is, therefore, very useful to become familiar with the notation, even though it is not common in modern structural analysis. We may use the notation as a useful shorthand to describe fabrics for determining strains.

We have seen that fabric, whether mesoscopic or microscopic, appears as sets of repetitive, parallel structural elements, which may be linear or planar. Where planar elements are present, these are labeled *s*; all *s* surfaces are **foliation** surfaces. Linear elements (lineations) are labeled *L* (formerly *B*). *L* may be directly observable, as a linear structure; or it may be an axis around which other elements appear to have been rotated. It is always normal to a plane of bilateral symmetry in the fabric, and may be so defined. Thus, determining *L* may amount to finding a plane of bilateral symmetry. In monoclinic fabrics, *L* is uniquely determined. If a preferred planar direction *s* is present, *L* will lie in *s*.

We may now define a set of three orthogonal axes of reference to describe a rock fabric. Let *L*, which lies in *s*, define a *direction*, *b*; let *a* be the direction in plane *s* which is perpendicular to *b*; and let *c* be perpendicular to *s*, that is, to both *a* and *b*. Then *a*, *b*, and *c* are the **fabric axes** or coordinates of fabric symmetry for a given point (or homogeneous domain) in a rock fabric which comprises both *s*-planes and lineations. Figure 9.7 shows fabric axes for an orthorhombic fabric; figure 9.8 shows fabric axes for a monoclinic fabric.

Surfaces and planes that intersect the fabric axes are identified by using the symbols *h*, *k*, and *l*, as in crystallography, where *h*, *k*, and *l* denote intersection at any angle with *a*, *b*, and *c*, respectively. (*h k l*) is a surface, or set of surfaces, that intersect all three axes. (*h 0 l*) is a set of surfaces parallel to *b*, or **tautozonal** about *b*. The symbols (*0 k l*) and (*h k 0*) (as in figure 9.7) refer to tautozonal surfaces about *a* and *c*, respectively. Also, *s*, which contains *a* and *b*, is known as the *ab* plane; and the symmetry plane perpendicular to *b*, which contains *a* and *c*, is known as the *ac* plane.

A fabric may, of course, comprise several lineations *L* and several *s* planes. These are usually identified by subscripts: $L_1$, $L_2$, $s_1$, $s_2$, etc. Direction *b* will be chosen along $L_1$, and *ab* will be $s_1$. Note that *L* and *s* are *identifiable* fabric elements, while *a*, *b*, and *c* are *abstract* directions. The most accurate two-dimensional representation of structure is the cross section normal to *b* (i.e., parallel to *ac*); this is the *tectonic profile* (see p. 125).

**Figure 9.7**
Fabric coordinates *a*, *b*, *c*, and related planes in orthorhombic symmetry. (From Dennis, 1967.)

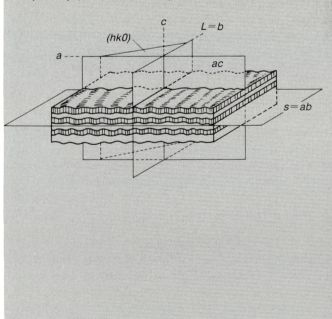

**Figure 9.8**
Monoclinic fabric symmetry expressed in mesoscopic fabric. (From Wilson, 1982.)

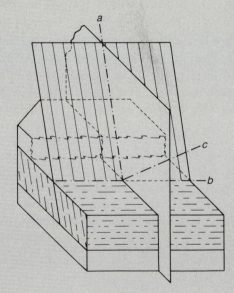

### Derived Axes

It is also convenient to make use of some derived axes. These are axes that are defined by geometric construction alone. An intersection between related *s* planes, or a statistical mean of such intersections, is a *β*-axis. Construction of *β* may help us to discover lineations that would otherwise remain undisclosed. In large-amplitude folds, finding the *β* of several bedding attitudes is a good way of approximating the axis of folding. The bedding surfaces, usually symbolized *ss* or *s$_o$*, are plotted on a stereographic projection as great circles, and each intersection of great circles is a *β*. The *β*-axis is the statistical mean of the attitudes of all construction intersections. This construction is a **β-diagram** (figure 9.9).

Instead of plotting *s* planes as great circles, we may plot them as their poles. The pole of an *s* plane is known as *π*. If all the *π*-poles are concentrated in a girdle along a great circle, the axis of that girdle is known as the *π*-axis of the family of *s* planes (figure 9.9). It is usually an *L*-axis. The construction is known as a **π-diagram** (see also figure 9.16).

*π*-diagrams are more reliable than *β*-diagrams. Many *β* concentrations, at first sight, look better than they really are, and some intersections are completely meaningless, such as intersections between unrelated *s* planes. The *π*-diagram selects related *s* planes, since only planes with a common *π*-axis will fall on a common great circle.

### Limitations

The *a, b, c* axial cross is meaningful only for monoclinic and orthorhombic fabrics. In orthorhombic fabrics, there may be no unique way of distinguishing *a* from *b*. We may select *b* as lying along the most prominent (and possibly the only) lineation *L*. Whatever the choice, the reasoning behind it must be explicit, even though the choice may seem obvious.

Sander (1930, 1948) and Schmidt (1932) defined **kinematic reference axes** in terms of postulated displacements during deformation; these axes Sander also labeled *a, b,* and *c*. The concept has proved confusing (Flinn 1962, Friedman and Sowers 1970), and older literature containing that notation must be read with care, to avoid misunderstandings. Many authors failed to indicate whether their reference axes were for fabric (acceptable) or "kinematic" (potentially misleading).

**Figure 9.9**
*π*-diagram and *β*-diagram. The mean position of all great circle intersections marks the attitude of the axis, in this case a fold axis. A great circle can be drawn through the girdle of bedding poles (black dots); the normal to the plane of this particular girdle is the axis. Note that there is less uncertainty in the positioning of the *π*-axis than of the *β*-axis. Normally, as here, *π* and *β* should coincide (large black dot, near top of stereogram).

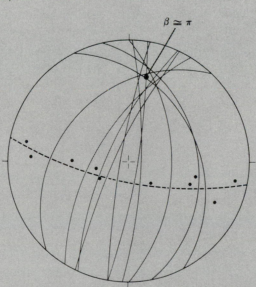

$\beta \simeq \pi$

**Figure 9.10**

Location of grains in gastropod shell (see figures 9.11 and 9.12). (From Friedman and Conger, 1964.)

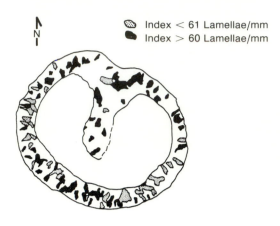

Index < 61 Lamellae/mm
Index > 60 Lamellae/mm

**Figure 9.11**

Distribution of normals to 100 sets of twin lamellae in calcite grains of shell in figure 9.10. Contours, 1, 2, 4, and 6% per 1% area; 12% maximum. (From Friedman and Conger, 1964.)

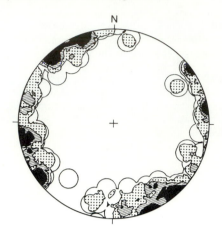

**Figure 9.12**

Approximate orientations of two $e$ twin planes and optic axes ($c_v$) in calcite crystals at four points within the gastropod shell of figure 9.10. With respect to an assumed general east-west greatest principal stress, grains along the north and south sides of the shell are favorably oriented for twinning because of the sense of shear on the twin planes, whereas those on the east and west sides are unfavorably oriented for twinning. (From Friedman and Conger, 1964.)

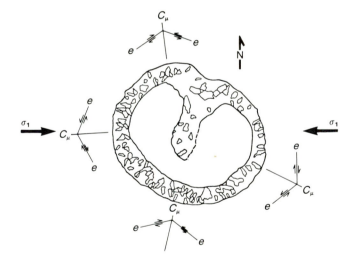

## Dynamic Interpretation of Fabrics

Past experimental and observational data enable us to relate preferred orientation in fabrics to the orientation of the related strain ellipsoid. In mesoscopic fabrics, such interpretation is easiest: for instance, cleavage forms normal to $Z$, extension fractures form normal to $X$ (see chapter 5). In some instances, both cleavage and fracturing may be discernible only on the microscopic scale, but interpretation is just as straightforward. Statistical flattening of grains, following Riecke's principle, suggests that schistosity is normal to $\sigma_1$, and hence to the $Z$ direction of the strain ellipsoid.

It is also possible to make use of the orientation of crystallographic elements. Friedman and Conger (1964) studied calcite crystals in a gastropod shell cross section from the Lower Cretaceous Kootenai formation. These crystals grow at a fixed orientation to the shell walls, so that a complete cross section (figure 9.10) gives a 360° coverage of orientations. Two opposing segments of arc were profusely twinned, and in these, preferred orientation of twin planes (figure 9.11) yielded a statistical orientation of the stress axes (figure 9.12), in accordance with the criteria previously established by Turner(1953) (see p. 102).

## Figure 9.13

Representative patterns (crossed girdles) of preferred orientation of c-axes in quartzites bordering Papoose Flat pluton; contours, 1, 2, 3, and 5% per 1% area. $L$ = lineation, $S$ = foliation. (From Sylvester and Christie, 1968.)

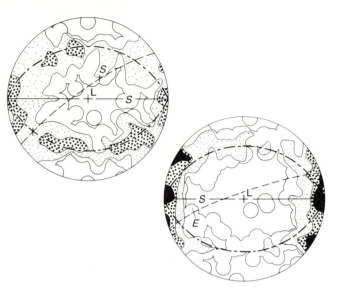

## Figure 9.14

Diagrammatic representation of the relationship between the foliation ($S$), lineation ($L$), and preferred orientation of quartz and calcite, and the long ($X$), intermediate ($Y$), and short ($Z$) axes of the strain ellipsoid (see figure 9.13). The principal stresses ($\sigma_1 > \sigma_2 > \sigma_3$, compressive stress positive) that are probably associated with the principal strains are shown in parentheses. (From Sylvester and Christie, 1968.)

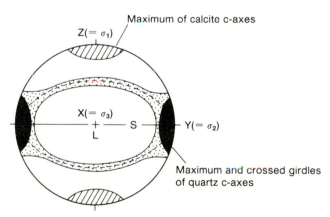

The Papoose Flat pluton in the Inyo Mountains of California is an intrusive granitic pluton that has stretched its wall rocks during emplacement. As far as is possible to ascertain, this is the only strain the wall rocks have undergone, and their flattening parallel to the contact at the west end of the pluton is amply confirmed in the mesoscopic fabric. Sylvester and Christie (1968) analyzed c-axis (optic axis) orientation of quartz crystals in wall rock quartzites and found a consistent pattern of two girdles crossing one another at nearly right angles (figure 9.13). This crossed-girdle pattern is symmetrically related to the strain ellipsoid as determined from mesoscopic fabric elements. The consequent probable relationship to the stress axes is summarized in figure 9.14. Away from the strained contact area of the pluton, quartz grains in the same quartzites are not significantly oriented.

**Figure 9.15**
Folds on the bank of Usken Lake, Central Sweden. Note discordant
lineation in hinge zone, and concordant minor folds (see figure 9.16).
(Courtesy H. J. Koark.)

Figure 9.15 shows minor folds in the hinge zone of a larger fold. The common axial direction and the symmetry are evident in the $\pi$-diagram of figure 9.16.

By means of these and similar analyses, we may obtain the orientation of stress axes for a given event throughout a selected rock unit or structure, and this enables us to plot a paleo-stress field over a given structure or structural element. Such analyses can be of tremendous importance in attempts to reconstruct the evolution of structures, and to understand the very nature of the deforming processes involved. We must remember that our analysis applies only to a given phase in the evolution of the structure. But it helps in charting the deformation path.

## Conclusion

We are now in a good position to evaluate the internal fabric of rocks in relation to the total deformation plan. We shall bear this relationship in mind as we look at some natural rock fabrics.

## Review Questions

1. Define and explain "rock fabric."
2. Distinguish between *isotropic* and *homogeneous* fabric.
3. What elements in a rock may become oriented in a preferred direction, and thus provide fabric data?

**Figure 9.16**

$\pi$-diagram of bedding attitudes (crosses) and fold lineations (dots) in figure 9.15. The large dot marked "$\pi$" is the axial direction obtained from the $\pi$-circle (dashed). (Courtesy H. J. Koark.)

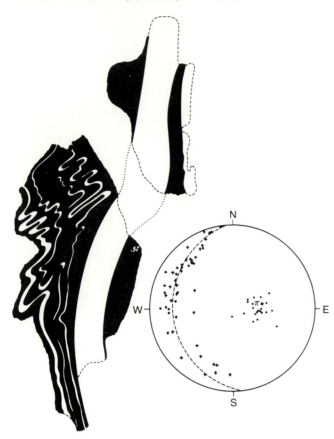

## Additional Reading

Fairbairn, H. W. 1949. *Structural Petrology of Deformed Rocks*. Reading, Mass.: Addison-Wesley.

Friedman, M. 1964. Petrofabric techniques for the determination of principal stress directions in rocks. Pages 12–37 in: *State of Stress in the Earth's Crust*. New York: Elsevier.

Kamb, W. B. 1959. Petrofabric observations from Blue Glacier, Washington, in relation to theory and experiment. *J. Geophys. Res.* 64: 1908–1909.

Knopf, E. B., and Ingerson, E. 1938. Structural petrology. *Geol. Soc. Am. Mem.* 6. 270 pp.

Oertel, G. 1983. The relationship of strain and preferred orientation of phyllosilicate grains in rocks—A review. Pages 413–447 in: M. Friedman and M. N. Toksöz, eds. *Continental Tectonics: Structure, Kinematics and Dynamics. Tectonophysics,* 100.

Paterson, M. S., and Weiss, L. E. 1961. Symmetry concepts in the structural analysis of deformed rocks. *Geol. Soc. Am. Bull.* 72: 841–882.

Turner, F. J., and Weiss, L. E. 1963. *Structural Analysis of Metamorphic Tectonites*. New York: McGraw-Hill. 545 pp.

Wenk, H. R., ed. 1985. *Preferred Orientation of Deformed Metals and Rocks: An Introduction to Modern Textural Analysis*. Academic Press, 610 pp.

4. What processes may give preferred orientation to fabric elements in rocks?

5. Illustrate different patterns of orientation on the stereonet.

6. Illustrate different fabric symmetries on the stereonet.

7. Define fabric axes *a, b,* and *c.* Apply them to a given example.

8. How would you obtain the attitude of a fold axis from a $\beta$-diagram, and from a $\pi$-diagram?

9. How does rock fabric reflect stress orientation at the time of its formation?

10. On differently shaped folds (open, tight, well-rounded hinges, angular hinges), measure at least 25 bedding attitudes and plot a $\pi$-diagram for each. Comment on the results. (If you cannot find suitable fold forms in your vicinity, make cardboard models.)

# 10

# Rock Fabrics— Field Relations

Early in the nineteenth century, geologists (e.g., Bakewell 1813) became aware that slate splits along well-defined, smooth parallel planes which are usually not parallel to bedding. This preferred direction of splitting has become known as *rock cleavage*. Adam Sedgwick in Wales and Arnold Escher in Switzerland were among the first to recognize that a relationship exists between cleavage and rock deformation. A great deal of attention focused on that subject in the mid-nineteenth century. In 1847 Daniel Sharpe explained that cleavage is caused by flattening of mineral grains; in 1853 H. C. Sorby attributed it to rotation of *originally* platy grains. Both came to the conclusion that cleavage forms perpendicular to compression. In 1849 C. F. Naumann listed all the fabric elements he had observed in the field, and classified them in the two categories of "platiness" and "stretching." He was referring to what today we should call *planar* and *linear* fabric elements, more specifically, *foliation* and *lineation*.

## Planar Fabric Elements (Foliation)

We may redefine **foliation** as referring to all planar fabrics, primary as well as secondary, with the exception of fractures. Foliation implies mechanical anisotropy, which manifests itself in potential, but not necessarily actual, parting. **Cleavage** is a secondary foliation that is commonly an expression of strain in rocks. Most cleaved rocks are metamorphic; that is, they have undergone partial or complete recrystallization. However, there are exceptions.

We recognize two broad categories of rock cleavage: *continuous* and *spaced*. In *continuous cleavage*, (figures 10.1–10.4), all observed platy mineral grains are statistically parallel, at least at the scale of observation. In *spaced cleavage*, domains of a statistically parallel fabric alternate with domains of an older, discordant fabric (figures 10.5–10.7), again, at the scale of observation. *Schistosity* is continuous cleavage due to complete recrystallization of the rock. Figure 10.1 shows a

**Figure 10.1**
Slaty cleavage in Aquidneck shale, Newport County, Rhode Island. Note bedding at large angle to cleavage. This is evidently the hinge area of a fold. (Subject area is approximately 4 m from left to right.) (Photography by J. B. Woodworth, U.S. Geological Survey.)

typical slate outcrop. The bedding, revealed by a faint banding, is steep; the cleavage is very prominent, dipping gently toward the right. As we shall learn below, the large cleavage/bedding angle indicates the hinge area of a fold.

## Continuous Cleavage

Figure 10.2 shows a photomicrograph of a slate. The mineral grains clearly have a statistically parallel orientation. Figures 10.3 and 10.4 show alternations of slate and graywacke. In figure 10.3 the cleavage is nearly vertical and transects both rock types without deflection. It is certainly the same feature in both rocks. The micrograph of another, similar occurrence (figure 10.4) reveals micaceous grains threading their way along grain boundaries of weakly oriented quartz crystals. This is true continuous cleavage also, because all the *platy* grains are statistically parallel, even though quartz crystals may form unoriented, or weakly oriented domains.

**Figure 10.2**
Continuous cleavage, Mariposa phyllite from the Sierra Nevada foothills, Caifornia. Note preferred alignment of grains throughout. This cleavage is also *zonal* (zones of contrasting texture). Natural light, ×160 magnification. (Thin section courtesy of A. L. Ehrreich.)

## Spaced Cleavage

Figures 10.5 and 10.6 are photomicrographs of phyllites. In each we see, superimposed on an older schistosity, a set of closely spaced cleavage traces that appear to have offset the earlier schistosity. These cleavage traces resemble the cleavages shown in figures 10.1–10.4 in that they are repetitive and parallel, and that no rupture has taken place along them. However, the later cleavage traces in figures 10.5 and 10.6 are discretely spaced and reveal an older fabric between them.

## Some Common Characteristics

In the field, cleavage must be distinguished from other common planar fabrics, especially bedding and closely spaced fractures. One needs a little experience for this in some cases, but the following criteria should help.

1. Cleavage normally consists of closely spaced parallel planes of weakness or potential parting; there is no rupture along these planes (except incidentally, as shown in figure 11.9). This characteristic usually distinguishes cleavage from fracturing, however closely spaced, for fractures are definite surfaces of rupture. Cleavage gives a marked anisotropy to rocks, which therefore tend to fracture preferentially along cleavage planes. Thus, jointing parallel to cleavage is common, but in such cases the cleavage always precedes the jointing.

**Figure 10.4**
Continuous cleavage passing from slate into graywacke; from Lower Carboniferous, Eastern Thuringia, Germany (thin section, natural light; magnification: ×70). (From Schroeder, 1966.)

**Figure 10.5**
Thin section of phyllite, central Vermont. Later (spaced) cleavage, which deforms earlier (continuous) cleavage, is much subdued along the fold limb and in the granular (quartz-rich) layers. The spaced cleavage is crenulation cleavage. Note its disappearance in the granular layers.

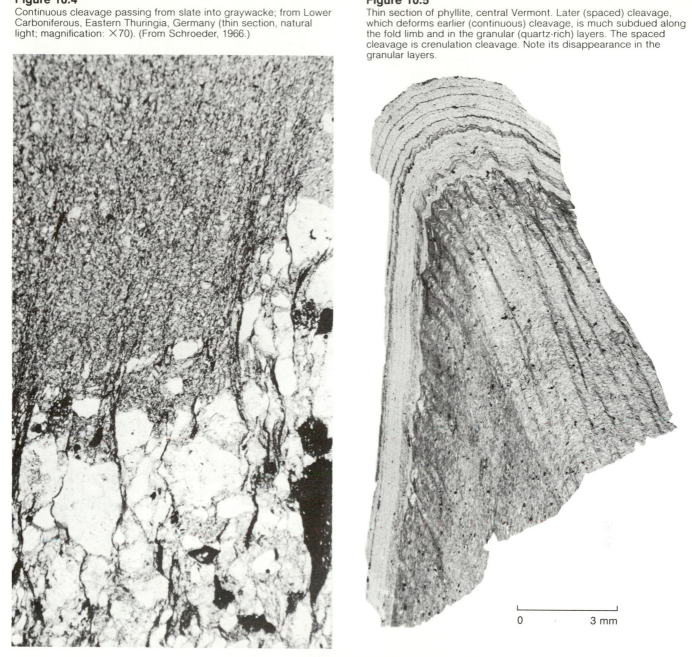

0          3 mm

**Figure 10.6**
Spaced cleavage, Ordovician Missisquoi Formation, Vermont (4 times natural size, direct (negative) projection to retain detail, by R. L. Harris, Jr.). Note incipient metamorphic differentiation along the spaced (late) cleavage planes.

2. The direction perpendicular to cleavage (the fabric *c*) lies close to the direction of greatest shortening in the strain directly related to the cleavage.

3. Cleavage may be distinguished from bedding in ambiguous cases by changes in texture or composition in the direction perpendicular to the foliation concerned. In cleavage, such changes are rare and are always repetitive on a small scale. In bedding, they tend to be less regular and on a larger scale. But there are exceptions!

## Descriptive Classification of Cleavage

The most useful classification of cleavage is descriptive, that is, a classification and nomenclature according to significant observable criteria. Older literature contains genetic cleavage terms whose meaning and applicability have led to misunderstanding and disagreement. Several of them are explained in the Glossary. Table 10.1 gives a modern descriptive classification based on Powell (1979). Let us define the most important terms.

**Rock cleavage** refers to all types of secondary planar fabric elements (excepting, by convention, coarse schistosity in crystalline schists) which impart mechanical anisotropy to the rock without apparent loss of cohesion. **Continuous cleavage** (figures 10.1–10.4) results from dimensional parallelism of all platy minerals present. It is continuous throughout the affected domain. It can be fine grained, as in slates, or coarse grained, as in schists. In rocks consisting almost entirely of small, platy, usually micaceous mineral grains, continuous cleavage tends to be fairly smooth and is thus distinguished as *smooth cleavage* (e.g., figure 10.1 and figure 10.3, *bottom*).

Where granular minerals make up an appreciable proportion of the rock, but all the platy mineral grains are statistically parallel, the cleavage has a "rough" appearance (e.g., figure 10.4, *bottom,* and figure 10.3, *top*); it is appropriately called *rough cleavage*. This is true continuous cleavage, because all platy minerals are statistically parallel (in contrast to spaced cleavage, defined below). In figure 10.3 the rough cleavage is evidently a continuation of the slaty cleavage in the slate band. By strict definition, **schistosity** is synonymous with continuous cleavage. As commonly used, however, it refers to a coarsely foliated continuous cleavage, in contrast to finer-grained slaty cleavage.

**Slaty cleavage** is the cleavage characteristic of slates: very smooth, very closely spaced cleavage planes. It can be either continuous or spaced (as revealed at high magnification). In very weakly metamorphosed slates, the cleavage is marked by spaced domains of recrystallized illite or sericite, with intervening domains of unaltered and unstrained rock. With increasing intensity of metamorphism, more argillaceous minerals become recrystallized so that eventually a schistose fabric results, consisting largely of well-oriented sericite. The scanning electron microscope has revealed the fabric of some very fine-grained weakly metamorphosed slates (figure 10.7).

**Spaced cleavage** includes all types of cleavage which, at the scale of observation (visual or microscopic) have discretely spaced surfaces of discontinuity separating largely undeformed intervening domains (**"microlithons"**), such as the examples shown in figures 10.5 and 10.6. They range from wide spacing (10 cm and more) to spacing barely discernible under the microscope. **Crenulation cleavage** is a spaced cleavage along which an earlier continuous cleavage has been deformed, so that the continuous cleavage surfaces have become crenulated (figures 10.5 and 10.6). There are two main types of crenulation cleavage: *zonal* and *discrete*. In zonal crenulation cleavage, the cleavage "planes" are actually zones of oriented micaceous minerals, and it is possible to trace the preexisting schistosity right through the cleavage zones (figure 10.6). In *discrete crenulation cleavage,* the cleavage surfaces are very thin and abrupt (figure 10.8); preexisting schistosity cannot be traced through them. Discrete crenulation cleavage is most common in very fine-grained rocks. *Disjunctive cleavage* consists of thin, distinct discontinuities in the preexisting rock fabric, but the intervening domain is totally unaffected, so that the cleavage planes tend to look like fractures, or microfaults (figure 10.8). However, this appearance is misleading for, in contrast to true fractures, there is no loss of cohesion along the cleavage planes.

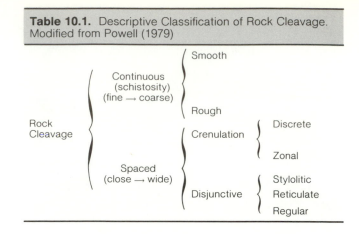

**Table 10.1.** Descriptive Classification of Rock Cleavage. Modified from Powell (1979)

| Rock Cleavage | | | |
|---|---|---|---|
| Continuous (schistosity) (fine → coarse) | | Smooth | |
| | | Rough | |
| Spaced (close → wide) | Crenulation | Discrete | |
| | | Zonal | |
| | Disjunctive | Stylolitic | |
| | | Reticulate | |
| | | Regular | |

**Figure 10.7**
Scanning electron micrograph of first-generation cleavage in Middle Devonian phyllite from the Rheinische Schiefergebirge, West Germany. Sample is from a fold limb. Bedding is parallel to scale bar (*bottom right*). Length of scale bar = 20 μ. (From Weber, 1981.)

**Figure 10.8**
Spaced cleavage resulting from selective dissolution, Devonian slates, Rheinische Schiefergebirge, West Germany. Reconstruction of original shape and attitude of fold limb, based on the assumption that the entire dissolved material has been carried away. (From Plessmann, 1965.)

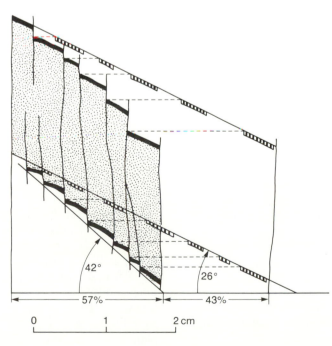

**Figure 10.9**
Anastomosing cleavage. (a) Longitudinal view showing traces in
Middle Devonian slates near Nuttlar, West Germany (60% natural
size). (Photograph by G. Langheinrich.) (b) cross section m-n in (a).

Disjunctive cleavage can be further subdivided according to several morphological types. Thus, *stylolitic cleavage,* seen in some deformed carbonate rocks, resembles a system of stylolites; indeed, as we shall see, there is a kinship between cleavage and stylolites. *Reticulate cleavage* is along surfaces that branch and rejoin in an anastomosing, network fashion (figure 10.9). *Regular disjunctive cleavage* consists of reasonably smooth, relatively closely spaced parallel planes, as in figure 10.8.

### Cleavage and Bedding—Rules of Geometry and Nomenclatures

Certain classical rules of thumb relate cleavage geometry to fold geometry. These rules apply only to ideal cases and to cases that represent an acceptable approximation to the ideal.

**Transverse cleavage** is any cleavage that transects known older foliation, usually bedding. Transverse cleavage tends to be parallel to axial planes of related folds. Where this relationship can be directly observed, the name **axial plane cleavage** (figure 10.10) is used. In many occurrences, cleavage planes are not strictly parallel to axial planes throughout the fold, but tend to *fan* about the axial hinge surface (figure 10.11). Such fanning cleavage surfaces are tautozonal about the *b*-axis of the fold. Furthermore, cleavage-bedding intersections are axial lines; in other words, they are parallel to the axes of related folds.

**Parallel cleavage,** usually a schistosity, is continuous schistosity that has developed along preexisting foliation. Parallel schistosity that follows bedding is also called **bedding schistosity.** Here, schistosity (cleavage) planes may be axial planes to occasional isoclinal subsidiary folds whose limbs are parallel to the dominant,

**Figure 10.10**
Axial plane cleavage showing cleavage–bedding relation in
overturned folds. In overturned limbs, the cleavage dip of axial plane
cleavage is less than the bedding dip.

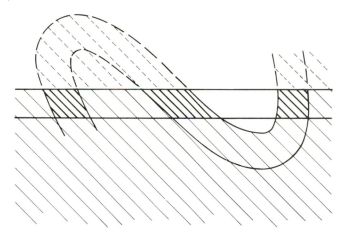

**Figure 10.11**
Cleavage refraction and fanning, syncline in Paleozoic phyllites,
southern Appalachians. (Photograph by C. D. Walcott, U.S.
Geological Survey.)

**Figure 10.12**
Intrafolial folds. (a) Intrafolial fold of feldspathic gneiss in biotite gneiss near Turoka, Kenya. (b) Rootless intrafolial fold of calc-silicate layer in marble near Weldon, California. (From Turner and Weiss, 1963.)

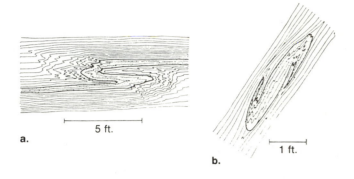

a.

5 ft.

b.

1 ft.

layer-parallel schistosity. Such folds, literally squeezed between layers, are called **intrafolial folds** (figure 10.12).

Cleavage may change attitude as it crosses a contact between rocks of different ductilities; this is **cleavage refraction** (figures 10.11 and 10.13). When this occurs, cleavage is normally deflected away from axial surfaces in the less ductile layers, and toward axial surfaces in the more ductile layers. As a corollary, cleavage tends to fan more in the less ductile layers, and to keep closer to ideal axial plane attitude in the more ductile layers.

It is clear that, in overturned folds, the dip of axial plane cleavage would be intermediate between that of the normal and that of the overturned limbs. This relationship furnishes a simple criterion for diagnosing overturning in the field, as demonstrated in figures 10.10 and 10.14.

**Figure 10.13**
Refraction of cleavage in Ardnoe Beds, Argyll, Scotland. The light-colored beds are quartzites, the darker beds are phyllites. Note constancy of cleavage attitude in each lithology. (U.K. Crown copyright, British Geological Survey photograph.)

**Figure 10.14**
Overturned syncline in Paleozoic slates, southern Appalachians.
Note dip of cleavage and its relation to the dip in each limb of the
fold. (Photograph by A. Keith, U.S. Geological Survey.)

**Figure 10.15**
Two models for compression of graptolites and the enclosing rock perpendicular to cleavage, where the bedding is perpendicular to cleavage (*top*), at a moderate angle to cleavage (*middle*), and parallel to cleavage (*bottom*). The right column shows the effect of constant-volume strain: the bedding plane and graptolite are greatly extended if cleavage and bedding are parallel, and somewhat extended if the cleavage/bedding angle is moderate. In the left column, the volume loss model, the bedding and graptolite are not extended significantly even where cleavage is parallel to bedding. The graptolites are shortened even where they lie on a bedding plane at a moderate angle to cleavage. (From Wright and Platt, 1982.)

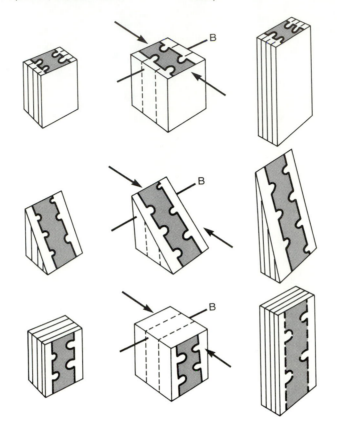

## Cleavage and Strain

Cleavage is a manifestation of strain in rocks. Much evidence, mainly from slates, has shown that cleavage tends to be perpendicular to the $Z$-direction of the strain ellipsoid associated with cleavage formation. This evidence includes strain markers such as conglomerate pebbles, originally spherical discoloration spots, and distorted fossils (figures 5.7, 10.15, and 10.16). The strains so measured vary over a wide range. Wood (1973) found the mean deformation ellipsoid from the Taconic slate belt to have axial ratios of 1.7:1:0.17, or an average shortening of 75%.

Cleavage attitudes may deviate from being strictly perpendicular to the short axis of the associated strain ellipsoid. For instance, where strain is not homogeneous, the orientation of the principal stresses and that of the principal strains may not coincide. In that case, a shear component would act along cleavage planes, resulting in slip along them. This is comparatively rare. But Dieterich (1969), in a computer-generated fold (chapter 8, p. 140), found resolved shear stresses acting parallel to cleavage directions (figure 10.17). This theoretical pattern is in remarkable agreement with some natural fabrics (as in figure 10.18).

In the course of progressive deformation, cleavage planes, once initiated, may well rotate out of their original attitude normal to the short axis of the strain ellipsoid. Or, in folded rocks, cleavage may not begin to form until late in the folding process. So the strain axes at the time of cleavage initiation may not be those that determine fold geometry. Thus, the relation of cleavage to the final fold form may sometimes appreciably deviate from the ideal. Also, some strain indicators (such as crystallization in strain shadows, figure 6.16) may form later than the cleavage, and thus may not reflect the true relationship between cleavage and strain. In any such deviations, a shear component of strain may form along cleavage planes, adding simple shear to dominant pure shear (flattening). But simple shear is then subsidiary. On the other hand, there are instances of dominant simple shear in planar fabrics: shear zones and mylonite fabrics. These fabrics are related to faulting, and we shall deal with them in chapter 12.

## Figure 10.16
Stretched belemnite. (a) Longitudinal view showing gaps between segments filled with calcite. Frette de Sailles, Switzerland (60% natural size). (b) Cross section m–n in (a). (From Heim, 1878.)

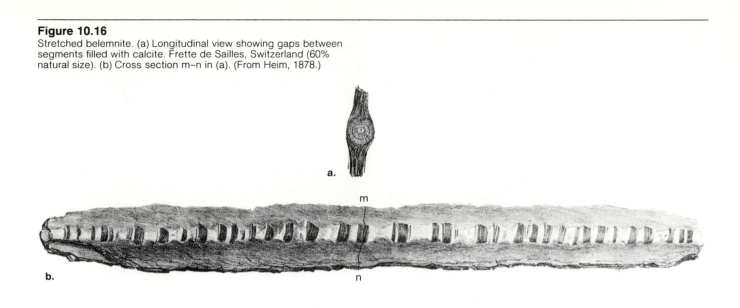

## Figure 10.17
Computer model of folded layer in less viscous matrix. Average lateral contraction of 100%. Arrows indicate sense of shear stress where the principal stress and strain axes are mutually inclined at large angles. (From Dieterich, 1969.)

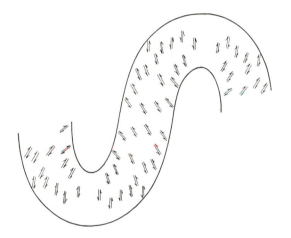

## Figure 10.18
Sketch of quartzite microfold in phyllite. Foliation in quartzite layer consists of preferentially oriented muscovite and chlorite. Foliation in the matrix is crenulation cleavage. (From Dieterich, 1969.)

1mm

**Figure 10.19**
Photomicrograph of Devonian slate, Rheinische Schiefergebirge,
West Germany, showing *Styliolina* dissolved along cleavage planes.
Scale: Width of micrograph = 850 $\mu$. (Photograph by W. Plessmann.)

**Figure 10.20**
Scanning electron micrograph of Devonian slate, Rheinische
Schiefergebirge, West Germany, showing *Tentaculites* partially
dissolved along cleavage plane. Scale: Width of micrograph = 500 $\mu$.
(From Weber, 1976.)

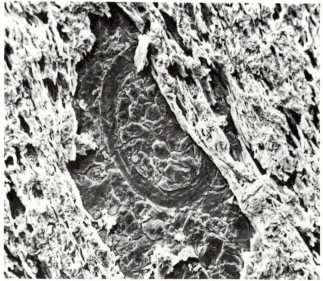

## Processes of Cleavage Formation

Three basic processes may form rock cleavage, acting either separately or in conjunction. They are
(1) rotation of mineral grains or buckling of a preexisting foliation; (2) pressure solution; and (3) recrystallization. A fourth process, discrete ductile shear,
may sometimes accompany one or more of the principal processes.

### Rotation and Buckling

Thin-section examination of some incipient cleavages
shows progressive rotation of platy mineral grains,
mostly illite, into nascent cleavage planes. Buckling as
a cleavage forming process is best exemplified by crenulation cleavage. Here cleavage develops along aligned
limbs of stacked microfolds (e.g., figure 10.5). This
buckling gives rise to crenulation of the folded surface,
hence the name.

### Pressure Solution

In some examples of spaced cleavage, the cleavage surfaces appear to occupy zones of "lost" volume, after the
manner of stylolites (p. 59). For instance, cleavage surfaces may truncate fossils (figures 10.19 and 10.20): a
part of the truncated fossil has evidently been dissolved
along cleavage surfaces. This process results in shortening perpendicular to the cleavage, as shown in figure
10.8. The lost volume may be documented not only by
truncated fossils, but also by offset of veins or other late
features that intersect the cleavage at an oblique angle.
The soluble material "lost" along the cleavage surfaces
may be redeposited along extension veinlets at a large
angle to the cleavage, and also in pore spaces in the
undeformed intercleavage lamellae. The cleavage surfaces themselves are commonly lined with insoluble
residues, in a manner reminiscent of stylolite seams. The
cleavage shown in figure 10.21 is also pressure-solution
cleavage, with rather wide spacing. It is not uncommon
in deformed limestones. Here the borderline between
pressure-solution cleavage and stylolites becomes indistinct.

### Recrystallization

Recrystallization is a common process in cleavage formation. In crenulation cleavage, micas may recrystallize along cleavage planes during and after rotation. In
such cases the grains will grow with (001), that is, with
mica cleavage flakes perpendicular to the short axis of
instantaneous strain—in other words, parallel to the
rock cleavage. Early crystallized grains will rotate into
final cleavage planes. If metamorphism leading to recrystallization lasts sufficiently long or is sufficiently
rapid—as in a wet environment—the whole rock will
recrystallize, resulting in a homogeneous schistosity.
Schistosity is always perpendicular to $\sigma_1$ at the time of
crystallization. In all cases of recrystallization, growth

**Figure 10.21**
Stylolitic cleavage in Turonian limestone, foreland of Harz mountains,
West Germany (irregular cleavage planes dip gently to the left, i.e.,
south). (a) General view. (b) Enlarged view of central portion.

a.

b.

and orientation of new platy mineral grains are controlled by stress, not by the orientation of preexisting crystal grains, except in isotropic stress: textural studies have shown that *mimetic schistosity,* that is, schistosity controlled by a preexisting foliation, rather than contemporaneous strain, must be extremely rare. Hence, *bedding schistosity* normally is due to shortening perpendicular to bedding. Most examples of bedding schistosity appear to have been initiated when the layering was relatively undeformed. The inference would be that early bedding schistosity is the result of crystallization under vertical load. While this can certainly lead to shortening perpendicular to bedding, there would also need to be a net loss of volume. Recrystallization of clay minerals to minerals of the mica family does

involve loss of water, so that volume loss can be accounted for. Bedding cleavage and schistosity may also form in limbs of recumbent and isoclinal folds.

Recrystallization is sometimes accompanied by metamorphic differentiation along cleavage planes: thus, cleavage-parallel lamellae may become alternately enriched in granular and in micaceous minerals: this is *chemical transposition.* (In contrast, *mechanical* transposition is the squeezing of contrasting lithologies into parallelism with cleavage in extreme strain.) Transposition gives a striped appearance to the rock. The common tendency toward chemical transposition in crenulation cleavage suggests that recrystallization has an important role in its formation.

## Discrete Ductile Shear

We have seen (p. 178) that simple shear along cleavage planes can be documented for some cases. The displacement surfaces represent true cleavage only if they are nonruptural. By definition, cleavage planes are planes of *potential* parting. Hence, the formerly used term "fracture cleavage" was misleading. Sets of cleavage-like, closely spaced fractures are not cleavage, but are true fractures; they are end members of a series grading from pervasive shear through discrete ductile shear to brittle fracture, as will be shown in chapter 13. On the other hand, fractures may form along preexisting cleavage planes (figure 11.9). Such fractures are not related to the stress field that formed the cleavage; it merely provides a convenient direction of weakness in the rock.

## Environmental Factors

In what environments are these different processes active? Buckling alone is a low-temperature process, and it needs a preexisting foliation. Rotation, similarly, is most evident in lowest-grade metamorphic rocks. Pressure-solution processes act over a wider range of temperatures. We have seen that these processes affect unmetamorphosed limestones. In that setting, cleavage planes are very widely spaced. But pressure solution also forms spaced cleavage under conditions of greenschist facies metamorphism. Chemical transposition (p. 181), for instance, is caused by pressure-solution processes. Its most common setting is in crenulation cleavage: granular minerals are depleted by dissolution from the limbs of microfolds and are enriched in fold hinges. Gray (1979) estimated the relative importance of buckle shortening and dissolution shortening in crenulated Silurian slate. He obtained the following ranges for the two processes; buckle shortening, 43–51%; dissolution shortening, 13–19%; and total shortening, 51–59%.

Pressure-solution processes in crenulation cleavage formation involve very short range migration of material, at the scale of a microlithon (cleavage lamella). In recrystallization diffusion occurs throughout the rock. Thus, recrystallization is a higher-temperature phenomenon. This is well illustrated by cleavage in the Willoughby arch of eastern Vermont, which consists of metamorphic rocks mantling gneiss domes. Near the crest of the arch, the whole fabric of the mantling rocks is recrystallized into a coarse schistosity. On the flanks, the same schistosity grades into crenulation cleavage, much of it exhibiting metamorphic differentiation. Thus, pressure solution evidently was active in the cooler flanking rocks, whereas the rocks closer to the hotter crest underwent complete recrystallization.

**Figure 10.22**
Kink bands. (After T. B. Anderson, 1964.)

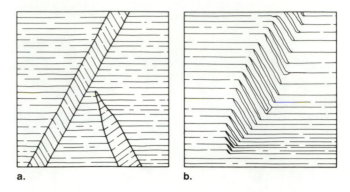

a.       b.

## Kink Bands and Kink Folds

In some schists and phyllites, schistosity is sharply deflected or kinked along tabular zones. These zones are bounded by kink surfaces in sets of two (figures 10.22, 10.23, and 10.24), and deflection is confined within the tabular zone or **kink band** between the kink surfaces. Kink surfaces may also form at more or less regular intervals, as axial planes to certain chevron folds called **kink folds** (see figure 7.9). Whereas "chevron fold" is a purely geometric, descriptive term, "kink fold" has a definite genetic implication: a chevron-type fold formed by the mechanism of kinking in thin, even layering or in a schistosity. The kink surfaces in kink folds may be part of a spaced schistosity in axial plane attitude.

The kink in these bands is quite sharp, and remarkably similar to analogous features in crystals. The kink band represents a shear zone, and it always belongs to a set of such zones. Two sets may form conjugate pairs, resulting in fold patterns that have been called *conjugate folds*. They have been reproduced in the laboratory, where they may precede the formation of regular shear folds (figure 10.24).

Where the shear sense of kink bands follows the shear sense of the foliation in which it is generated, the kink band is said to be **synthetic,** in analogy to similarly classified normal fault displacements. Where the shear sense of kink bands is opposed to the shear sense of the foliation, the shear sense is **antithetic.**

Following Dewey (1965), with some modifications, we may distinguish four types of kink bands: discrete, segregation, depletion, or shear. **Discrete kink bands** are synthetic kink bands in thinly foliated rocks, bounded by a pair of parallel kink surfaces (figure 10.22 *left*).

## Figure 10.23

Kink bands in Lewisian amphibole schist, Loch Hourn, Scotland.

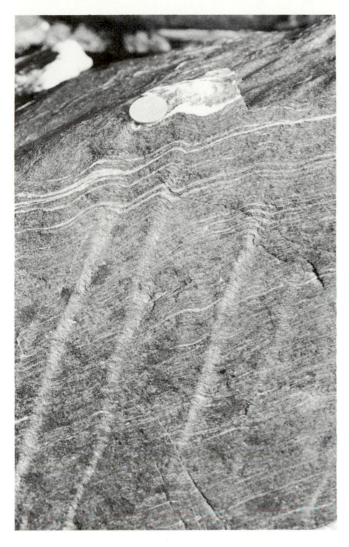

## Figure 10.24

Experimentally produced kink folds. Phyllite shortened approximately 20% parallel to the foliation, in a thick brass jacket, showing folds formed at the intersection of kinks (thin section; magnification: ×65). (From Paterson and Weiss, 1962, reprinted by permission from *Nature*, copyright © 1962 Macmillan Journals Limited.)

**Segregation kink bands** are discrete kink bands in which the bounding surfaces are weakly developed. There is extension normal to the foliation planes within the kink zone, and this results in segregation of soluble minerals between foliation surfaces (figure 10.22 *right*, and figure 10.23). **Depletion kink bands** have kink zones impoverished in quartz relative to the host rock by migration of silica. **Shear kink bands** have foliation deflected along penetrative axial planes, resulting in simple shear within the kink zone. Shear kink bands may grade into crenulation cleavage and kink folds.

## Tectonic Setting of Cleavage

We use the terms **pre-, syn-, and postkinematic** to refer to structures or rocks formed before, during, and following strain, respectively, thereby indicating a sequence of tectonic events. It is also possible to establish certain sequential relationships of cleavage and metamorphism, mainly by studying the relationship between cleavage and porphyroblasts developed in the same rock.

**Figure 10.25**
Relations between relict schistosity in porphyroblasts and external
schistosity. (1a, 1b) Prekinematic metamorphism, no internal
schistosity. (2) Early synkinematic metamorphism. (3a) Development
of internal schistosity in porphyroblast. (3b) Deflection of external
schistosity during subsequent flattening. (3c) Rotation due to simple
shear along cleavage planes. (4) Late kinematic metamorphism,
more intense crenulation in rim of crystal. (5a) Preservation of
originally weakly crenulated cleavage $s_1$ in crystal; intense
crenulation of external schistosity, resulting in spaced cleavage $s_2$.
(5b) Relic of $s_1$ in crystal, external $s_1$ obliterated by $s_2$. (From Zwart,
1961.)

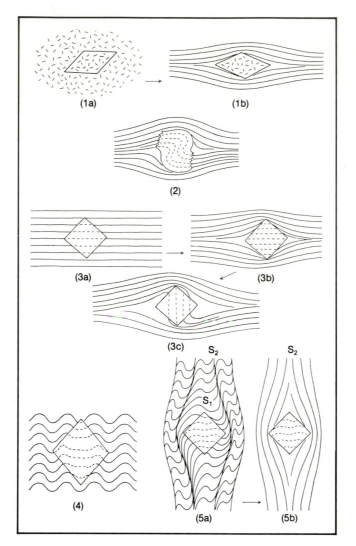

Briefly, if cleavage relics in porphyroblasts are un-
deflected (as in 3a of figure 10.25) then the porphyroblast is younger than the cleavage, or is postkinematic.
If cleavage is deflected around the porphyroblast, then
this cleavage is younger and the porphyroblast is prekinematic. Evidence of porphyroblast rotation (as in 2 and
3 of figure 10.25) indicates that prophyroblast growth
may have accompanied the rotation (as in 2) or preceded it (as in 3). A number of other related situations
are illustrated in 4 and 5 of figure 10.25.

## Cleavage in Orogenic Belts

Except for a very few localized occurrences, cleavage
is confined to orogenic belts, where its pattern expresses the flow of rocks. A surprisingly large proportion of known, well-developed occurrences of cleavage
are in Paleozoic orogenic belts. There, as a general rule,
it is predominantly bedding plane cleavage in its earlier
manifestations, and at deeper tectonic levels; it is predominantly axial plane cleavage in most later phases,
and at higher tectonic levels. Where folds are overturned in one direction, vergence and flow expressed by
cleavage are generally toward a cratonic foreland.

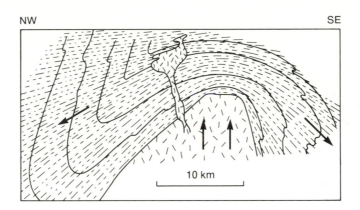

### Doming Cleavage

Rising diapirs compress mantling rocks in the direction
of their advance. Where the mantling rocks are metamorphic, as in plutonic diapirs, cleavage and schistosity
may form a draping pattern over the diapir (figure
10.26). This reflects strain during diapir ascent. Subsidiary folds formed in this process face *away* from the
crest of the diapiric fold, as illustrated in figure 10.26.
This contrasts with the more common case of compressive folding where subsidiary folds face *toward*
main fold hinges (e.g., figure 7.1).

Overall, cleavage expresses flow of rocks, and regional analysis of cleavage can reveal the flow pattern—an important step in reconstructing the strain
history and dynamics of regional deformation.

## Lineation

The first recognition of linear parallelism in rocks came
from the study of volcanic rocks. Early in the nineteenth century, von Buch and Scrope noted the arrangement of long axes of phenocrysts in lavas, and
found them aligned in the direction of flow. Scrope also
noted that gas bubbles were drawn out in the same direction. Soon afterward, he and others found minerals
in many metamorphic rocks aligned in a definite direction, usually along cleavage or foliation planes. Naumann (1833, 1849), in particular, studied linear features
of this kind, and he and Scrope independently came to
regard the linear parallelism of minerals in lavas and
in metamorphic rocks as of similar mechanical origin.
They believed that these structures were due to flowage
of the rock, and marked a direction of "stretching."
These early observers clearly understood that "linear
parallelism," or lineation, is penetrative and reflects deformation of a rock body as a whole. Evidence suggested that some lineations represented elongation in
the direction of lineation, that some marked the intersection between sets of planes (such as cleavage and
bedding), and that some were the result of folding about
parallel axes.

All pervasive linear rock structures, whatever their
extent or scale, are called lineations. They may either
lie within a foliation or constitute an independent fabric
element.

### Primary Lineation

The best-known and most commonly observed primary
lineations are aligned phenocrysts and elongated inclusions in igneous rocks. The alignment can be readily
correlated with the direction of flow of magma. Aligned
crystals in igneous rocks, therefore, form what are
known as **flow lines.**

### Secondary Lineation

The great majority of secondary lineations, lineations
that have developed in a rock body after its emplacement, are in metamorphic rocks. There are five processes by which consolidated rocks may develop
lineation: elongation; growth of minerals; folding; intersection of s planes; and slippage along s planes.

### Lineation from Elongation

A common example of elongation lineation is provided
by elongated conglomerate pebbles. Of course, strong
currents may impose a statistical orientation on long
pebbles at the time of deposition, but such orientation
is controlled by bedding planes. Tectonically oriented
conglomerate pebbles are elongated within cleavage
planes, and the elongation may be many times that of
naturally oblong pebbles. Common tectonic elongations are 1:2 to 1:3; elongations of 1:10 and more are
known. Evidently, the direction of elongation of pebbles
is a direction of elongation of the whole rock. Such
elongation may be parallel to axes of minor folds, an
interesting point that we shall consider later.

**Figure 10.27**

Deformation of oolite in pure shear, resulting in elongation lineation, and flattening in *ab*. (a) Block diagram of undeformed oolite. (b) Deformation 50%. (c) Deformation 100%. The block is now twice as wide and half as high. Bedding shows intense crenulation and has also doubled in thickness. (From E. Cloos, 1947a.)

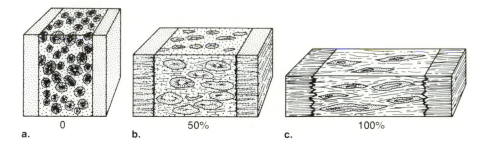

a. 0
b. 50%
c. 100%

**Figure 10.28**

Four cross sections through the western slope and crestal portion of the South Mountain fold, Maryland. Black representative strain ellipses are constructed from ooid measurements. (From E. Cloos, 1947a.)

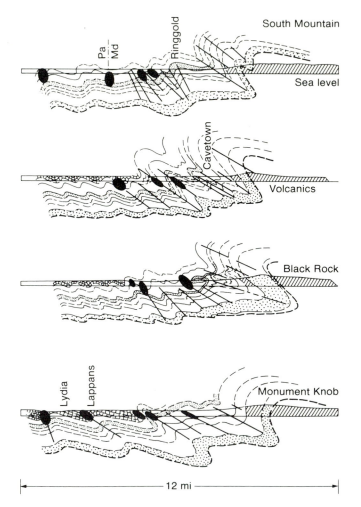

Ooids in oolitic limestone may become elongated into ellipsoids. Such instances are extremely valuable strain indicators, since the ooids represent almost perfect strain ellipsoids (figure 10.27). One of the best illustrations of this is the South Mountain anticline in Maryland, investigated in a classic study by E. Cloos (1947). It is a fold overturned to the west in oolitic Cambrian Conococheague limestone (figure 10.28), and has a primary fabric of tiny spherical ooids. It also carries a somewhat fanning axial plane cleavage. As a result of deformation, the ooids have become deformed into ellipsoids, true strain ellipsoids. They are flattened in cleavage planes, and are elongated at right angles to the fold axis. Clearly, the cleavage planes are parallel to $XY$ in the overall strain, and the lineation defined by the strained ooids is parallel to $X$ (elongation).

### Lineation from Crystal Growth

Some minerals, such as hornblende and other amphiboles, may grow during metamorphism with their long (*c*) axes in parallel alignment (figure 10.29). Other mineral grains may assume dimensional orientation not necessarily related to crystal form. Many minerals crystallizing under stress may assume a statistical orientation by lattice, as we have seen in chapter 9.

Mesoscopically visible lineations from crystal grain orientation are parallel to other types of lineation, and serve to reinforce them. This orientation by preferred direction of mineral growth is controlled largely by anisotropic permeability for mineral solutions. (*Wegsamkeit* is Sander's German term for this, which has no English equivalent) and by Riecke's principle (p. 96).

**Figure 10.29**
Lineation by crystal growth. Amphiboles in Tremola schist, southern
border of Gotthard massif, Switzerland.

## Fold Lineation

Folds and crenulations are linear structures. Platy minerals, especially micas, may also occur in a fabric in such a way that they all have one direction in common: their poles plotted on a stereogram ($\pi$-diagram) will form a girdle. The axis of this girdle is a lineation (see figure 9.3). Such a lineation may not be immediately apparent mesoscopically, but other features, such as parallel minor folds, may indicate it.

*Mullion and rodding structures* are lineations developed in strongly metamorphosed rocks. **Mullions** are columnar structures bounded by long cylindrical surfaces (figures 8.16 and 10.30). The surfaces may be polished, covered by mica films, or striated longitudinally. They may be any size, from a few centimeters to a few meters in cross section. Some are the right size for fence posts, and, according to Wilson (1961 and 1982), are so used.

Wilson distinguished four types of mullions.

1. **Fold mullions** possess regular, curved cylindrical surfaces corresponding to bedding or preexisting *s* surfaces. Most of these are largely detached hinges of minor folds, and identifiable bedding surfaces within fold mullions are conformable with the external mullion surface.
2. **Bedding mullions,** in Wilson's terminology, are structures like those originally named "boudinage" in the Ardennes (see figure 8.16). Some geologists prefer to restrict the term "mullion" to this type.
3. **Cleavage mullions** are formed in cleavage, rather than in bedding surfaces; they tend to be more angular in cross section.
4. **Irregular mullions** are the most common type. They have the typical cylindrical mullion surfaces, but in cross section the surfaces interlock irregularly, and internal bedding surfaces are only locally comformable with the mullion surfaces.

Mullions furnish the same geometric indications as cylindrical folds. Minerals such as hornblende are oriented parallel to the long axes of some mullions, suggesting possible longitudinal extension.

**Rodding** is somewhat similar to mullion structure, but the material of rods is monomineralic, generally quartz. Most rods are also much thinner than mullions. They may be detached hinges of smeared-out intrafolial folds in quartz or other veins, but in many cases it seems that the quartz has crystallized in fold hinges *after* most of the deformation has occurred.

**Figure 10.30**
Mullion structure in Lewisian gneiss, Poolewe, Scotland. (U.K. Crown copyright, British Geological Survey photograph.)

Some minor folds have their axes parallel to maximum extensional strain ($X$). This seeming paradox is well illustrated in diapirs. In salt domes, fold axes of bedding surfaces are steep to vertical, apparently following the flow direction, the rise of the salt (see figure 6.26). Secondary foliation planes are parallel to axial planes of folds; they are flow planes. Thus, the steep folds in salt diapirs have axes of folding parallel to the direction of flow. In the famous stretched pebble conglomerate at Bygdin, Norway, fold axes are parallel to the direction of pebble elongation (Hossack 1968). In figure 10.31, E. Cloos shows one possible mechanism for achieving this relationship.

## Intersection Lineation

Different sets of planar fabric elements that are symmetrically related, such as cleavage and bedding in simple folds, or *s* surfaces formed in the same act, intersect along traces that form a lineation. We know that many intersection lineations are parallel to fold lineations of the same fabric. In phyllites and schists, these intersection lineations become apparent as crenulations on schistosity; hence the name crenulation cleavage. The presence of more than two intersecting surfaces may be revealed by two or more sets of intersecting crenulations on an *s* surface (figure 10.32).

## Figure 10.33

(a) Flexural folding of an early lineation $L_1$ on a surface $S_1$.
(b) Passive shear folding of an early lineation $L_1$ on a surface $S_1$.
(From Wilson, 1982.)

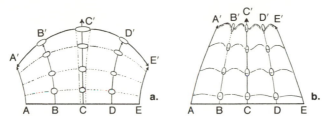

## Figure 10.32

Intersection of two sets of cleavage forming intersecting crinkles in the earlier surface; Northfield formation, near Hardwick, Vermont. The steeper lineation, locally more pronounced, is older.

**Figure 10.33**
(a) Flexural folding of an early lineation $L_1$ on a surface $S_1$.
(b) Passive shear folding of an early lineation $L_1$ on a surface $S_1$.
(From Wilson, 1982.)

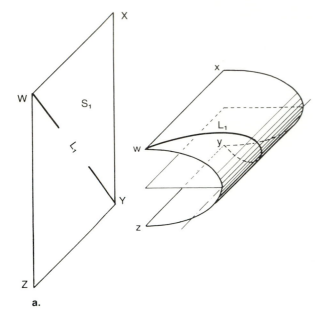

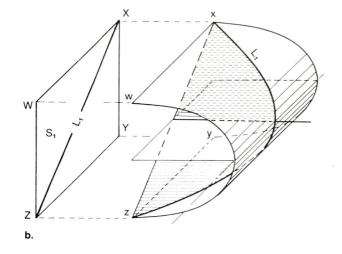

**Figure 10.34**
Geometry of rotation of a linear structure ($L$) by later folding.
(a) Flexural slip folding, $\beta = L$ moving along small circle about $\beta'$.
(b) Passive (shear) folding, with $L$ moving along a great circle. (From Weiss, 1959.)

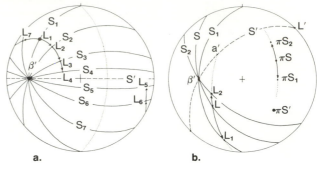

### Slippage Along Surfaces

Slickensides can always be distinguished from true penetrative lineation by the very fact that slickensides are confined to surfaces or, at most, thin layers along fractures, whereas true lineation is pervasive in three dimensions. Slickensides as such do not concern us here. They are discussed in chapter 12.

### Folded Lineations

Where linear structures are folded, including fold hinges, the resulting patterns may give useful clues as to deformation history. Figure 10.33 a and b shows the fate of a lineation in an originally straight surface, after flexural and passive folding. In mesoscopic folds, it may be possible to trace such lineations onto a piece of paper directly from the fold. If on unfolding the paper, the lineation trace becomes a straight line, then folding was flexural. If the folded lineation projects into a straight line (i.e., if every part of it lies along a straight plane), then folding was passive, by either simple or pure shear. Deformed lineations on larger folds may be plotted on a stereonet (see Weiss 1959). A flexurally folded lineation will plot along a small circle. A lineation folded passively, in simple or in pure shear, will remain within a straight plane, and hence will plot as a great circle (figure 10.34).

## Tectonic Significance of Lineation

Lineation is a preferred alignment in the fabric of rocks. It is a valuable clue for two distinct parts of tectonic analysis.

Lineation can be a point of departure in *describing the fabric,* most conveniently with reference to the fabric axes with which we became acquainted in chapter 9. Lineation is, by definition, parallel to the *b* fabric coordinate. This applies only to the domain and scale of observation, however, and difficulties arise as soon as we attempt to extrapolate. For instance, in the case of South Mountain, Maryland, described earlier (Cloos 1947a), the lineation marks *b* of the cleavage fabric, but is parallel to *a* of the fold. It seems advisable to retain the fabric axial cross only for small-scale homogeneous fabrics, and not to extrapolate, since this really serves no useful purpose and may actually lead to misunderstanding.

Many lineations are *strain indicators,* and their orientation may help us to determine the orientation of the strain ellipsoid. Elongation of the ooids in South Mountain evidently represent the *x*-direction of the finite strain ellipsoid. Intersection and folding lineations, on the other hand, develop in the *y*-direction along which the change in length has been minimal. We have seen, however, that minor folds along *x* are not altogether uncommon, and this is particularly well illustrated by steeply plunging minor folds in diapirs. Also, we have seen that deformed conglomerates may be associated with minor folds aligned along the direction of pebble elongation. Great care is therefore necessary when proceeding from fabric analysis to strain analysis.

Fossils are convenient closed strain indicators, provided we know their unstrained shapes. In figure 5.7b, the deformed trilobite indicates strain: elongation is parallel to the strong lineation in the matrix, and shortening is perpendicular to the cleavage surface shown. This again illustrates the useful rule that normally the maximum (elongation) and intermediate strain axes lie within planes of cleavage or schistosity associated with the deformation; the minimum strain (shortening) axis is perpendicular to the cleavage.

## Conclusion

The fabric imposed on rocks after their deposition reflects strain. While opportunities for determining relative lengths of strain axes are comparatively rare, the fabric at least gives us the *orientation* of the strain ellipsoid, and superposed fabrics give clues as to strain history. We may thus reconstruct the movements of a portion of the earth's crust, an important step toward tectonic synthesis.

In chapter 11, we shall begin our study of fractures in rocks, and their role in the deformation of the earth's crust.

## Review Questions

1. How would you identify rock cleavage in the field?
2. What is the difference between "cleavage" and "schistosity"?
3. What is the difference between continuous cleavage and spaced cleavage? Give examples of each.
4. By means of sketches, explain:
   (a) cleavage refraction;
   (b) relationship between cleavage, bedding, and fold axis; and
   (c) attitude of cleavage as a guide to overturning of beds in overturned folds.
5. What evidence exists to link cleavage formation to solution processes?
6. Explain the main processes of cleavage formation. Under what conditions does each of them operate?
7. What is a kink band? (Illustrate your answer.)
8. (a) Define "lineation" in rocks. (b) Give examples for three different types of lineation.
9. By means of sketches, refer the South Mountain fold, Maryland, to (a) the *a, b, c* reference axes of folding; (b) the *a, b, c* reference axes of its fabric; and (c) strain ellipsoid.
10. What are mullions?
11. Show how lineations may be used (a) in fabric description; and (b) as strain indicators.

## Additional Reading

Alvarez, W., Engelder, T., and Lowrie, W. 1976. Formation of spaced cleavage and folds in brittle limestone by dissolution. *Geology* 4: 698–701.

Borradaile, G. J. 1978. Transected folds: A study illustrated with examples from Canada and Scotland. *Geol. Soc. Am. Bull.* 89: 481–493.

Borradaile, G. J., Bayly, M. B., and Powell, C. McA. 1982. *Atlas of Deformational and Metamorphic Rock Fabrics.* Heidelberg: Springer. 551 pp.

Cloos, E. 1971. *Microtectonics Along the Western Edge of the Blue Ridge, Maryland and Virginia.* Baltimore: Johns Hopkins Press. 234 pp.

Crook, K. A. W. 1964. Cleavage in weakly deformed mudstones. *Am. J. Sci.* 262: 523–531.

Dale, T. N. 1896. Structural details in the Green Mountain region and in eastern New York. *U.S. Geol. Surv. 16th Annu. Rep.* p. 549–580.

Engelder, T. 1982. A natural example of the simultaneous operation of free-face dissolution and pressure solution. *Geochim. Cosmochim. Acta* 46: 69–74.

Ghosh, S. K. 1982. The problem of shearing along axial plane foliation. *J. Struct. Geol.* 4: 63–67.

Gratier, J. P. 1980. Deformation pattern in a heterogeneous material: Folded and cleaved sedimentary cover immediately overlying a crystalline basement (Oisans, French Alps). *Tectonophysics* 65: 151–180.

Groshong, R. H. Jr. 1976. Strain and pressure solution in the Martinsburg slate, Delaware water gap, New Jersey. *Am. J. Sci.* 276: 1131–1146.

Harker, A. 1886. On slaty cleavage and allied rock structures. *Br. Assoc. Adv. Sci. Rep. 1885,* 813–852.

Hobbs, B. E. Means, W. D., and Williams, P. F. 1982. The relationship between foliation and strain: An experimental investigation. *J. Struct. Geol.* 4: 411–428.

Lebedeva, N. B. 1978. Significance of mechanical heterogeneities of rocks for formation of flow cleavage. *Tectonophysics* 54: 61–79.

Lister, G. S., and Williams, P. F. 1983. The partitioning of deformation in flowing rock masses. *Tectonophysics* 92: 1–33.

Nickelson, R. P. 1966. Fossil distortion and penetrative rock deformation in the Appalachian Plateau, Pennsylvania. *J. Geol.* 74: 924–931.

Powell, C. McA., and Vernon, R. H. 1979. Growth and rotation history of garnet porphyroblasts with inclusion spirals in a Karakoram schist. *Tectonophysics* 54: 25–43.

Ragan, D. M. 1969. Structures at the base of an icefall. *J. Geol.* 77: 647–666.

Rosenfeld, J. L. 1970. Rotated garnets in metamorphic rocks. *Geol. Soc. Am. Spec. Pap.* 129. 105 pp.

Siddans, A. W. B. 1976. Deformed rocks and their textures. *R. Soc. London, Philos. Trans.* A283: 43–54.

Simpson, C., and Schmid, S. M. 1983. An evaluation of criteria to deduce the sense of movement in sheared rocks. *Geol. Soc. Am. Bull.* 94: 1281–1288.

Weiss, L. E. 1968. Flexural-slip folding of foliated model materials. *Geol. Surv. Can. Pap.* 52: 294–357.

Williams, P. F. 1972. Development of metamorphic layering and cleavage in low grade metamorphic rocks at Bermagui, Australia. *Am. J. Sci.* 272: 1–47.

Wood, D. S. 1974. Current views of the development of slaty cleavage. *Annu. Rev. Earth Planet. Sci.* 2: 369–401.

Woodland, B. G. 1982. Gradational development of domainal slaty cleavage, its origin and relation to chlorite porphyroblasts in the Martinsburg Formation, eastern Pennsylvania. *Tectonophysics* 82: 89–124.

Part Three

# Discontinuous Structures

# 11

# Description of Rock Fractures

Fractures pervade all rock outcrops. They are the most universal of geological structures and have important roles in many geological processes. They are also of great practical importance. For instance, fractures contribute to the permeability of rocks. This influences groundwater flow, petroleum migration and accumulation, geothermal "plumbing," and storage of all kinds of fluids, including water, petroleum, natural gas, and toxic wastes. Fractures also allow the passage of hydrothermal fluids, some of which may carry valuable metals, and they may serve as traps for mineral deposits. As pointed out by Mead (1925), "A fractured rock body expands and attracts any available fluid into its fracture-formed voids." Fractures also guide such geological processes as igneous intrusion, volcanism, and

weathering. Fractures are essential in mining and quarrying operations, where they guide ground-breaking operations. Civil engineers find fractures important in their determination of the suitability of rocks for foundations and excavations. Some fractures become faults, which will be discussed in succeeding chapters.

In order to describe, classify, and analyze fractures, we need to note several important geometric characteristics. These include *attitude* (dip and strike); any regularity in grouping or *pattern,* including degree of preferred orientation; regularity and density of *spacing;* and their *extent* and relative *age.* We also need to note the nature of their surfaces, their filling, and wall rock alterations, if any.

**Figure 11.1**
Prominent jointing in Entrada sandstone, near Moab, Utah. This pattern is present only in the Entrada formation, not in the underlying and overlying rocks. (Air photograph by U.S. Geological Survey.)

## Some Definitions

A **fracture** is a surface along which material has lost cohesion. Geological fractures along which movement has been negligible or absent are commonly called **joints.** (Miners imagined that the rocks were "joined" along these surfaces like building blocks.) Geological fractures along which there has been appreciable movement are **faults** (chapter 12). Fractures whose walls have moved apart are **fissures.** Fissure fillings—dikes or veins—may have been deposited as the fracture opened, or the filling may result from replacement of wall rock. Empty, gaping fissures are near-surface phenomena that do not exist at greater depths.

A group of related parallel fractures forms a **set** (figures 11.1 and 11.2). Two or more sets that intersect at fairly constant angles form a **system.** A system of joints or fractures consisting of two or more sets that appear to have formed simultaneously is a **conjugate system.** Joints that form sets or systems or are otherwise related in a geometric pattern are called **systematic joints.** If penetrative at the scale of observation, they constitute a fabric element, as defined in chapter 8. **Nonsystematic joints** usually have appreciably curved surfaces, and their distribution is somewhat irregular. They tend to be subsidiary to systematic sets. They meet but do not cross older joints or bedding surfaces. The angle between two joints or joint sets is known as the **dihedral angle.** It is measured in a plane at right angles to the intersection between the joint planes.

**Pervasive** fractures are evenly distributed over an extensive domain while **localized** fractures are restricted to relatively narrow zones.

**Figure 11.2**
Joint faces in Entrada sandstone near Moab, Utah. Note fringe zone at base of sandstone. These are typical orthogonal joint sets in near-horizontal layered rocks.

# Descriptive Classification of Fractures

We may classify fractures according to the pattern in which they occur. **Radial fractures** (figure 11.3a) are found near volcanic centers and domal uplifts, and many are filled with dikes. As their name implies, they are more or less radially arranged around a present or former center of (usually) igneous activity. Fractures that form a cylindrical pattern around a central area are termed **ring fractures** (figure 11.3a). Parallel fractures that are successively offset in a constant direction or along an arc are said to be arranged *en echelon*, a French term meaning "in ladder-rung fashion" (figures 11.3b and 13.10). The word "staggered" is an acceptable English equivalent.

We may also classify fractures by their geographic orientation. Thus, one speaks of north-south, east-west, 120° fractures, and so on. Or we may classify them according to their relations to other structures. **Longitudinal fractures** have their traces aligned more or less parallel to regional structural trends. **Cross fractures,** or **transverse fractures,** transect the regional structure at large angles. **Oblique fractures,** or **diagonal fractures,** have attitudes intermediate between longitudinal and cross fractures. **Bedding joints** are parallel to bedding. The strike of **strike joints** is parallel to the strike of country rock; the strike of **dip joints** is parallel to the dip direction of country rock.

**Sheeting** is the name given to a set of joints that lie subparallel to the topographic surface (figure 11.4). Sheeting is most common in plutonic rocks, but it also exists in areas of fairly massive sedimentary or metamorphic rocks. Sheeting may be associated with an orthogonal system perpendicular to it, and together these joints then divide the rock mass into rectangular slabs.

**Figure 11.3**
Outcrop pattern of (a) radial and ring fractures and (b) *en echelon* fractures.

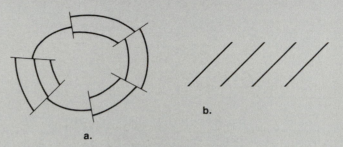

a.

b.

**Figure 11.4**
Horizontal sheeting in granite, Yosemite National Park, California. (Photograph by C. A. Nelson.)

## The Fundamental Joint System

One system of joints seems so basic and universal that it has been called the **fundamental joint system** (Nickelsen and Hough 1967). It consists of two more or less vertical sets that are very nearly at right angles to one another; since most occurrences are in flat-lying layered rocks, they are also perpendicular to the bedding. A third set of joints, parallel to the bedding (bedding joints), serves to divide the rock body into orthogonal blocks. Hence, the system is also called the **orthogonal** joint system (figure 11.5a). In massive rocks, a flat-lying joint set completes the orthogonal system. The two vertical sets need not be alike; one of the sets—the **dominant** set—tends to be better developed (see figure 11.5b). The other set—the *subsidiary* set—is less well developed.

## Fractures Related to Preexisting Fabrics

Fractures tend to form parallel to foliation, such as bedding, layering, and cleavage. They also form perpendicular to such foliations, as **normal joints** (figures 11.5 and 11.6). Another very common and prominent fracture type is normal to strong linear structures, especially to fold axes and to flow structures. Fractures so oriented are known as **cross joints,** or *ac* joints, since they tend to lie in the *ac* fabric plane (figure 11.7; see also figure 9.7). Such joints are common loci for vein and dike fillings. The fabric of folds may also control fractures (e.g., figures 11.8 and 11.9). All such fractures are pervasive.

**Figure 11.5**
The fundamental joint system: (a) KK1 = master joints; KK2 = cross joints; SS = bedding; (b) and (c) fundamental joint systems represented by Hodgson; (d) traces of a fundamental joint system on a bedding plane. (a. From Bock 1976a; b. and c. from Hodgson, 1961; d. surveyed by A. Rautenberg; from Bock, 1980.)

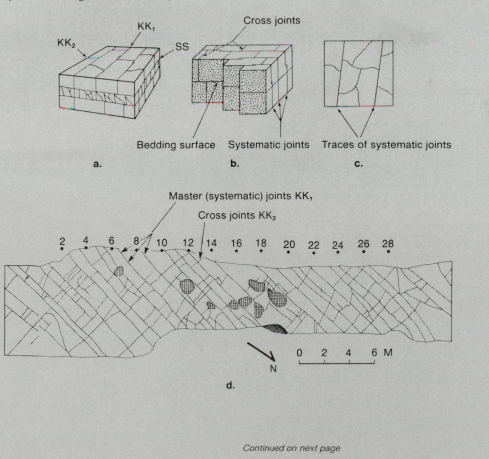

*Continued on next page*

**Figure 11.6**
Normal jointing in an anticline. Interbedded Upper Silurian sandstone and slate, near Hancock, Maryland. (Photograph by C. D. Walcott, U.S. Geological Survey.)

**Figure 11.7**
Extension fractures perpendicular to fold axes at Rillage Point, North Devon. (After a photograph by Bull, 1922; from Wilson, 1961, 1982.)

**Figure 11.8**
Small northwest-facing anticline in graywacke and schists near Altenahr, Rhineland, West Germany. Shaded area represents bedding surfaces with slip striations. White area represents mainly *ac* joint faces, showing traces of quartz veins along normal *bc* joints, which have been displaced in steps along some of the bedding surfaces; displacement is toward the hinge on both sides of the hinge surface. In the core of the fold, note traces of axial plane cleavage. (From H. Cloos, 1948.)

**Figure 11.9**
Fabric-controlled joints in Waits River Limestone, near Albany, Vermont.

**Figure 11.10**
Aerial photograph of part of the Duncan Lake area, Northwest Territories, Canada. Scale: 1:14,500. (Courtesy of the Royal Canadian Air Force.)

## Lineaments

Anyone who flies over extensive shield areas (where basement is exposed), such as the Canadian shield, or who studies aerial or satellite photographs of these areas, cannot fail to notice surprisingly regular, mostly straight lines in the topography, marking slight linear depressions or scarps from differential erosion (figure 11.10). Such lines are called **lineaments.** They are particularly striking on satellite imagery. Some lineaments continue into areas where the basement is covered by layers of sedimentary rock. Most lineaments mark traces of steep, usually vertical fractures, along which no offsets can be detected. Most of these are not faults.

Many sets form orthogonal patterns. Curiously, lineaments do not have random strikes: in any given area, only a very limited number of sets exists. Lineaments are defined as "alignments of topographic features" and thus may mark geological features other than fractures. However, the term "lineament" customarily has been restricted to conspicuous fracture traces; the name "linear" has been limited to any discernible straight or slightly curved line seen from the air or in satellite imagery, without genetic implication. These may include valleys, drainage patterns, distinct vegetation changes not caused by human intervention, and any kind of scarp (e.g., figures 11.10 and 11.11). They may result from

**Figure 11.11**
Santa Elena Canyon, Rio Grande, Texas (left of the canyon) and Mexico (right of the canyon), looking southeastward. The Rio Grande here is an antecedent (prefaulting) stream. The scarp in the left background is a fault scarp. The scarps on the right are cuesta scarps, with subsequent stream valleys controlled by the bedding strike. Streams on the plateau in the foreground and center of the picture are controlled by an orthogonal joint pattern. Vegetation differences reveal bedding on many of the hill slopes. (Photograph by John S. Shelton.)

**Figure 11.12**

The Cameroons Lineament. Volcanics in black. (After Illies, 1970, cartography by M. Hewson.)

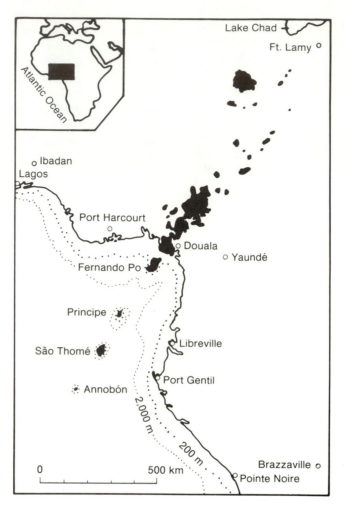

**Figure 11.13**

The Monteregian Hills, Quebec, and associated alkalic plutons in New England (Brighton Line). These plutons (black areas) are along an alignment that appears to mark a major tectonic break at depth and to line up with the Kelvin seamounts in the Atlantic Ocean. M, Montreal; B, trend of Brighton Line expressed in local rock fabric. (Cartography by M. Hewson.)

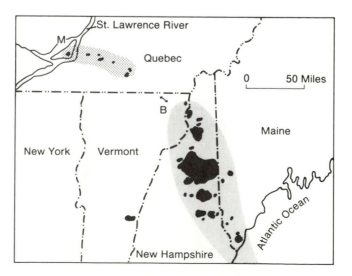

Joint face structures in Lower Carboniferous rocks near Steinach, Thuringia, Germany. (a) Main joint face in slate (*Mj*), clearly bounded by a sharp shoulder (*K*), which marks the beginning of fringe zone *Rk–Z₂*. The main joint face carries a plume, fringe faces *Rj,* a fringe zone ($RK–Z_1$) immediately adjoining the shoulder and rib structures (R), which can also be seen on $RK–Z_2$. (Hammerhead, for scale, at top right.) (b) Secondary plumes in graywacke developed from branches of the main plume. (From Bankwitz, 1966.)

a.

b.

fractures or from contacts between rocks with contrasting resistance to erosion. Some lineaments consist of alignments of plutons or volcanoes, implying feeder fractures at depth (figures 11.12 and 11.13). Some are revealed by linear magnetic anomalies. Many lineaments are of economic importance, for they may act as channels for ore fluids, and tracing them may lead to ore deposits, where suitable traps exist.

Observations in platform areas have shown that old lineaments become reactivated on different occasions throughout their history and may propagate through younger, overlying rocks. This persistence is very characteristic, and reveals the mechanical segmentation of the continental crust. Many joint patterns are directly related to the larger-scale lineaments in their region, and joint patterns may "trace through" from basement to overlying rocks.

## Structures on Fracture Surfaces

Some fracture surfaces display delicate relief patterns. Although Woodworth (1896) gave a good account of these, they were generally neglected until both Hodgson (1961) and Roberts (1961) independently drew attention to them once more. The most thorough accounts to date are by Bankwitz (1965, 1966).

The essential structures in the ideal case (shown in figures 11.14 and 11.15) are the **main joint face,** carrying the so-called **hackle marks** that tend to form a featherlike pattern called the **plume;** the **fringe,** a system composed of small subsidiary **fringe faces;** and a set of **rib marks** that tend to form annular patterns. These structures may occur in any combination, and any of them may be missing.

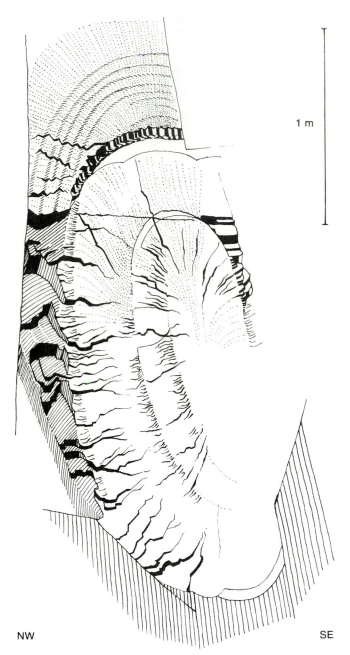

1 m

NW                                    SE

The plume, or plumose structure, has a featherlike outline, with a central axis and diverging branches. The plume axis in normal joints in bedded or foliated rocks tends to be parallel to the trace of the bedding or foliation. Ideally, two plumes, aligned along a common axis, diverge from a common point, the origin. The featherlike branches point, arrowhead fashion, toward the origin. The surface relief becomes more pronounced away from the origin. The plume terminates at the fringe or fringe zone, where it may be replaced by subsidiary plumes (see figure 11.14). Joints that terminate against other surfaces, such as older joints or bedding surfaces, have fringes along the intersection with the older surface. Thus, many fringe zones are rectangular in outline. Where the joint remains unimpeded by crosscutting joints, the fringe tends to be elliptical (figure 11.15). Younger joints, of course, traverse older plumose structures without affecting them. On a freely developed joint, annular rib marks may form concentrically with the fringe (figure 11.15). These do not appear to form on joint faces that terminate against other surfaces, but only on joint faces that have propagated freely.

Plumose structure may be used to determine the relative age of joint sets. At joint intersections, if one joint has a fringe zone along the intersection, it belongs to a younger set. Also, a joint that cuts a plume of another joint is younger than the intersected joint. These are more reliable criteria than relative displacement, since such displacement is not necessarily proof of age difference.

The formation of joint-surface markings depends largely on the material properties of the rock. For instance, the more cohesive and fine-grained rocks, such as argillite and mudstone, seem to develop the best plumose structures. We shall see, in chapter 13, that joint-surface markings are important clues in the genetic interpretation of joints.

## Representation of Fracture Orientation

Apart from measurement of selected fracture attitudes and their representation on geological and tectonic maps, much useful information can be obtained from statistical diagrams that show attitudes of many different fractures within a given domain. What at first sight may appear to be a meaningless jumble in many attitudes may resolve itself into significant groupings on statistical diagrams. The principal function of statistical joint diagrams is to enable the geologist to establish and to compare patterns. We shall briefly review some common representations.

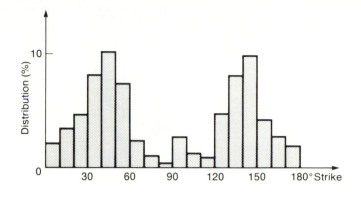

## Histogram

The abscissa represents bearings from 0° to 180°; the ordinate is proportional to the number of fracture-strike measurements, to any convenient scale (figure 11.16). Only strikes can be represented.

## Rose Diagram (Polar Histogram)

Instead of representing bearings along an abscissa, we may construct a polar representation (figure 11.17) and show true bearings, plotting each fracture trace along its strike. The length of each line entered is then proportional to the number of measurements made in a stated interval, usually 5°. For better visual impact, ends of lines are usually joined (inner rose of figure 11.17). Instead of using radial lines, we may outline the whole interval used as a sector (outer rose of figure 11.17). The area so outlined is usually shaded, and different shadings may be used for different domains, as in figure 11.17.

Rose diagrams are superior to histograms because they show strikes directly in relation to compass directions. They may be entered at critical locations on geological or tectonic maps, where they represent a stated domain. Like histograms, they do not show dips and therefore do not indicate true attitude. They are useful only where fracture strikes are significant. This is true where almost all of the attitudes represented are nearly vertical, or where dips are simply not obtainable, as in most air and satellite photographs. However, most fractures read from remote sensing imagery are steeply dipping.

**Figure 11.17**
Rose diagram of fracture strikes in a quarry near Treuchtlingen, Bavaria. Different shadings represent different domains. (From Bannert, 1969.)

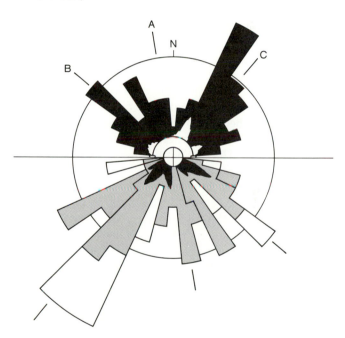

**Figure 11.18**
Stereogram of *ac* joints in coal of the Ruhr district, Germany. The stereogram brings out the southwesterly plunge of the *b*-axis, a result unobtainable from either the histogram or the rose diagram. (From Bolsenkötter, 1955.)

**Figure 11.19**
Combination of statistical fault stereograms and rose diagrams; Arnspitzstock, Bavarian Alps. This type of representation gives more complete information on fracture distribution. Note maxima on the rose diagram and corresponding maxima on the stereograms. (From Schneider, 1953.)

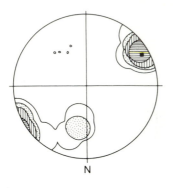

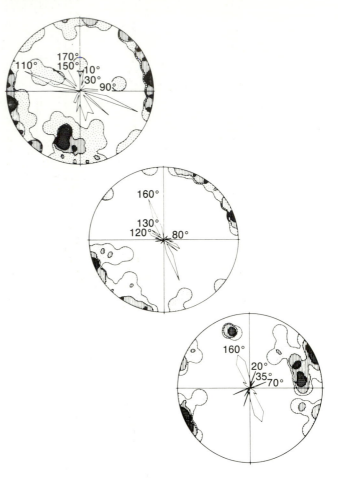

## Stereographic Projection

The most convenient way to represent planes stereographically is by their poles (chapter 3). The poles of fracture surfaces in stereographic projection give a true representation of their attitudes in space (figure 11.18). For point diagrams we may use the Wulff net, but if we wish to contour point densities to better portray the pattern, we must use the equal-area (Lambert) projection of the Schmidt net (figure A.38b). We may superimpose such representations on maps. Although they give more information than rose diagrams, they have one drawback: because planes are shown as poles, their projections become separated from those of any lines (e.g., striations) that lie within them. The stereogram and the rose diagram may be combined to emphasize strikes, as in figure 11.19.

The equatorial plane of stereographic fracture diagrams is usually the horizontal. Where necessary for special purposes, however, we may choose an equatorial plane in any attitude. If we do so, we must then rotate joint readings to fit the new reference plane. Examples of such a situation include excavations in bedrock, where the attitude of fractures with respect to sloping or vertical bedrock faces is more directly significant than standard orientation by dip and strike.

Concentration of fracture poles on a stereogram indicates a preferred attitude of fracture surfaces (figure 11.18). Because poles represent normals to planes, concentrations around the equator of the stereographic projection mark dominantly vertical joints, and concentrations at the center mark dominantly horizontal joints. Boundaries of the individual domains of each stereogram must be carefully selected to reduce scatter to a minimum and to increase statistical validity. There are no rules for fixing these boundaries; individual judgment, experience, and common sense are the only guides. It is best, in the beginning, to keep domains small. Finally, two important geometric properties of fracture systems—their density (frequency) and the angles they make with the principal stress axes—depend on the material properties of the host rock. Statistical diagrams, therefore, should discriminate between lithologies.

## Conclusion

One of the most significant aspects of rock fractures is their pattern in the earth's crust (and in the moon's, and in that of the solid crust of other planets, for that matter). Fracture patterns are clues to the stress history of rocks. Chapter 13 will have more to say on this, following the description of faults in chapter 12.

## Review Questions

1. Classify joints according to (a) relationship to other structures, and (b) pattern.
2. Define *set, system,* and *conjugate system* of fractures.
3. Distinguish between the *fundamental* joint system and *fabric-related* joint systems.
4. What is a lineament? Give examples.
5. Make a labeled sketch showing the principal joint face markings: plume, fringe, and rib marks.
6. Discuss the advantages and disadvantages of different methods of representing fracture patterns in statistical diagrams.
7. Measure the attitude of at least 50 joints in an outcrop in your vicinity, and make a stereographic plot of the joint poles. Comment on the resulting pattern.

## Additional Reading

Bahat, D., and Engelder, T. 1984. Surface morphology on cross-fold joints of the Appalachian Plateau, New York and Pennsylvania. *Tectonophysics* 104: 299–313.

Friedman, M. 1975. Fracture in rock. *Rev. Geophys. Space Phys.* 13: 352–389.

Gay, S. P. 1979. Pervasive orthogonal fracturing in earth's continental crust: Review and update. Page 318 in: M. H. Podwysocki and J. L. Earle, eds. *2d Int. Conf. Basement Tecton. 1976 Proc.*

Kelley, V. C., and Clinton, N. J. 1960. *Fracture Systems and Tectonic Elements of the Colorado Plateau.* University of New Mexico: Publications in Geology No. 6. 104 pp.

Kohlbeck, F. and Scheidegger, A. E. 1977. On the theory of the evaluation of joint orientation measurements. *Rock Mech.* 9: 9: 25.

Kutina, J. 1976. Relationship between the distribution of big endogenic ore deposits and the basement fracture pattern—Examples from four continents: *1st Int. Conf. New Basement Tecton. 1974 Proc.* 565–593.

Lattman, L. H., and Matzke, R. H. 1961. Geological significance of fracture traces. *Photogramm. Eng. Contrib.* 60–94, pp. 435–438.

Nickelsen, R. P. 1976. Early jointing and cumulative fracture patterns. *1st Int. Conf. New Basement Tecton. 1974 Proc.* 193–199.

O'Leary, D. W., Friedman, J. D., and Pohn, H. A. 1976. Lineaments, linear, lineation—some proposed new standards for old terms. *Geol. Soc. Am. Bull.* 87: 1463–1469.

Parker, J. M. 1942. Regional systematic jointing in slightly deformed sedimentary rock. *Geol. Soc. Am. Bull.* 53: 381–408.

Pincus, H. J. 1951. Statistical methods applied to the study of rock fractures. *Geol. Soc. Am. Bull.* 62: 81–130.

Pohn, H. A. 1981. Joint spacing as a method of locating faults. *Geology* 9: 258–261.

Williams, E. 1967. Joint patterns at Dalrymple Hill, northeast Tasmania. *Geol. Mag.* 104: 240–252.

Wise, D. U. 1976. Sub-continental sized fracture systems etched into the topography of New England. *1st Int. Conf. New Basement Tecton. 1974. Proc.* 416–422.

# 12

# Description of Faults

**Figure 12.1**
Steeply dipping normal fault in sandy shale, 1 mile southwest of
Little River Gap in the Chilhowee Mountains, Blount County,
Tennessee. Note differential "drag" and weak brecciation along fault
trace; also note penecontemporaneous deformation in some of the
sedimentary layers. (View is to the southeast.) (Photograph by A.
Keith, U.S. Geological Survey.)

Miners in the past often faced the problem of finding the continuation of a faulted ore body, so the investigation of faults became an early concern of applied geology. It remains important to the present-day structural and economic geologist. This chapter will explain the geometric relationships of fault displacement and will provide the descriptive basis for the discussion in chapters 13 and 14 of the role of faulting in rock deformation.

## Definitions

A **fault** is a fracture surface or zone along which rocks have moved relative to each other (figure 12.1). Several attempts have been made to systematize fault nomenclature, which derives from mining usage. Here we shall conform with the *International Tectonic Dictionary,* which follows recommendations of a committee of the Geological Society of America (Reid et al. 1913), with some modifications.

Where a fault has formed from a single fracture, the **fault surface** is a surface along which fracture and displacement have occurred. A **fault scarp** is a scarp that has formed at the earth's surface as a direct result of fault displacement (figure 12.2). A **fault-line scarp** is the result of differential erosion of different rock types brought into juxtaposition by faulting (figure 12.2).

The country rock laterally bounding a fault, fracture, or fracture filling is the **wall,** "block," or "side." "Wall" refers to the bounding surface of a block at the fault. The upper wall of an inclined fault is the **hanging wall** (the wall on which the miner hangs his lamp). The lower wall is the **footwall**—the wall on which the miner rests his foot (see figure 12.3).

The trace of a fault on the surface of the earth or on any artificial or imaginary surface, such as a mine level, is a **fault trace** or **fault line.** A zone of sheared, crushed, or foliated rock, in which numerous small dislocations have occurred, adding up to an appreciable total offset of the undeformed walls, is a **fault zone. Displacement** is a useful general term for relative change in position along a fault of formerly adjacent features. The term "movement" includes the rate and the path of displacement and is not appropriate if we wish to consider only the geometric result of movement.

## Figure 12.2

Physiographic expression of faults. In cases of negligible erosion (*left*), scarp height represents throw (see p. 219). Where the erosion rate exceeds relative uplift of the rising unit, a scarp can form only by differential erosion. The intermediate condition is most widespread; the rising unit deposits an alluvial fan on the subsiding unit. Note truncated spurs between canyons. After cessation of fault movements and erosional planation, evolution may be along two directions (*bottom*): renewed sedimentation followed by renewed, posthumous, faulting and fault scarp formation (*left*); or differential erosion, resulting in a fault-line scarp facing either the same way as the original fault scarp, or in the opposite direction (*bottom right*). (From H. Cloos, 1936.)

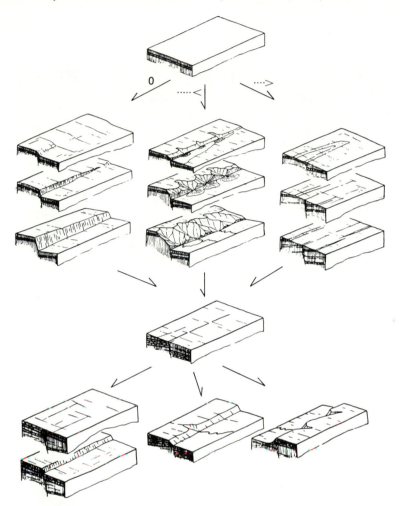

## Figure 12.3

Hanging wall and footwall. Originally used for the wall rocks of an ore body (some of which are emplaced along faults), these terms refer to the wall from which a miner hangs his lamp ("hanging wall"), and the wall on which he rests his foot ("footwall").

**Figure 12.4**
Topographic expression of the Perry Fault, Vermont. (North is at top.) Drag along the fault is clearly expressed to the south of the valley, which marks the fault trace. The curved linears are bedding traces in the Devonian Gile Mountain formation, which consists of quartzites and quartz-rich schists. Southwest of the large pond in the valley, a set of southwest-striking linears probably marks pinnate fractures. Vertical fractures (linears not deflected by topography) south of the fault are parallel to the fault and are almost certainly related to it. Very little of this information can be obtained on the ground, where exposures are scarce. (Photograph by Fairchild Aerial Surveys, Inc.; from Dennis, 1956.)

**Figure 12.5**
Feather (pinnate) joints, showing movement sense of associated
fault. (From Dennis, 1967.)

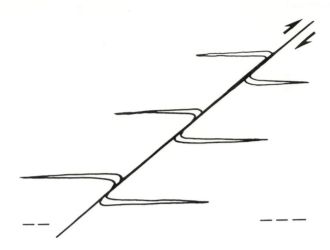

## Field Setting and Recognition of Faults

A fault is most readily recognized where it separates
once-continuous rock bodies or structures. This may
become evident in any observed or constructed section
through the structure in surface outcrop or in drill holes.
In either case, the immediate observation may be a repetition or omission of layers. Observed displacement
may need confirmation by identification of an actual
fault, for flexures may also displace structures.

Truncation of structures may indicate faulting, but
unconformities also truncate structures (see chapter 4);
and metasomatic structures may end against fractures
without displacement (figure 4.13). Repetition and
omission of rock units also indicate faulting.

Discordant alignments of topographic features may
indicate fault traces. Preferential erosion along some
faults results in linear depressions, many of which are
evident in the topography and on aerial photographs
(figure 12.4). However, preferential erosion may also
occur along fracture zones that are not faults, or along
less resistant strata. Aligned springs and anomalies in
kind or density of vegetation mark many fault traces.
Scarps may form along, and retreat from, fault traces,
but they must be distinguished from other types of erosional scarps. In addition to topographic evidence, we
also need evidence for displacement.

Many movement surfaces carry sets of polished,
sometimes fibrous, striations called **slickensides.** These
are classic indicators of direction and possibly of sense
of displacement, but they must be used with extreme
caution. When the fingers are run over the fibrous surface of slickensides in the direction of the striations, they
feel rough in one sense and smooth in the opposite sense.
There is no agreement as to which sense is that of
movement along the fault, but most geologists believe
that movement of the hand in the smooth sense indicates relative displacement sense. In any case, the direction of the slickensides indicates only the last
displacement; previous displacements may have been
in other directions.

Deflection of layers ("drag") along fracture surfaces as in figure 12.1 is usually considered a good indication of relative displacement along a fault and of
sense of displacement. Also, short and, in many instances, vein-filled gash fractures, set out in an *en echelon* pattern along the fault walls (figure 12.5) normally
indicate fault displacement. The mechanical interpretation of these **pinnate** or **feather joints** is given in
chapter 13. Their acute angles with the main fracture
point, arrowhead-fashion, in the direction of relative
displacement of their respective side of the fault.

All these features, where present, are useful clues
for the recognition of faults. However, they are certainly not universal attributes of all faults; in fact, the
presence of even one or two of these clues is usually
considered a lucky find.

**Figure 12.6**
Translational fault displacement. The displacement vector between
two separated points is the net slip revealed by the linear body *AB*
(separated at the fault along *SS'*). Actual movement may have been
along any path between *S* and *S'*.

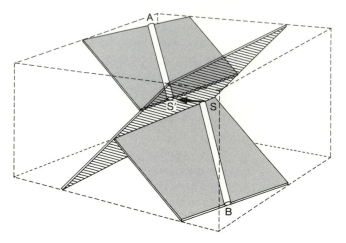

**Figure 12.7**
Components of slip. Arrows indicate relative displacement of
hanging wall. *OP* = net slip; *OS* = strike slip; *OD* = dip slip;
*OV* = vertical component of net slip; *VP* = horizontal component of
net slip.

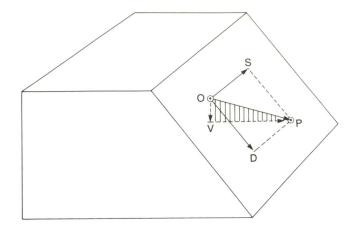

## Analysis of Displacement

Faulting normally occurs parallel to the fault surface
or zone. Any component perpendicular to the walls (e.g.,
gaping) is not a component of faulting. If straight traces
that were parallel before displacement remain parallel
afterward, the displacement is said to be **translational**
(figure 12.6). If formerly parallel straight traces on opposite sides of the fault are no longer parallel after
faulting, the displacement is **rotational**. A displacement that is the result of both translational and rotational components is essentially rotational, because a
translation added to a rotation results in a rotation with
a different (greater) radius than that of the original
component rotation. For most practical purposes, however, local displacements may be treated as purely
translational.

Whenever the terms "up" and "down" are used in
faulting, they refer to displacement of blocks *relative*
to each other. It is important to remember that observed vertical displacement alone does not show the
*absolute* movement sense of fault blocks with respect
to the earth's surface.

### Slip

In 1913, the Reid committee defined **slip** or **net slip** as
"the distance, measured on the fault surface, between
two originally adjacent points situated, respectively, on
opposite sides of the fault. It would be represented by
a straight line in the fault surface connecting these two
points after displacement." (Reid et al. 1913) This relationship is shown in figure 12.6. Note that the measured finite slip vector need not coincide with the actual
movement path from *S* to *S'* composed of incremental
slip vectors; it could deviate from *SS'* and even reverse
itself before finally reaching S'. Reactivation of old
faults may further complicate the process.

### *Components of Slip*

It is often convenient to resolve slip into components
along certain preferred directions (as in figure 12.7).
The component along the fault strike is the **strike slip;**
that along the fault dip is the **dip slip.** Where it is necessary to refer to components outside the fault surface,
it is best to use self-explanatory composite terms, such
as in "vertical component of slip," and "horizontal
component of slip." Net slip may become modified by
"drag." In figure 12.8, *AA'* is the net slip; *BB',* the true
relative displacement of the two blocks, is called the
**shift.**

### Separation

As stated by Reid, et al. (1913) and by Gill (1941), the
**separation** of any recognizable geological surface, such
as bedding that has been offset along a fault, is the *distance between the displaced parts measured in any
specified direction.* The emphasis here is on "specified." The term "separation" has geological meaning
only if the direction of measurement is clearly specified
by a qualifying adjective. We can measure separation
from field, map, or cross-section data of displaced
structures alone, without any knowledge of the net slip.

Figure 12.9 shows commonly measured separa-
tions. **Strike separation** is along the fault strike, either
in the field or on a map; **dip separation** is along the line
of dip of the fault. **Vertical separation** is that seen in
vertical boreholes, shafts, and vertical cross sections.
**Horizontal separation** or **offset** is measured at right angles to the two separated horizontal traces of a faulted
surface (such as that seen on maps or on subsurface
levels). **Perpendicular separation** is the shortest, or perpendicular, distance between the two separated parts of
the faulted surface; if the faulted surface is a bedding
surface, this separation is the **stratigraphic separation.**

**Figure 12.8**
Net slip reduced by drag. Assuming net slip = dip slip, *AA'* is the net slip, but it does not represent total relative displacement of the two walls, given by *BB'*. This latter quantity is sometimes referred to as the **shift.**

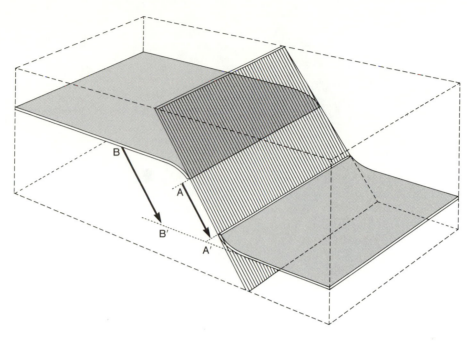

**Figure 12.9**
Separations. *H* = offset or horizontal separation; *S* = strike separation; *D* = dip separation; *V* = vertical separation. Perpendicular or stratigraphic separation is perpendicular to the separated surface. (Based on Gill, 1941.)

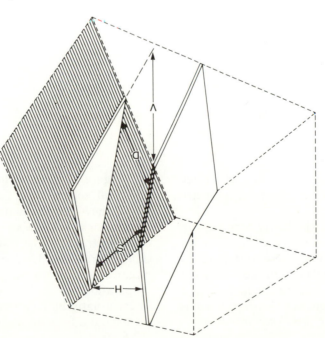

**Figure 12.10**
Different slips leading to the same separation. *SA* = strike separation; *UE* = dip separation; *OA, OB, OC, OD, OE* are all possible net slips. For *OB*, net slip = strike slip; for *OD*, net slip = dip slip. Displacement *OA* would be a reverse fault (left reverse slip); *OC, OD,* and *OE* would be normal faults.

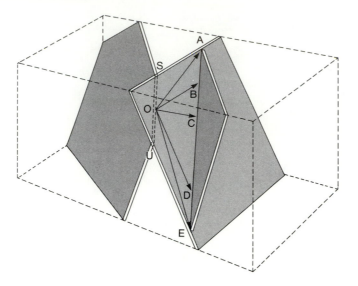

**Figure 12.11**
Strike-slip fault. Strike slip *EG* results in normal separation at *A*, reverse separation at *B*. This relationship is sometimes erroneously interpreted as a "scissors fault." (After Gill, 1935.)

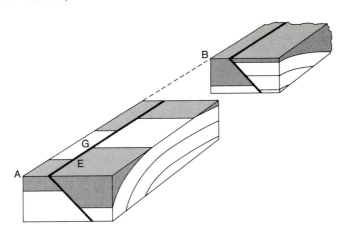

It is commonly measured (especially in subsurface studies) to specify the thickness of strata that are missing or repeated by faulting.

It is important to distinguish clearly between slip and separation. Slip refers to the *actual* relative displacement of the two walls of a fault. Separation refers to the *apparent* relative displacement of the traces of a tabular body or surface separated by the fault. It is a local geometric effect. We might say that separation is a *geometric* term, whereas slip is a *kinematic* term.

Figure 12.10 shows how different slip may result in identical separations. Conversely, figures 12.11 and 12.12 show that the same net slip will impart different separations to surfaces with different attitudes. Net slip that is parallel to the trace of a faulted surface on the fault is **trace slip** (figure 12.13). Note that trace slip, regardless of its length, produces no separation at all. Separation can be observed and measured directly; slip must usually be computed or constructed from separation data, for situations such as that shown in figure 12.6 are rare.

**Figure 12.12**
Displacement of veins along a fault, Maggia gneiss, Swiss Alps.
Note different separations and try to explain the reasons for the
differences.

## Heave and Throw

**Heave** (horizontal displacement component perpendicular to the fault strike) and **throw** (vertical displacement component) are old miners' terms. In the past, they have been used by some to refer to slip components, by others to separation components of displacement. The latter usage would be more appropriate, but heave and throw are rarely used now, because they lack precision. Nevertheless, "throw" is still a convenient term when referring to displaced morphology, such as alluvial or erosional surfaces or to displaced horizontal beds.

**Figure 12.13**
Trace slip. No separation of faulted surface (gray), but linear
structure LL' is displaced and indicates net slip AA'.

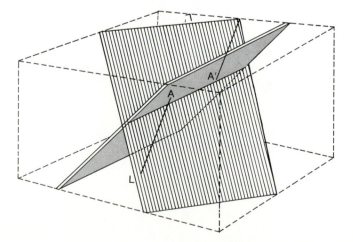

## Geometric and Kinematic Classification of Faults

Faults may be classified according to the following categories:

1. Pattern of sets and systems
2. Relation to regional structure or topography
3. Direction of relative displacement

In categories 1 and 2, classification is identical to that used for fractures in chapter 11 (p. 198). In category 3, it is by relative displacement along the dip and along the strike of the fault.

### Displacement Along the Dip

A fault whose predominant component of displacement is along the line of dip is a **dip-slip fault,** which can be either *normal* or *reverse*. The traditional terms normal fault and reverse fault are miners' terms. According to Blyth (1952):

> The custom of describing faults as normal or reversed originated in English coal-mining practice. Faults which were inclined towards the downthrow side were met most commonly. When a seam which was being worked ran into such a fault, it was necessary to continue the heading a short way, and then sink a shaft to recover the seam; this was the usual or normal practice.

A seam that was offset the opposite way was the reverse of the usual conditions, and so the fault was called "reversed."

In other words, a **normal fault** dips toward the block that seems relatively lowered (e.g., figure 12.1); a **reverse fault** dips toward the block that seems relatively raised (e.g., figure 12.5). These terms apply regardless of the magnitude of any strike-slip component present. A fault with a 90° dip (vertical fault) evidently dips in no direction, but is traditionally classified as a normal fault. Some faults change from normal to reverse (figure 12.14).

Where a dip separation can be determined, but net slip is unknown, we may use the terms normal and reverse to qualify the dip separation. For instance, a fault shows **normal dip separation** where the hanging wall trace of the separated surface lies below its footwall trace (see figure 12.10).

### Displacement Along the Strike

A fault whose predominant slip is along its strike is a **strike-slip fault.** In a **right-slip fault,** the block across the fault from an observer has been offset toward the

right. In a **left-slip fault,** the block across the fault from an observer has been offset toward the left. Common partial synonyms of strike-slip fault are **wrench fault,** and **transcurrent fault.** In an **oblique-slip fault,** displacement has large components along both strike and dip (figure 12.15). Most fault slip is oblique to some extent. Figure 12.16 is a classification diagram according to slip vectors.

Where only the separation effect on some surface has been observed, we may use the terms right separation and left separation to qualify the strike separation, just as normal and reverse were used to qualify the dip separation.

### Classification Problems

Fault classification has been subject to much debate. The main point at issue has been whether separation is a valid criterion for classification (Gill 1935 and 1971; Hill 1947 and 1959).

**Figure 12.14**
Effect of changing fault dip. The whole fault cannot be classified as either reverse or normal, but it can be characterized as "facing east."

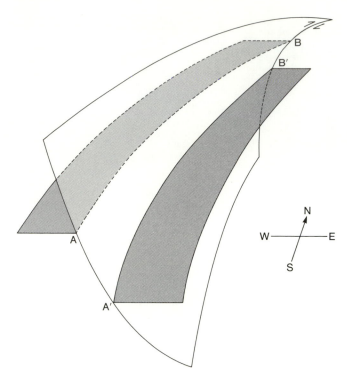

**Figure 12.15**

**Figure 12.15**
Block diagram of an oblique-slip fault showing net slip (with respect to the footwall), drag, and effect on river valley crossing the surface trace. Note spring (s). (From King, 1978.)

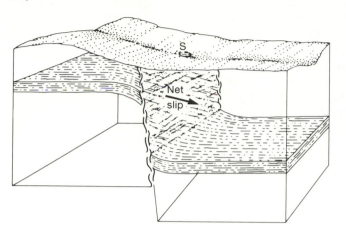

**Figure 12.16**
Classification diagram, according to slip vectors, for translational faults. (a) Dip-pitch triangular grid; pitch is measured along line representing the dip of the fault plane. (b) Fault categories on dip-pitch triangles. Dip-slip categories are ruled; strike-slip categories are dotted; oblique-slip categories are blank; *f is an example of a right-normal-slip fault. (Slightly modified from Rickard, 1972.)

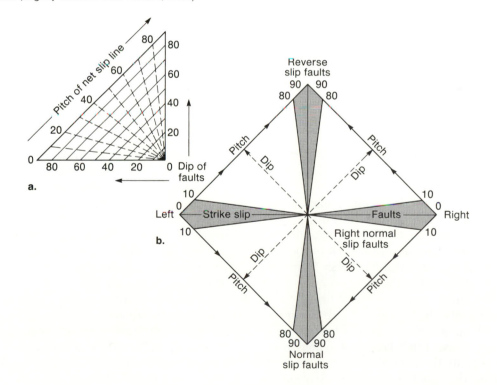

Table 12.1 is an integrated classification that takes into account both slip and separation criteria.

In practice, "slip" and "separation" are rarely added to fault terms, but the distinction is important. In oil-field and some mineral exploration, most faults are characterized by separation, because that is usually all the information that is available. So, implicitly, fault classification in such cases is by separation. However, slip sense often can be determined by means of a displaced linear feature, by drag features, or by circumstantial evidence. In such cases, classification is implicitly by slip. It might be better, however, to make the distinction explicit. Admittedly, confusion is usually unlikely. However, figures 12.9 and 12.10 show that a normal separation fault may well turn out to be a reverse-slip fault. Strike slip may become apparent by juxtaposition of different facies or thicknesses, as in figure 12.17.

Different cases of separation in a number of standard views are illustrated in Appendix C.

## Thrusts and Overthrusts

Most, but not all, thrusts have low dips, and many have large displacements (of the order of tens of kilometers). Most bring older rocks to lie over younger, and that is why they have traditionally been classified with reverse faults. Their prefaulting relationships usually cannot be constructed directly by the simple geometric methods set out earlier. Elucidation of many thrusts hinges on genetic hypotheses, and observation must be clearly separated from preconceived notions about the movement and mechanics involved. We shall devote chapter 15 to a fuller discussion. An explanation of basic descriptive terminology follows.

### Thrust Terminology

Thrust and overthrust are terms that are difficult to define. Thrust has been used as a synonym for low-angle reverse fault. Both terms have been used for any fault that brings older rocks to rest on younger ones. They have also been used for any kind of low-dip ("low-angle") fault. To confuse matters further, these terms have been and are being used both descriptively and genetically; genetically, they imply a "push from behind," crustal shortening, or both.

It is now becoming common for "thrust" and "overthrust" to be used only descriptively. To minimize confusion, we shall adopt definitions that follow fairly widespread usage. Thus, **thrust fault** is a synonym of reverse fault (especially those with low dip). **Thrust** or **overthrust** is a fault that has low dips over a large area and along which there is a large overlap of rocks that are not in their original site of emplacement. It commonly implies a local shortening of section.

| Table 12.1 Fault Classification | |
|---|---|
| **Classification Where Slip Sense Is Known (Classification indicates slip sense)** | **Classification by Geometric Effect on a Specified Surface (Classification indicates separation)** |
| **Strike-slip faults**<br>Right-slip fault<br>Left-slip fault | **Strike-separation faults**<br>Right-separation fault<br>Left-separation fault |
| **Dip-slip faults**<br>Normal-slip fault<br>Reverse-slip fault | **Dip-separation faults**<br>Normal-separation fault<br>Reverse-separation fault |
| **Oblique-slip faults**<br>Right-normal-slip fault<br>Left-normal-slip fault<br>Right-reverse-slip fault<br>Left-reverse-slip fault | (Most faults produce both strike and dip separations on surfaces they displace) |

Note: From Dennis, J. G., *International Tectonic Dictionary, English Terminology, Memoir 7.* © 1967 American Association of Petroleum Geologists. Reprinted by permission.

Strictly speaking, since thrusts were originally considered a class of reverse faults, they should bring older rocks over younger ones. But in many settings, such as nonfossiliferous rocks, it may not even be possible to determine which is the older and which is the younger sequence. To designate low-dip faults, E. B. Bailey suggested **slide,** and Bailey Willis suggested **tectonic discordance;** but these terms, useful and logical though they seem, have not found general acceptance. **Detachment fault** and its French equivalent, **décollement,** imply detachment and shear of a layer or sheet along a bounding surface below it. **Overthrust** has been used for thrusts with large displacement; a common but arbitrary criterion is a displacement of 5 km or more. Overthrust has also been used as the antonym of **underthrust.** In overthrusts, the upper unit or hanging wall is the assumed active unit; in underthrusts, it is the lower unit. Notwithstanding contrary opinion (e.g., Lovering 1932), it is not possible to determine objectively which unit moved actively unless there is good evidence for movement by gravity. This does not rule out, in specific cases, certain geotectonic hypotheses postulating active movement of either the upper or the lower unit.

"Thrust," "slide," and similar terms refer to the movement *surface.* These terms have also been applied, incorrectly, to the unit that overlies the movement surface. This latter unit is properly named a **thrust sheet, nappe, or allochthone.** Of these, "thrust sheet" is the most commonly used English term. *Nappe* is French for "tablecloth," and by extended meaning, it refers (in French) to any spread-out geological bodies such as groundwater pools, oil pools, and lava sheets; as an abbreviation for *nappe de recouvrement* (cover sheet), it is used in the sense of a large, sheetlike geological body covering lower rocks tectonically (figure 12.18). It thus includes thrust sheets, detachment sheets, and large recumbent folds.

**Figure 12.17**
Segments of contrasting thickness juxtaposed by faults indicate strike-slip: the unit concerned evidently thickens along strike, so strike-slip faulting brings together segments of different thickness. Miocene Puente Formation, Southern California. (Redrawn from an original compilation by A. Maher.)

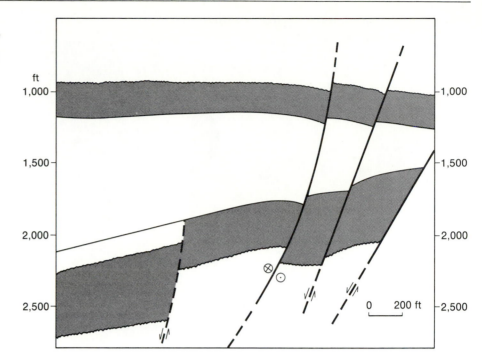

**Figure 12.18**
Drusberg nappe, overlying Mürtschen nappe of the Helvetic zone in eastern Switzerland, at the western end of Churfirsten range above the Walensee. Movement was toward the left and back of the picture (northwest). Trace of slide surface outlined in black. (Photograph Swiss National Tourist Office.)

**Figure 12.19**
Klippe (*K*) and window (*W*) in map and in cross section.

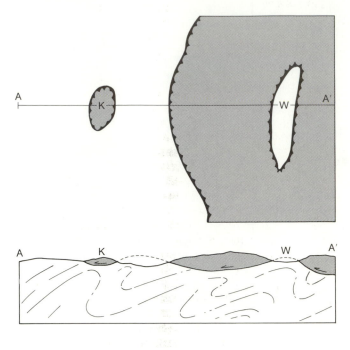

Rocks that have been moved from their original emplacement site and overlie others with tectonic contact are **allochthonous.** The unit above the tectonic contact is the allochthone. In contrast, rocks that have their original roots in their present site are **autochthonous.** The term "upper plate" is common in western North America and refers to the tectonic unit above a thrust. It is unsatisfactory, because the term "plate" wrongly conveys an impression of rigidity. (Also, the lower "plate" may extend to the center of the earth!) A **tectonic outlier** (*K* in figure 12.19), in analogy to a stratigraphic outlier, is an isolated outcrop of an allochthone surrounded in map pattern by tectonically lower rocks. It is also commonly called a **klippe.** A **tectonic window** (or *fenster*) is an isolated outcrop of a tectonic unit entirely surrounded in map pattern by tectonically higher rocks (*W* in figure 12.19). The general term **tectonic unit** is useful for denoting any rock unit defined by tectonic contacts. A succession of low-dip fault slices is an **imbricate** or **schuppen zone** (e.g., figure 15.6).

## Klippes

A klippe is an isolated rock unit, large or small, that overlies another unit along a tectonic contact (figures 12.19 and 12.20). Isolation of a klippe may come about by erosion or through detachment and independent movement of the klippe rocks. Both of these mechanisms may act in conjunction, so that a klippe may become detached from a main nappe by erosion, and then may move independently of the main sheet, along the same surface.

In the Carpathians, large units of isolated rocks have been found to lie in matrices of younger rocks. In this situation, they are rather like oversized breccia blocks. By definition, these, too, are klippes; in fact, they were the first features so named (figure 12.21). *Klippe* is a German word meaning "reef," in the sense of a protrusion of rock in shallow water. By analogy, it was originally applied to protrusions of older erosion-resistant rock in softer, younger surrounding rock. Later it became apparent that these "protrusions" were, in fact, rootless, but the term *klippe,* though no longer apt, continued to be used. It gained wide currency when it was applied to Alpine tectonic outliers, which resemble the Carpathian klippes morphologically, but which are more directly related to thrust sheets than the Carpathian klippes.

**Figure 12.20**
Klippe of Precambrian and Paleozoic Monte Cristo Formation on
Tertiary fanglomerate, Shadow Mountains, Mojave Desert, California.

**Figure 12.21**
Group of Pienine Klippes, Czechoslovakia. Klippe rock is Malm
limestone, surrounding rock is Senonian marl. (Photograph courtesy
of D. Andrusov.)

## Fault Rocks

Shear along faults sometimes forms distinctive rocks called **fault rocks.** These may aid in the recognition of faults in the field. We shall briefly review important types of fault rocks with reference to table 12.2.

*Incohesive* fault rocks are generated near the surface, forming an aggregate that lacks cohesion. They may later become cemented by secondary processes, but our classification applies to rocks as they are generated at the time of faulting. If a rock consists of fragments, largely angular, that are visible to the naked eye and make up over 30% of the rock, it is a **fault breccia.** If it is largely a claylike paste, it is a **fault gouge.**

*Cohesive* fault rocks are generated at depths where confining pressures are high. Rocks in this class consist of both cataclastic (broken) and recrystallized grains, and are classified according to the predominance of one or the other of these constituents. Many cohesive fault rocks have a characteristic wavy or "flowing" foliation ("fluxion structure"), usually visible at the microscopic scale and, quite commonly, at the mesoscopic scale (figures 12.22 and 12.23). Its presence or absence distinguishes rocks in column 1 of table 12.2 from those in column 2. **Crush breccia** is an intensely fragmented cohesive breccia in which the matrix makes up less than 10% of the rock; all fragments are unoriented. **Cataclasite** is a structureless cohesive fault rock with few or no fragments visible to the naked eye. Very fine, optically isotropic crush powders are **ultracataclasites.**

**Pseudotachylite** has been so intensely sheared that it has melted and looks like volcanic glass. True tachylite is a basaltic glass, and pseudotachylite looks somewhat like it, even assuming intrusive, veinlike relationships with the country rock.

**Table 12.2**  Classification of Fault Rocks.

|  | Random Fabric | Foliated |
|---|---|---|
| **Incohesive** | Fault breccia<br>Fault gouge |  |
| **Cohesive (grain growth subordinate)** | Crush breccia<br>Cataclasite | May be foliated<br>Mylonite (including protomylonite and ultramylonite) |
| **Special Matrix Textures** | Pseudotachylite (glossy or devitrified) | Blastomylonite (pronounced grain growth)<br>Phyllonite (pronounced mica fabric) |

Note: From Sibson, R. H., Fault Rocks and Fault Mechanisms'' in *Geological Society of London Quarterly Journal, 133,* pp. 191–214, 1977. Copyright © 1977 Geological Society of London. Reprinted by permission.

**Figure 12.22**
Mylonite at base of Pachalca Thrust, Mojave Desert, California.

**Figure 12.23**

Photomicrograph of mylonite from the Santa Rosa, California, mylonite zone, cut normal (perpendicular) to stretching lineation, showing feldspar porphyroclasts. (Diameter of large porphyroclast is about 3 mm.) Fine matrix of quartz, biotite, feldspar, and other minerals associated with the granodiorite protolith. Note the fold above the large porphyroclast on the left (uncertain origin). Also note the segregation of quartz into discrete bands. (Courtesy Bradley Erskine.)

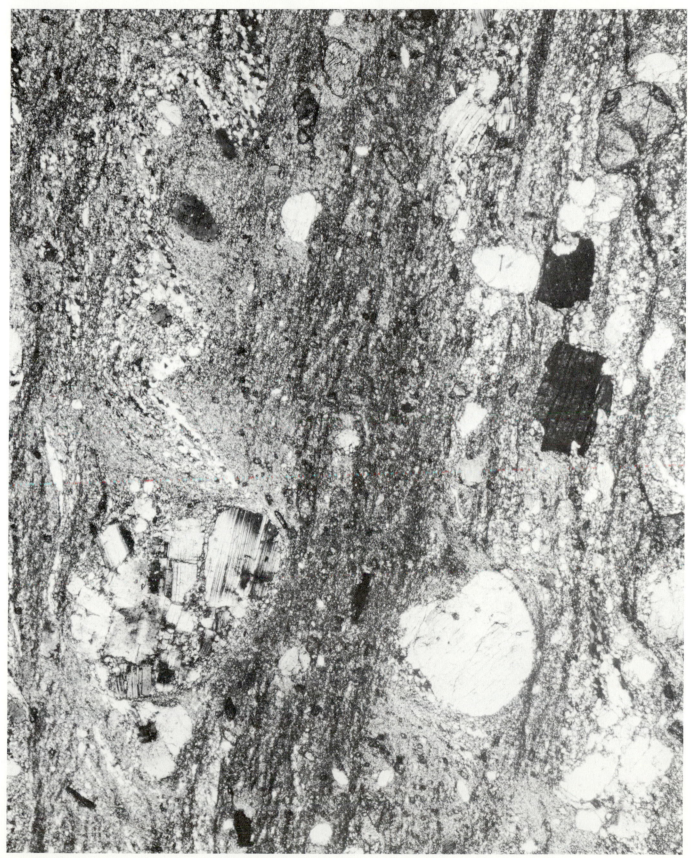

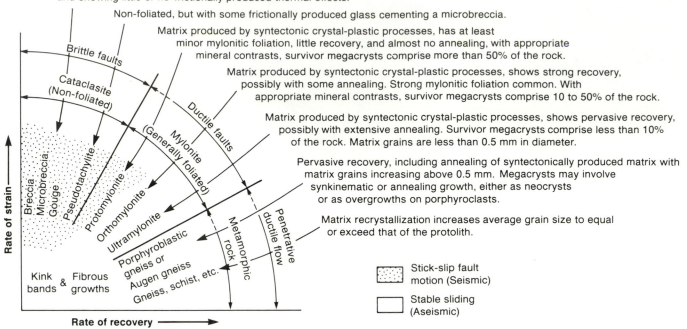

**Figure 12.24**
Terminology of fault-related rocks, as proposed by Wise et al. (1984)
Horizontal and vertical scales are variable, depending on
composition, grain size and fluids.

Coherent but unfoliated rocks produced by micro- and/or macro-fracturing
and showing little or no frictionally produced thermal effects.

Non-foliated, but with some frictionally produced glass cementing a microbreccia.

Matrix produced by syntectonic crystal-plastic processes, has at least
minor mylonitic foliation, little recovery, and almost no annealing, with appropriate
mineral contrasts, survivor megacrysts comprise more than 50% of the rock.

Matrix produced by syntectonic crystal-plastic processes, shows strong recovery,
possibly with some annealing. Strong mylonitic foliation common. With
appropriate mineral contrasts, survivor megacrysts comprise 10 to 50% of the rock.

Matrix produced by syntectonic crystal-plastic processes, shows pervasive recovery,
possibly with extensive annealing. Survivor megacrysts comprise less than 10%
of the rock. Matrix grains are less than 0.5 mm in diameter.

Pervasive recovery, including annealing of syntectonically produced matrix with
matrix grains increasing above 0.5 mm. Megacrysts may involve
synkinematic or annealing growth, either as neocrysts
or as overgrowths on porphyroclasts.

Matrix recrystallization increases average grain size to equal
or exceed that of the protolith.

Brittle faults

Cataclasite
(Non-foliated)

Ductile faults

Mylonite
(Generally foliated)

Rate of strain

Breccia,
Microbreccia,
Gouge

Pseudotachylite

Protomylonite

Orthomylonite

Ultramylonite

Porphyroblastic
gneiss or
Augen gneiss

Gneiss, schist, etc.

Metamorphic
rock

Penetrative
ductile flow

Kink bands & Fibrous growths

Stick-slip fault
motion (Seismic)

Stable sliding
(Aseismic)

Rate of recovery ──────▶

Foliated metamorphic fault rocks are varieties of the rock called **mylonite** (figures 12.23 and 12.24). There has been some disagreement in the past about what exactly constitutes a mylonite. A Penrose conference of specialists (Tullis et al. 1982) has agreed on the following characteristics: (1) grain size reduction; (2) occurrence in a relatively narrow planar zone (the qualification "relatively" is important here, as mylonite zone widths range from a few millimeters to several kilometers); (3) foliation that is more intense than in adjacent rock, usually a result of strain concentration.

These are strictly descriptive, nongenetic features. Mylonites are generally assumed to be the result of ductile deformation under metamorphic conditions, whereas cataclasites are the result of relatively high level cataclastic deformation. Mylonites are most common along certain overthrusts, such as the Moine thrust in Scotland, some Caledonian thrusts in Norway, and some Sevier belt thrusts in the western United States. They probably form early in the faulting process, possibly even before the fault break has fully developed, and some are associated with shear zones rather than with faults (see chapter 13).

A **protomylonite** is a rock in the early stages of mylonitization. Fragments make up about 50% of the rock. Protomylonites may resemble conglomerate or arkose on weathered surfaces. An **ultramylonite** has been sheared beyond the stage of a mylonite *sensu stricto*. Most fragments have been reduced to streaks, and those that remain are smaller than 0.2 mm in diameter. Some recrystallization is present, but to the naked eye these rocks are aphanitic and could be mistaken for quartzite or felsic volcanic rock. **Blastomylonites** have been largely recrystallized during or after mylonitization. In many of these rocks, recrystallization is so advanced that they are difficult to identify as fault rocks. This is particularly true of **mylonite gneisses,** which look like augen gneisses and can easily be mistaken for true augen gneisses of regional metamorphism. **Phyllonite** is a mica-rich mylonite that has the mesoscopic appearance of a phyllite. Some fault rocks contain relatively large fragments of mineral grains or aggregates of grains in a finer groundmass; these larger fragments are called **porphyroclasts.**

Recognition of fault rocks in the field can be difficult, and positive identification may only be possible in thin section in some cases. Field criteria include obvious association with faults, gradation into unaltered rock, and recognition of microbreccia fragments. Fault rocks may range in thickness from less than a millimeter to hundreds of meters. These thicknesses vary considerably, even along the same fault. Cohesive fault rocks tend to be darker than the unaltered rock. Most of them belong either to the cataclasite or to the mylonite series. Table 12.2 and figure 12.24 give a useful classification and terminology.

## Conclusion

Recognition of faults and measurement of displacement along them are important tasks for the structural geologist. These tasks are rarely easy, because many clues tend to be hidden. We have discussed some guidelines for finding and using them; but there is no substitute for field experience.

## Review Questions

1. Distinguish between fault scarp and fault-line scarp.
2. Describe field criteria that could help you in identifying faults. Discuss their validity.
3. Differentiate between "slip" and "separation" as used with respect to fault displacement.
4. Assuming a sheet of paper is a fault surface, draw a horizontal line representing the strike of the fault, and also a line at 60° to it, sloping left, representing the direction of net slip. Next, draw two lines sloping right at 35° to the strike line, and spaced 10 cm apart along the strike; they represent the hanging wall trace (left) and the footwall trace (right) of a geological surface separated by the fault.

    On this diagram enter in red: net slip, strike slip, dip slip. In blue enter strike separation and dip separation.

    Can any other separation be shown on this diagram?

5. Explain why some known faults are classified by slip, whereas some have to be classified by separation.
6. Distinguish between thrust, detachment fault, thrust sheet, and nappe.
7. Distinguish between cataclasite and mylonite.
8. A vertical drill hole traverses a dipping fault. Draw vertical cross sections illustrating situations for which (a) a displaced horizon is encountered twice; and (b) a displaced horizon is completely missed by the drill hole. Do this for both normal and reverse separation.

## Additional Reading

Bell, T. H., and Hammond, R. L. 1984. The internal geometry of mylonite zones. *J. Geol.* 92: 667–686.

Blackwelder, E. 1928. The recognition of fault scarps. *J. Geol.* 36: 289–311.

Brown, C. W. 1961. Comparison of joints, faults and airphoto linears. *AAPG Bull.* 45: 1888–1892.

Crowell, J. C. 1959. Problems of fault nomenclature. *AAPG Bull.* 43: 2653–2674.

House, W. M., and Gray, D. R. 1982. Cataclasites along the Saltville thrust, U.S.A. *J. Struct. Geol.* 4: 257–269.

Kupfer, D. H. 1960. Problems of fault nomenclature. *AAPG Bull.* 44: 501–505.

Simpson, Carol. 1986. Determination of movement sense in mylonites. *Jour. Geol. Ed.* 34: 246–261.

Simpson, C., and Schmid, S. 1983. An evaluation of criteria to deduce the sense of movement in sheared rocks. *Geol. Soc. Am. Bull.* 94: 1281–1288.

# 13

# Fracturing in Rocks

In this chapter we shall discuss the relationship between fractures and the stress fields that cause them. Fractures reflect stress in the lithosphere. They are ubiquitous, and many form very early in the history of their host rocks. Experience with engineering materials, rock deformation experiments, and results of geological model experiments have provided much of our knowledge of the nature of the fracturing process.

There are two distinct types of failure in solids: brittle failure and ductile failure. Brittle failure occurs with a sudden loss of cohesion across a plane, following elastic deformation. Ductile failure occurs beyond the elastic limit in the form of ductile deformation; fracture during ductile deformation may be regarded as a velocity discontinuity in the material whereby one portion moves relative to the other until cohesion is lost.

A broad mechanical classification of fractures recognizes two kinds: **extension fractures,** resulting from outward separation, however minute, of two formerly contiguous surfaces; and **shear fractures,** resulting from displacement, however small, *along* the fracture surface. (This latter displacement is, in fact, faulting.)

## The Principle of Effective Stress

The pores of rocks (except those close to the earth's surface) normally carry a pore fluid, usually water with some dissolved mineral matter, but occasionally with hydrocarbons. The pore fluid is under isotropic pressure. This *pore pressure (p)* is a component of the *total stress (s)* acting on the rock according to the equation

$$S = p + \sigma_e \text{ or } \sigma_e = S - p \tag{13.1}$$

where $\sigma_e$ is the *effective stress*. The pore pressure produces no shear stress, and hence no shear deformation. It acts in a sense opposite to that of the overburden load: if $p$ increases, both $\sigma_1$ and $\sigma_3$ decrease, but the stress difference $\sigma_1 - \sigma_3$ does not change. In other words, increasing $p$, an isotropic stress, will move Mohr circles (see chapter 5, p. 68) toward the origin (figure 13.1), because both $\sigma_1$ and $\sigma_3$ have then decreased by the same amount.

Pore pressures may rise above the normal (hydrostatic) level in rocks by several geological mechanisms, principally: (1) *rapid loading* by sediment deposition or thrusting (see chapter 15), so that rate of loading considerably exceeds rate of escape of pore fluids during compaction of rocks of low permeability, such as shales; (2) *tectonic stress;* (3) *thermal expansion* of pore fluids; and (4) *dewatering* of hydrous minerals.

**Figure 13.1**
Diagram showing the effect of increase of fluid pressure on the position of the Mohr stress circle, assuming that total principal stresses remain constant. $-\sigma_t = T$, the tensile strength, $\Delta\sigma = -\Delta p$, the incremental change in fluid pressure. The bold parabolic curve joins points in $\sigma - \tau$ space at which failure by rupture takes place. (Slightly modified from Secor, 1965.)

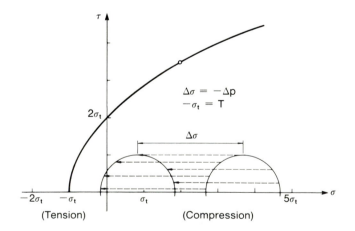

## Failure Criteria

At rupture, $\sigma$ and $\tau$ are related by an equation of the form

$$\tau = f(\sigma)$$

Experiments have shown that, for materials with no cohesive strength, such as soils, the relationship is

$$\tau = \sigma \tan \phi \qquad (13.2)$$

where $\phi$, the slope angle of the line (figure 13.2), is known as the *angle of internal friction.* The term "internal friction" of loose, noncohesive materials describes a material property of slip resistance along the fracture: the steeper the slope of the line, the greater this resistance.

In cohesive materials, frictional resistance and cohesive strength (in shear) have to be overcome before rupture can occur; the relationship then becomes

$$\tau = \tau_0 + \sigma \tan \phi \qquad (13.3)$$

where $\tau_0$ is the shear strength of the material. Here, no friction comes into play before actual rupture takes place, so it may be misleading to call $\phi$ an angle of "internal friction" in cohesive materials.

Equations 13.2 and 13.3 are forms of what is commonly known as **Coulomb's Law** (Coulomb 1773), after the eighteenth century French physicist who first proposed the relationship. It is not truly a "law"; rather, it gives the condition or *criterion* for failure in shear. It is also known as the *Coulomb-Navier criterion,* after Navier, the early-nineteenth century French mathematician who later developed the concept.

Otto Mohr (1900) showed that the failure criterion is not necessarily a straight line, as in equation 13.1, but has the more general form, $\tau = f(\sigma)$. This curve (shown in figures 13.1 and 13.3) constrains Mohr circles for a given material. Circles below the curve represent states of stress in real material. Any circle that touches the curve at point $P$ fails along planes oriented at an angle $\theta$ to $\sigma_3$ where $2\theta$ is the angle between the tangent to the circle at $P$, and the $\sigma$-axis. No part of any circle can go above the curve: that region does not represent real stress states. So, if we can determine a number of values of $\sigma_1$ and $\sigma_3$ at failure for any given rock, we shall be able to plot a family of Mohr circles for states of stress that cause failure in shear in a given material. Mohr drew tangents to such families of circles and found that, instead of Coulomb's straight lines,

**Figure 13.2**
Coulomb failure envelope. $T_o$ = cohesive shear strength; $O'$ = center of Mohr circle for failure in tension; $O$ = center of Mohr circle for failure in compression.

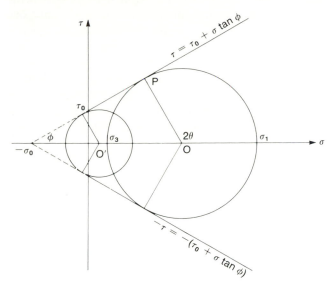

**Figure 13.3**
Mohr failure envelope for unaltered Mineral Mountains granite. (From Bruhn et al., 1982.)

| TEST # | $\sigma_3$ (MPa) | $\sigma_1 - \sigma_3$ (MPa) |
|---|---|---|
| I | 0 | 1.21 |
| II | 34 | 4.55 |
| III | 69 | 5.38 |

Tensile strength
$T_o = 5.0 \pm 2.4$ MPa
Cohesive strength
$\tau_o = 20$ MPa

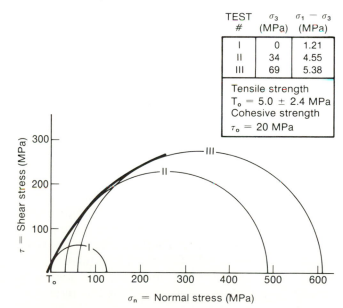

he always obtained curves, now known as **Mohr envelopes,** slightly concave toward the $\sigma$-axis, as in figure 13.3. This is the **Mohr criterion of failure.** Note that tangency of the circles to the Mohr envelope fixes the orientation of the fractures that will form: only one angle $2\theta$ is possible at failure. Note also that, as in chapter 5, we are ignoring $\sigma_2$, the intermediate stress. In fact, the magnitude of $\sigma_2$ rarely has any significant influence, and it would unduly complicate an elementary discussion. Additionally, tangency with negative $\sigma$ value (left of the $\tau$-axis) represents failure in extension, and tangency with positive $\sigma$ values represents failure in shear.

The curvature of the Mohr envelope results from change in material properties with increasing confining pressure, reflecting an increase in ductility. For perfectly ductile material, the envelope would be a straight line parallel to the $\sigma$-axis, for in that case equation 13.3 reduces to $\tau = \tau_0$.

Figure 13.4 shows that $\tau$ can also be negative, yielding a Mohr envelope that is the mirror image of the envelope for positive $\tau$. For an appropriate sign convention, see chapter 5 (p. 69).

## Extension Fractures

Rocks fracture in extension when the numerical value of $\sigma_3$ exceeds that of the tensile strength $\sigma_t$ of the rock. Figure 13.2 shows that in that case both these values are negative, and that only cohesive materials can have tensile strength.

The classical theory of brittle rupture in extension was developed by Griffith (1921) while working on glass. Griffith sought to explain why values of strength actually observed are appreciably lower than those predicted by solid-state theory. He postulated the existence of submicroscopic flaws, "Griffith cracks," throughout the material, in the shape of extremely eccentric, flattened ellipsoids. Tensile stresses will concentrate at the highly curved ends of the cracks, and the material will fail there, at an overall stress difference well below the theoretical tensile strength of the material. The greater the effective length of the cracks, the lower the tensile strength of the material. The Griffith criterion for brittle failure, by postulating that cracks exist from the start, makes it unnecessary to consider how rupture begins in unflawed material. Extension fractures form by propagation and coalescence of Griffith cracks, in planes normal to the least principal stress, and spaced at intervals that depend on the material and on external conditions. Griffith postulated

**Figure 13.4**
Ideal Griffith failure envelope (parabola). (From Ode, 1956.)

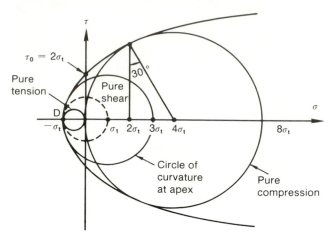

ellipsoidal primary cracks for homogeneous materials like glass. In rocks, grain boundaries may act as Griffith cracks.

The Griffith theory predicts a relationship between normal and sheer stress at failure represented by the parabola

$$\tau^2 - 4\,\sigma_t\sigma - 4(\sigma_t)^2 = 0 \qquad (13.4)$$

where $\sigma$ and $\tau$ are the normal and shear stresses, respectively, and $\sigma_t$ is the tensile strength. The parabola is shown in figure 13.4, where it represents a failure envelope. Although based on experimental criteria, the relationship is theoretical and only an approximation. It probably applies only to the tension side of the plot. For the compression (right-hand) side of the axis, Brace (1960) proposed the relationship

$$\tau = \mu\sigma + 2\sigma_t \qquad (13.5)$$

where $\mu$ is the coefficient of sliding friction, introduced here to account for postulated closure of Griffith cracks under compression. Equation 13.5 is a straight line similar to that in equation 13.3, except that $2\sigma_t$ replaces $\tau_0$, and $\mu$ replaces $\tan\phi$. This agrees with the common observation that $\tau_0$ equals about twice the tensile strength in most rocks. Thus, Brace combined equation 13.4 on the tension side of the $\tau$-axis and equation 13.5 on the compression side, as shown in figure 13.5. Note that circles that are tangent to the envelope on the tension (negative $\sigma$) side of this diagram represent failure in tension, and circles that are tangent to the envelope on the compression (positive $\sigma$) side represent failure in shear.

## Figure 13.5
Trace of quasiparabolic failure envelope. (From Ziony, 1966.)

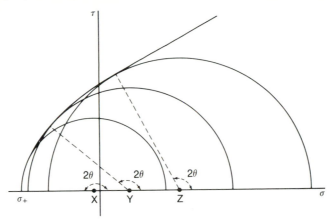

## Geological Setting of Extension Fractures

We have seen, in chapter 11, that we may distinguish fracture systems empirically according to association: those that are associated with local stress fields, and those that are pervasive over large areas.

### Localized Extension Fractures
Failure in extension may be localized. Fractures so formed tend to be filled with vein material. Some localized extension fractures occur in competent layers embedded in a more ductile matrix (figure 13.6); their setting is similar to that of boudinage (p. 142), but the ductility contrast in these fractures is higher than that for boudinage. Many fracture systems associated with local stress fields accompany faults: besides shear fractures, they include extension fractures associated with shear. We shall therefore discuss such extension fractures later together with shear fractures.

### Pervasive Extension Fractures
Many pervasive joints of fundamental joint systems carry plumose structures (p. 206). These structures are associated exclusively with extension, for their delicate patterns are destroyed by the slightest tangential displacement along the joint surface. Furthermore, the ideal complete plumose structure develops in a circle or ellipse: extension joints propagate from an origin, and their outer boundary moves outward radially in an ever-widening circle or ellipse, until the local stresses at the leading edge drop below the tensile strength of the rock (see figures 11.14 and 11.15).

## Figure 13.6
Deformation in Hopla quartz conglomerate and schist, north of Trondheim, Norway. In the conglomerate, extension has resulted in down-dip elongation of pebbles and in quartz-filled fractures perpendicular to elongation; schists have accommodated by ductile flow.

## Figure 13.7

Principal stresses at the initiation of the dominant joint set of the fundamental (orthogonal) system. (From Bock, 1980.)

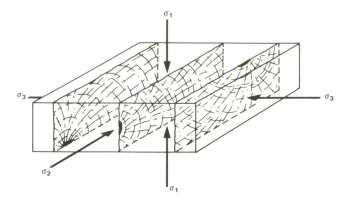

## Figure 13.8

Diagrammatic illustration of extension in a crustal segment raised from *AC* to *BD*. (From Price, 1959.)

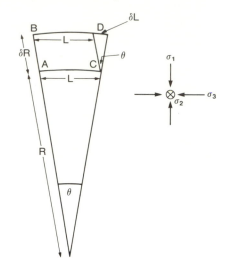

Thus, joint surface structures found on many pervasive joints are probably of extension origin. It is difficult to explain such joints, especially those of the fundamental system, as something other than extension fractures. How, then, do these pervasive extension fractures form?

### Origin of the Fundamental Joint System

The origin of the fundamental joint system has been the subject of much controversy. Bock (1980) lists three basic conditions for its initiation: (1) $\sigma_1$ vertical; (2) horizontal principal stresses $\sigma_2 \neq \sigma_3$ (the difference could be quite small); and (3) effective values of $\sigma_2$ and $\sigma_3$ close to zero, or even negative (necessary to open the joints, even by an infinitesimally small amount) (see figure 13.7).

The pervasiveness of fundamental and some fabric-controlled joints strongly suggests body forces. Possible candidates include volume decrease (e.g., dewatering or cooling), abnormal pore pressure, and overall volume increase through epeirogenic uplift. This last mechanism is not altogether supported by in-situ stress measurements, but it may have a contributing role and therefore merits discussion.

As a rock rises from depth to higher levels in the crust, where it becomes exposed, it occupies an increasingly large space (figure 13.8). Price (1959) suggested that horizontal tensile stresses thus developed may be significant. Even where the initial stress field is isotropic, these stresses may exceed the tensile strength of rocks. According to this concept, above a certain critical depth the two lateral principal stresses, $\sigma_2$ and $\sigma_3$, will be tensile. In isotropic rocks, a set of vertical extension fractures will develop normal to the least (algebraically) of these, $\sigma_3$, which is the greatest tensile stress. As soon as stress is relieved in the $\sigma_3$ direction, the original $\sigma_2$ becomes the greatest tensile stress. When the tensile strength is exceeded once more, a set of extension fractures perpendicular to the original set will form, generally somewhat less well developed than the first.

### The Role of Fluid Pressure and Hydraulic Fracturing

Secor (1965) demonstrated that the higher the ratio of fluid pressure to overburden weight, the greater the depth at which extension fractures can form. Where the ratio reaches 1:1, the fluid pressure is capable of "floating" the rock.

Abnormal pore pressure is sometimes created artificially, intentionally or otherwise. Petroleum producers use it to enhance the flow of oil to wells by hydrofracturing the rocks around them. The effect, expressed in terms of the Mohr diagram, is to move the local Mohr circle toward the tension area of the plot (figure 13.1), opening fractures that increase permeability and, hence, flow of carbohydrates. This is called hydraulic fracturing or hydrofracturing. The orientation of extension fractures so opened will conform to that dictated by the orientation of the ambient principal stresses, for change in pore pressure does not affect the orientation of the stress axes. Thus, in most oil fields, extension fractures opened by hydraulic fracturing are vertical ($\sigma_3$ horizontal) at depth, and horizontal ($\sigma_3$ vertical) at shallow levels. This was predicted by Hubbert and Willis (1957) on theoretical grounds.

### Shear Fractures

Figure 13.2 shows that two sets of fractures theoretically may form at failure in shear: one oriented at an angle $\theta$ with $\sigma_3$, the other at angle $-\theta$ with $\sigma_3$, such that their sum $2\theta$ is always obtuse. The corollary to this relationship is that $\sigma_1$ always bisects the acute angle between the two sets of fractures that theoretically form at failure. This important relationship (figure 13.9) has been verified by many experiments on real materials

## Figure 13.9

Diagram illustrating Hartmann's rule. The maximum compressive stress bisects the acute angle between conjugate surfaces of failure in shear. (a) Normal faults (cross section). (b) Reverse faults (cross section). (c) Strike-slip faults (map view).

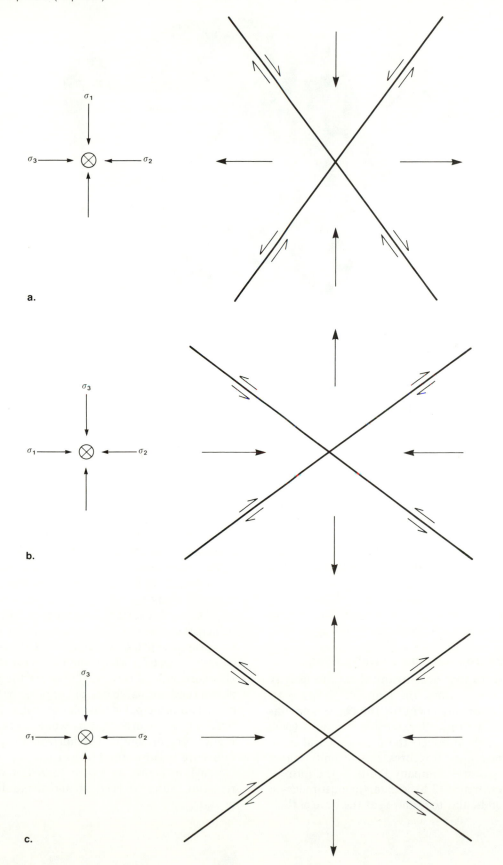

a.

b.

c.

**Figure 13.10**

*En echelon* "tension gashes" in dolomite. Compression was parallel to the joint surfaces; *en echelon* alignment is parallel to one of the potential conjugate shear fractures under this stress pattern. (Specimen in G. Wilson's collection at Imperial College, London.)

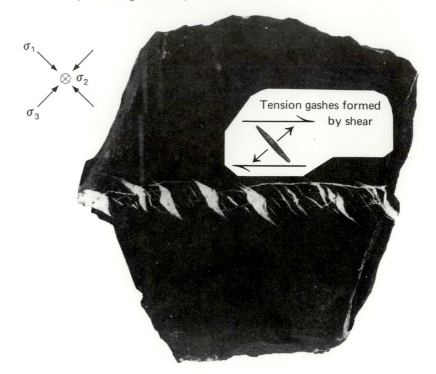

and is known as **Hartmann's rule,** from the name of the French metallurgist (Hartmann 1896). This rule was independently applied to rock fractures and faults by Anderson (1905, 1942), and specifically to joint systems by Bucher (1920, 1921).

Figures 13.10 and 13.11 show some practical applications of Hartmann's rule. Where orientation and sense of shear (fault slip) are known, we may deduce, within certain limits, a reasonable attitude for a set of possible conjugate shear surfaces, and also for the principal stress and strain axes. If we don't know the sense of shear, we may derive it from the attitude of any extension fractures that are clearly related to the observed shear, as in figure 13.10. Such extension fractures must be perpendicular to $\epsilon_1$, and hence to $\sigma_3$.

### Extension Fractures Associated with Shear

Extension fractures that accompany shear are usually arranged *en echelon,* along the plane of shear, as in figure 13.10. This arrangement fits the Hartmann diagram in that the attitude of extension fractures (perpendicular to $\epsilon_1$ and $\sigma_3$) bisects the acute dihedral angle between conjugate shear fractures. Extension fractures of this type may also accompany faults in the form of pinnate fractures (figure 12.5). Again, their attitude will always be perpendicular to $\epsilon_1$ and $\sigma_3$ at the time of their formation.

In this discussion we assume that $\epsilon_1$ and $\sigma_3$ are parallel. That may not always be strictly true, but it is an acceptable approximation in this context.

### Faulting

The basic dynamic principle of faulting is very simple. The states of stress illustrated in figure 13.9 will tend to cause shear fractures as shown, in accordance with the Mohr criterion. The potential relative displacement sense along the fractures becomes evident when we resolve the principal stresses along them; that is how faults will form. Note that conjugate shear fractures always intersect along $\sigma_2$.

The earth's surface influences the orientation of the principal stresses in upper crustal rocks. Parallel to the topographic, surface shear stresses are zero. Hence, by definition (p. 67), where the earth's surface is reasonably horizontal, it is parallel to one of the principal stress planes (and normal to one of the principal stresses). The other two principal stress planes would have to be vertical, which ensures that $\sigma_2$ would normally be either vertical or horizontal. For horizontal $\sigma_2$, the resulting faults are either normal ($\sigma_1$ vertical) or reverse ($\sigma_1$ horizontal), as shown in figure 13.9. For vertical $\sigma_2$, the resulting faults are vertical, and strike slip is either left or right.

**Figure 13.11**
Dolomite and black limestone inclusions in shaly matrix. Elongation of inclusions is parallel to the knife pictured. Extension fissures in inclusions are filled with calcite. Note prominent calcite-filled fissures, and orientation and symmetry of conjugate fracture system. (From Korn and Martin, 1959.)

## Sliding Along Existing Faults

What happens when stress is induced (or, more generally, the state of stress changes) in a rock that already is fractured? There must be conditions under which failure is more likely to take the form of sliding along preexisting fractures rather than opening new ones. Sliding along preexisting fracture surfaces does not have to overcome cohesion, only sliding friction. Thus, in such cases, the relationship between $\tau$ and $\sigma$ is that of equation 13.2 (line $OC$ in figure 13.12). A state of stress whose Mohr circle intersects line $OC$ will result in sliding along any surfaces oriented at angles $\theta$ with $\sigma_3$ that correspond to the intercept $AB$, made by the Mohr circle in figure 13.12. This relationship has some importance in practice, because it shows that increased pore pressure, which shifts Mohr circles to the left (see figure 13.1), may reactivate existing fractures.

**Figure 13.12**
Mohr diagram showing envelopes for failure in intact Weber sandstone (tangent) and for slip on preexisting faults in Weber sandstone (line $OABC$). Also shown is a Mohr circle tangent to the envelope for failure in the intact rock. (From Raleigh et al., 1972.)

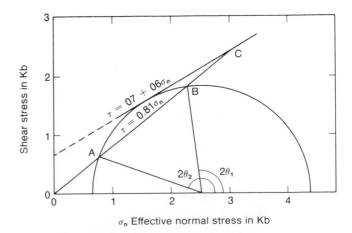

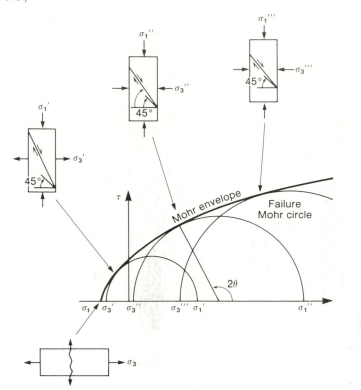

## Oblique Extension Fractures

Oblique extension fractures (that is, oblique to $\sigma_1$ and $\sigma_3$) in single or conjugate sets may take the place of the more common extension fractures normal to $\sigma_3$ (figures 13.13 and 13.14). In conjugate systems they are so disposed that the theoretical $\sigma_1$ direction at failure bisects the acute angle between them. They are typically very pervasive, and some carry plumose structures. How do they form?

The Mohr circles in figure 13.5 and figure 13.13 become tangent to the failure envelope at $\sigma_t$, and extension fractures parallel to the greatest normal stress $\sigma_1$ will result. Larger circles (representing a greater stress difference) may become tangent to the failure envelope between $\sigma_t$ and $\tau_o$, in cases where the envelope is continuous in that region (e.g., circle $Y$ in figure 13.5 or the smallest circle in figure 13.13). Such a configuration will result in failure along conjugate surfaces that enclose a comparatively small acute angle (Muehlberger 1961, Ziony 1966), which would be bisected by the $\sigma_1$ direction. Geometrically, this resembles the pattern predicted by Hartmann's rule; but in the Hartmann pattern the conjugate surfaces fail in shear under a compressive normal stress component. Oblique extension fractures, though parallel to planes carrying a resolved shear stress, result from oblique extension across the fracture surface. Displacement is therefore predominantly extensional at a large angle to the fracture surface, with a negligible shear component and no friction (Dennis 1969). These oblique extension fractures have a characteristically small dihedral angle as explained in figure 13.13 and illustrated in figure 13.14. They are typically pervasive and are not normally caused by localized stress fields.

## Trajectories

The Anderson-Hartmann relationship (figure 13.9), though extremely useful, is an oversimplification of the real world. It assumes a homogeneous stress field uniquely related to the earth's surface. The curvature of many, if not most, fault surfaces shows that many of the stress fields that give rise to faulting are not homogeneous. M. King Hubbert (1951) showed how the stress field curves in a crustal segment under compression (figure 13.15). The continuous lines are **stress trajectories,** and they represent lines of equal values for a stated stress component, in this case $\sigma_1$ and $\sigma_3$. We can, of course, construct a Hartmann relationship (p. 237) for any point of figure 13.15. The broken lines are traces of associated potential shear surfaces (slip lines). Note that true rigid body translation along a fault can occur only if the fault is perfectly planar (or, improbably, part of a spherical or cylindrical surface). All curved faults require straining of wall rock, usually in the hanging wall. Such strain can be accommodated by ductile flow, by subsidiary faulting, or both (see chapter 14).

## Figure 13.14
Jointing associated with lineation in Pennsylvanian schists, Shadow Mountains, California. These calcite-filled joints are in two sets: one (top left to bottom right) is prominent; the other (symmetrically disposed to the first about the *ac* plane of the lineation) is very subdued. This system evidently "substitutes" for the normally more common *ac* set. It is a system of oblique extension joints. Note the characteristic small dihedral angle.

## Figure 13.15
Stress trajectories in a rectangular slab under lateral compression. (a) Complete stress system. (b) Trajectories of principal stresses and potential fault surfaces (indicated by broken lines). (From Hubbert, 1951.)

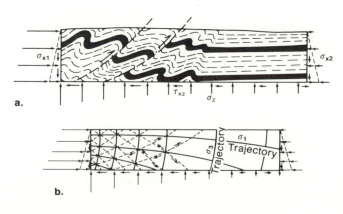

**Figure 13.16**
Schematic representation of the spectrum from brittle fracture to ductile flow, with typical strains before fracture and stress–strain curves for uniaxial compression and extension. The ruled portions of the stress–strain curves indicate variation within each case, and overlap between cases 3, 4, and 5. (From Griggs and Handin, 1960.)

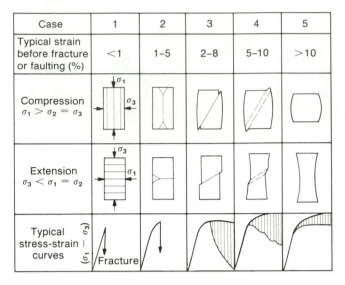

$\sigma_1, \sigma_2, \sigma_3$, are maximum, intermediate, and minimum principal stresses, respectively

## Ductile Shear Zones

Shear zones are tabular zones in rocks that localize shear strain. They range in width from infinitesimal, as in perfectly brittle faults, to hundreds of meters, as in some ductile shear zones in basement. Shear deformation occurs in a whole spectrum of styles, from clean, spaced shear fractures to continuous bulk strain (figure 13.16). Each of the fundamental deformation mechanisms discussed in chapter 6 (pp. 94–96) can result in localized *or* pervasive deformation of a rock. Localization and its manifestation and intensity depend on rock type and environmental factors. Figure 13.17 illustrates the relationship between the different deformation mechanisms and strain localization. It also implies that the contrast "brittle/ductile" is an oversimplification that may mislead (Rutter 1986).

In *ductile* shear zones, strain is concentrated along more or less widely spaced parallel zones. The strain becomes apparent in the rock fabric, generally by conspicuous, enhanced foliation (figure 13.18) and, in many cases, a lineation indicating the direction of greatest extension. Many ductile shear zones are characterized by mylonites and other fault rocks (see chapter 12). Some ductile shear zones are considered faults, some are not. Those in which it is possible to follow fabric elements from one boundary to another are not true faults (e.g., figure 13.18). Those in which prestrain fabric elements are so strained that they lose their identity must be considered faults.

As we follow a shear zone from the ductile to the brittle environment, we find that it narrows in width until eventually cohesion is lost along a fracture surface. Many faults go through a stage of initial ductile deformation, which imparts an S shape to suitably oriented markers. These S shapes give the appearance of "drag" along the fault after rupture. However, true drag along faults must be very rare because, once shear strain is taken up by a fracture surface, ductile deformation along adjacent rock is minimal.

**Figure 13.17**
Diagrammatic representation of deformation mechanism fields, for shear in rocks with and without localization of shear. (From Rutter, 1986.)

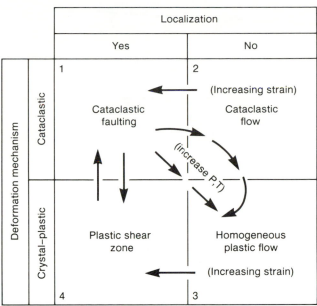

**Figure 13.18**
Shear zone in Maggia gneiss. (Photo by H. Masson.)

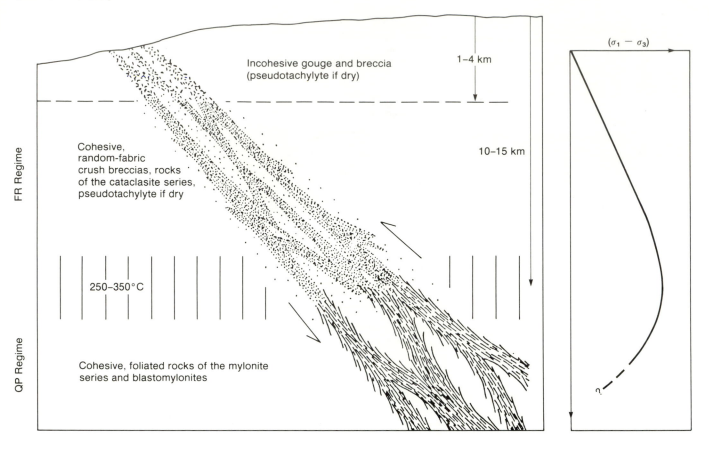

## Mechanisms of Movement Along Faults

Sibson (1977) has recognized two main shear regimes: the *frictional* or *FR regime,* and the *quasiplastic* or *QP regime.* The FR regime is characteristic of higher crustal levels than the QP regime (figure 13.19).

There are three fault mechanisms in the FR regime: brittle shear failure of intact rock, sliding on existing planes, and shear across a cataclastic shear zone.

### Brittle Shear Failure of Intact Rock

This is the mechanism of initial rupture in shear. The fracture must overcome the cohesive strength of the rock, which, in practice, means propagation and coalescence of microcracks and other flaws. It requires a relatively large amount of energy, part of which can be dissipated in seismic radiation and heat upon rupture, depending on material properties and loading conditions.

### Sliding on Existing Planes

Most fault movement in the FR regime occurs by sliding on existing planes, for once cohesion is overcome, fault movement is by sliding along the fracture surface. Such sliding is normally episodic. That is, stress accumulates until frictional resistance along the fault is overcome, and sudden displacement along the fault occurs until the stress drops to a value at which friction stops further movement. Only certain segments of a fault would normally move at any one time; in other words, movement is "patchy," both in time and in space. Such patchy movement is called **stick-slip.** However, some faults, or segments of faults, may move by steady displacement called **fault creep** (do not confuse this with ductile creep, defined on p. 76). Friction may be considerably reduced by pore pressure (see p. 239), so that FR regime sliding may operate at comparatively deep crustal levels.

### Shear Across a Cataclastic Shear Zone

The third mechanism in the FR regime is shear across a cataclastic shear zone. In this, a ductile mechanism, resistance is controlled by internal friction in the shear zone.

## Shear in the QP Regime

In the *QP regime,* shearing is accommodated by flow in tabular zones of finite thickness: these are ductile shear zones, as discussed on p. 242. Here failure and movement rate are governed by the yield strength of the material. The failure mechanism in the QP regime is plastic flow and therefore is thermally activated (see chapter 6). In fact, failure along shear zones is an intermediate mode between true fracture and ductile flow (figure 13.17). In the course of displacement along ductile shear zones, cohesion may be lost temporarily along discrete surfaces, giving rise to a transient ductile fracture.

In most fault movement in the deep crust (deeper than 20 km), the closing component of stress across faults is so large that sliding along fault surfaces would be inhibited but for special friction-reducing circumstances. Such circumstances include abnormal pore pressure (p. 239), and rock layers or tabular zones of low yield strength (which make friction irrelevant) (see also chapter 15).

### The Effect of Heat

In brittle deformation, episodic sliding friction at seismic rates produces transient, intensely localized heating. Where water is present, it inhibits friction melting because the first increments of frictionally generated heat during a slip event can boost fluid pressure to extremely high values; this could lower frictional shear resistance to almost zero. Where water is absent, enough heat may be generated by friction to melt the wall rock of the fault, resulting in generation of pseudotachylite (see p. 226). In steady fault creep, not enough heat is generated for the formation of pseudotachylite. Heat may also cause fracture by unequal thermal expansion of adjacent segments of rock, or by expansion of pore fluids, which will result in an increase of pore pressure.

Figure 13.19 summarizes some of the factors that affect and control faulting along a major fault zone.

## Development of Fractures

Development of fractures includes their initiation and their propagation.

### Initiation of Fractures

Experimental investigation of the microfabric of rocks has shown that joints and faults originate from the coalescence and propagation of microscopic openings. These may be preexisting flaws, as well as new cracks (*microcracks*) generated in the course of bulk strain, which always precedes failure. Pinnate or feather joints (p. 238) probably have a similar origin; that is, they

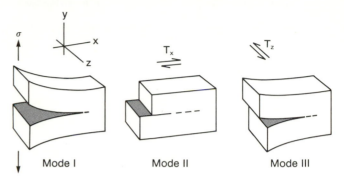

**Figure 13.20**
Diagrammatic representation of the three modes of subcritical crack propagation (modes I, II, and III). (From Paterson, 1978.)

*precede* movement along the fault. After initial rupture, the stress near the new fault is relieved, so that pinnate fractures are no longer likely to form.

The initiation of microcracks manifests itself in two well-known phenomena: dilatancy and acoustic radiation. *Dilatancy* is an increase in volume resulting from opening of microcracks. It seems logical that the opening of cracks would increase the volume of a rock by increasing its porosity, and this has been confirmed in rock-deformation experiments. *Acoustic emissions* have been detected in experiments, and it seems that swarms of microseisms in the crust signal microcrack opening and limited propagation. Microearthquakes are common along seismically active faults.

### Propagation of Fractures

Fractures form by propagation from initial crack tips. If the stress intensity exceeds a certain "critical" value (which varies in different rocks and under different environmental conditions), fracture propagation proceeds at or near the speed of sound. Below that critical value ("subcritical"), propagation from tips is slower and may proceed in any of three different ways or "modes" (figure 13.20): displacement normal to the crack plane and propagation as extension fractures along the plane of the original crack (mode I); displacement parallel to the crack plane and normal to the crack edge with propagation as shear fractures along it (mode II); and displacement parallel to the crack plane and to the crack edge with propagation as shear fractures ("tear" mode—mode III). The process of propagation is not yet fully understood, but subcritical propagation is aided by chemical reactions along crack tips. This is what is known as "stress corrosion," and it proceeds by circulation of solutions along networks of incipient and propagating cracks: opening fractures draw in available fluids.

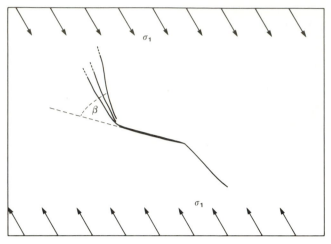

Fracture relieves stress. In the process, fracture uses up the mechanical energy manifested as stress and deformation. As a fracture propagates, energy is dissipated and the fracture finally stops. Waning energy near the periphery of a propagating fracture eventually tends to induce curvature of the fracture surface as well as branching into subsidiary fractures. Figure 13.21 shows how this happens along some faults. The fringe zones of plumose structures along extension joint surfaces (figures 11.14 and 11.15) are good examples of this tendency. On a larger scale, so are the famous "horsetail veins" at Butte, Montana, which carry rich copper ore (figure 13.22). The main fractures in that region are original shear fractures; the horsetail ends curve toward the standard attitude of extension fractures.

**Figure 13.22**
Plan of a portion of the 1,200 level of the Leonard Mine, Butte, Montana, showing "horsetail veins." The pattern is the result of fracture branching. (From Sales, 1914.)

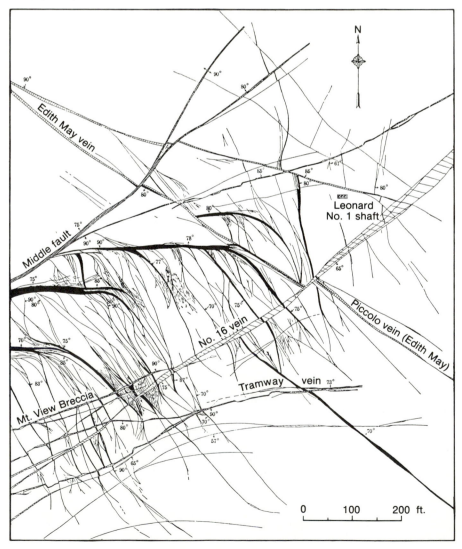

## Special Fractures

### Sheeting

Sheeting, as illustrated in figure 11.4, is usually parallel to the ground surface and confined to a restricted zone below it. The accepted explanation is that sheeting is the result of vertical extension following load release by erosion. It is most common in massive rocks.

### Columnar Jointing

Columnar jointing, a well-known form of fracture (figure 13.23) is characteristic of certain volcanic rocks. Polygonal columns form along shrinkage fractures by a process not unlike that which produces mud cracks.

As lava flow, ash flow, or dike rock cools, shrinkage occurs in a direction parallel to surfaces of equal temperature at any one time. If columns form, fractures bounding them propagate from the cool outside to the hotter inside of the body as it cools, at right angles to surfaces of equal temperature (isothermal surfaces). Hence, the attitude and pattern of the columns may give a clue as to contact configuration at the time of emplacement. Good columns form only under conditions of even cooling in a homogeneous rock.

Columnar joints are not necessarily the first joints to form in a cooling lava. Bankwitz (1978) documented many orthogonal fracture systems of the fundamental type (p. 199) that precede the initiation of columnar jointing.

**Figure 13.23**
Columnar jointing, Devil's Postpile, California. (Photograph by J. Crowell.)

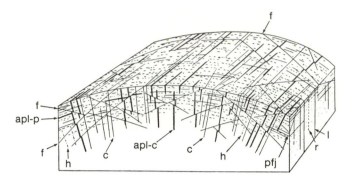

## Primary Fractures of Plutonic Rocks

Apart from fracture systems imposed long after emplacement, most plutons carry what Balk (1937) called primary joint systems, which are directly associated with the emplacement and with the fabric of the body. These joints may develop in four main systems (figure 13.24): cross joints, diagonal joints, longitudinal joints, and primary flat joints.

1. **Cross joints** perpendicular to lineation and foliation, the most conspicuous fractures, are analogous to *ac* joints of sedimentary and metamorphic rock fabrics. Mechanically, they are normal extension fractures. Many carry vein fillings, commonly of igneous differentiates.
2. **Diagonal joints** are steep fractures that form acute dihedral angles with the cross joints. They are oblique extension fractures.
3. **Longitudinal joints** are steep fractures parallel to lineation. They are *bc* (normal) extension fractures.
4. **Primary flat joints,** flat-lying fractures, are usually, but not necessarily, parallel to planar fabrics where these are also flat lying. Their overall attitude is governed by the fabric where suitably oriented. These pervasive fractures must not be confused with near-surface sheeting, discussed earlier.

Primary fracture systems in plutons develop best in flat-lying, sheetlike bodies, or in portions of plutons in which the fabric is flat lying. The steep walls of some plutons carry characteristic pinnate extension fractures associated with shear during emplacement.

## Cleats in Coal

Gangel and Murawski (1977) investigated **cleats,** which are joints in coal, and found that several fracture properties can be conveniently studied in cleats. Cleat spacing is quite close, of the order of millimeters and centimeters. By means of a scanning electron microscope, Gangel and Murawski were able to distinguish between extension fractures and shear fractures; extension fractures commonly show characteristic plumose patterns, whereas many shear fractures show striations resembling those of slickensides. These and other investigations (Murawski and Merkel 1983) have shown that a pervasive cleat system is orthogonal, just like the fundamental joint system (figure 13.25). It appears to have formed very soon after coalification. But in the neighborhood of faults, shear fracture systems complicate the pattern. The appearance and increasing frequency of such "shear cleats" in coal mining can act as a warning signal when faults are approached in mining operations. This phenomenon can also be observed in fracture systems in other rocks (figure 13.26).

**Figure 13.25**
Directional statistics of cleat pattern of a Paleogene coal from the southeastern part of the Manji dome, Japan. (a) Bedding traces of cleats. (b) Rose diagram of traces in (a). Dashed circle encloses radii representing 10% of fracture population. (From Murawski and Merkel, 1983.)

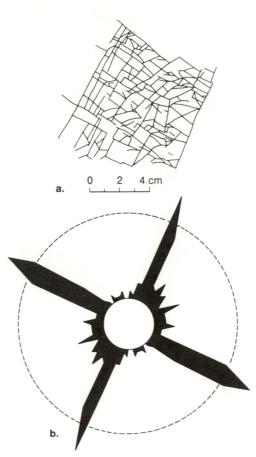

**Figure 13.26**
Schematic cross section of the Bonita fault zone in east-central New Mexico, showing fault-related conjugate shear fracture systems superimposed on the regional orthogonal system. (From Stearns, 1972.)

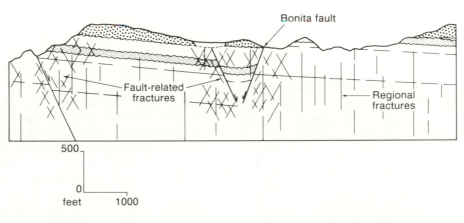

**Figure 13.27**
Fractures modeled in soft mud. (a) Extension fractures and faults in soft mud. Deformation experiments as described by Ernst Cloos (1955). Extension cracks open if the surface is liberally sprinkled with water. (b) Shear fractures develop if the clay surface is left "dry." (Photographs courtesy of E. Cloos.)

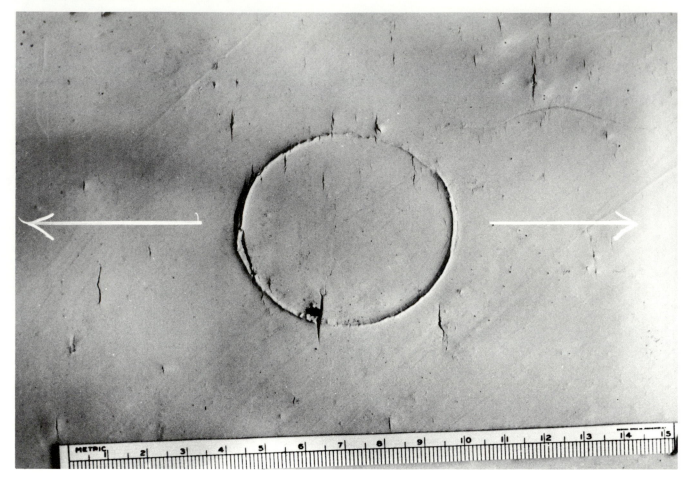

a.

**Figure 13.28**
Ernst Cloos' wire cloth experiments; (a) setup for nonrotational deformation (pure shear). The pull is in the diagonal of the square net. (b) Setup for rotational deformation (in this case a combination of simple shear and pure shear). The front is bolted to the table, and the rear can move from left to right and vice versa. (From E. Cloos, 1955.)

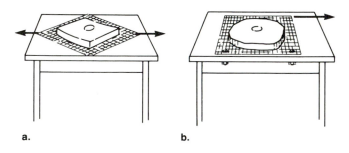

a.                                    b.

## Fracture Patterns in Clay

Noncohesive sediments have negligible tensile strength. The failure envelope is almost entirely in the compression region, and extension fractures will not normally form. Hence, most fractures in unconsolidated sediments are faults. The only exceptions are in clays, which possess some cohesion. That is also why clay is an excellent model material for the experimental investigation of fracture patterns. For, although experiments with rocks and engineering materials reveal much about the nature of fracture, they do not tell us much about large regional patterns. Hans Cloos and Ernst Cloos have been the pioneers in the technique of reproducing natural fracture patterns in clay models. A simple version, used by Ernst Cloos (1955), is very instructive for classroom use. It consists of a cake of clay that is placed on a wire cloth laid flat on a table (figures 13.27 and

**b.**

13.28). The mesh can be deformed in shear, and the resulting fracture patterns can then be studied in relation to known strain orientation. Relationships when fractures first appear are particularly critical. For shear fractures, the dihedral angle is less than 60° at first, but it increases by rotation. Faults then appreciably displace the ellipse outline, and the dihedral angle increases. More sophisticated (and far more expensive) arrangements for deformation control have been described by Oertel (1961) and Hoeppener et al. (1969).

## Conclusion

Most rock fractures fall into two clearly defined dynamic categories: extension fractures (both normal and oblique) and shear fractures. The best field identification of extension fractures is by the presence of plumose structure or vein fillings; the surest field identification of shear fractures and faults is by demonstrable primary slip. But neither of these indications may be present in any one case, so that field setting becomes the only reliable guide.

The following dynamic classification of rock fractures and faults emerges. *Normal extension fractures* form normal to the direction of least principal stress ($\sigma_3$), or the X direction of the strain ellipsoid. *Oblique extension fractures* tend to form a conjugate system, enclosing a small acute dihedral angle bisected by the direction of greatest principal stress ($\sigma_1$).

*Shear fractures* normally follow Hartmann's rule: in conjugate systems $\sigma_1$ bisects acute dihedral angles between pairs of sets. *Faults* are shear fractures or more or less coherent shear zones. They are nonpervasive; that is, they occur in well-defined deformed segments of the crust. In the next chapter, we shall consider the geological setting of some such segments.

## Review Questions

1. Explain the principle of effective stress and the role of pore pressure in rock fracturing.
2. What is the nature of resistance to slip along fractures?
3. Derive the equation for the Coulomb-Navier fracture criterion and explain its meaning.
4. What is the Mohr envelope? What information does it provide?
5. How does the Griffith criterion help in explaining brittle fractures?
6. What clues do plumose structures on fracture surfaces provide as to the origin and propagation of extension fractures?
7. Draw a Mohr's circle diagram to explain the principle of hydraulic fracturing.
8. What are oblique extension fractures? Explain by means of a diagram.
9. Apply Hartmann's rule to normal faults, reverse faults, and strike-slip faults.
10. What is a ductile shear zone? How could you assess displacement along such a zone?
11. Differentiate between fault mechanisms in the FR and QP regimes.
12. What are the three modes of subcritical crack propagation?
13. How does sheeting in near-surface rocks originate?
14. What does the orientation of columnar joints indicate about the cooling history of volcanic rocks in which they are found?
15. How are fractures in plutons related to emplacement of the pluton?

## Additional Reading

Atkinson, B. K. 1982. Subcritical crack propagation in rocks: Theory, experimental results and applications. *J. Struct. Geol.* 4: 41–56.

Ball, A. 1980. A theory of geological faults and shear zones. *Tectonophysics* 6: T1–T6.

Bles, J. L., and Feuga, B. 1986. The fracture of rocks. Amsterdam: Elsevier. 128 pp.

Bock, H. 1979. A simple failure criterion for rough joints and compound shear surfaces. *Eng. Geol.* 14: 241–254.

Corbett, M. K. 1979. Origin of pervasive orthogonal fracturing of the earth's crust. *2d Int. Conf. Basement Tecton. Proc.* 319–325.

Delaney, P. T., Pollard, D. D., Ziony, J. I., and McKee, E. H. 1986. Field relations between dikes and joints: Emplacement processes and paleostress analysis. *J. Geophys. Res.* 91: 4920–4938.

Engelder, T. 1982. Is there a genetic relationship between selected regional joints and contemporary stress within the lithosphere of North America? *Tectonics* 1: 161–177.

Engelder, T., and Geiser, P. 1980. On the use of regional joint sets as trajectories of paleostress fields during the development of the Appalachian Plateau, New York. *J. Geophys. Res.* 85: 6319–6341.

Fletcher, R. C., and Pollard, D. D. 1981. Anticrack model for pressure solution surfaces. *Geology* 9: 419–424.

Friedman, M. 1969. Structural analysis of fractures in cores from Saticoy field, Ventura County, California. *AAPG Bull.* 53: 367–389.

Gretener, P. E. 1969. Fluid pressure in porous media—Its importance in geology: A review. *Bull. Can. Pet. Geol.* 17: 255–295.

Griggs, D. T. 1960. Observations on fracture and a hypothesis of earthquakes. Pages 347–364 in: D. T. Griggs and J. Handin, eds. Rock Deformation. *Geol. Soc. Am. Mem.* 79.

Hafner, W. 1951. Stress distributions and faulting. *Geol. Soc. Am. Bull.* 6: 373–398.

Jaeger, J. C., and Cook, N. G. W. 1979. *Fundamentals of Rock Mechanics.* 3d ed. London: Methuen. 593 pp.

Muecke, G. K., and Charlesworth, H. A. K. 1966. Jointing in folded Cardium Sandstone along the Bow River, Alberta. *Can. J. Earth Sci.* 3: 579–596.

Nelson, R. A. 1981. Significance of fracture sets associated with stylolite zones. *AAPG Bull.* 65: 2417–2425.

Nur, A. 1982. The origin of tensile fracture lineaments. *J. Struct. Geol.* 4: 31–40.

Odé, H. 1960. Faulting as a velocity discontinuity in plastic deformation. *Geol. Soc. Am. Mem.* 79: 293–321.

Ramsay, J. G. 1980. The crack-seal mechanism of rock deformation. *Nature* 284: 135–139.

Shainin, V. E. 1950. Conjugate sets of *en echelon* tension fractures in the Athens limestone at Riverton, Virginia. *Geol. Soc. Am. Bull.* 61: 509–517.

Sibson, R. H. 1981. Controls on low-stress hydro-fracture dilatancy in thrust, wrench and normal fault terrains. *Nature* 289: 665–667.

Spry, A. 1962. The origin of columnar jointing, particularly in basalt flows. *Geol. Soc. Aust. J.* 8: 1295–1308.

Stearns, D. W., Couples, G. D., Jamison, W. R., and Morse, J. D. 1981. Understanding faulting in the shallow crust: Contributions of selected experimental and theoretical studies. Pages 215–229 in: N. L. Carter, M. Friedman, J. M. Logan, and D. W. Stearns, eds. *Mechanical Behavior of Crustal Rocks.* Washington D.C.: Am. Geophys. Union Geophysical Monograph 24.

# 14

# Geological Setting of Faults

**Figure 14.1**
Landslide in varved silt and clay near Grand Coulee Dam, Grant
County, Washington. Note the relation of the beds, and compare this
with figure 14.22. (Photograph by F. O. Jones, U.S. Geological
Survey.)

We now turn to the geological setting of faults. Their pattern, orientation, and displacement reflect the local mechanism of crustal failure. We shall deal with each of three main crustal environments in turn: extension, steeply dipping horizontal shears, and shortening.

## Crustal Extension, Thinning, and Subsidence

Normal faulting, commonly listric, is the brittle response to crustal thinning. Such thinning nearly always involves horizontal extension, and brittle zones of extension are the habitat of normal faults. This seems quite natural, because in normal faulting, $\sigma_3$ is horizontal (figure 13.9a). Like most faults, normal faults are not isolated structures, even though they occur in restricted zones. They are always members of a system.

## The Influence of Gravity

Normal faults are sometimes called gravity faults. This is intended to underline the importance of the influence of gravity in initiating many normal faults; but genetic terms may lead to misunderstanding, and the descriptive term "normal fault" is preferable. We shall see how gravity acts as an important factor in much normal faulting. To gain an intuitive understanding of this phenomenon, we shall first briefly discuss well-known, observable effects of gravity, such as landslides and other downslope movements of rock masses.

### *Landslides*

From the structural point of view, there are two types of slides: *slab slides* and *rotational slides*. Slab slides utilize preexisting surfaces of weakness, such as lubricated joint or bedding surfaces. Those that create their own sliding surfaces are called rotational slides, because in such cases slip is rotational and along cylindrical surfaces (figures 14.1 and 14.2).

**Figure 14.2**
Fissures in landslides, indicating extension in the upper part, but compression in the toe. 1, bedrock; 2, soil; 3, slide material. (From Ter-Stepanian, 1962.)

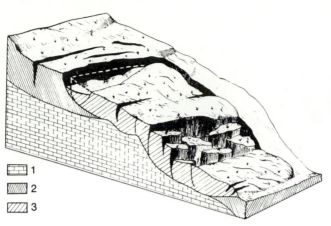

Slab slides have a ready-made slip surface and are usually of greater extent than rotational slides. They move rapidly and usually disintegrate during movement, coming to rest as a jumbled, brecciated mass. Rotational slides or slumps also detach themselves from the underlying mass, but the detachment surface is newly made, and concave upward. Most are smaller than slab slides, and many move with almost imperceptible slowness (although rapid rotational slides are known). They may retain some measure of coherence. Failure begins at the contour (head) cracks, which serve as the first danger signal in affected slopes (a and b in figure 14.2). Soon the surface of failure, or slip surface, propagates downward, then curves back up toward the free surface, producing the well-known spoon-shaped form.

### Collapse Along Flanks of Anticlines

Harrison and Falcon (1934) have described a classic example of collapse structures in southwest Iran. There, a succession of Cretaceous and younger marls and limestones has been folded and subsequently eroded. Comparatively more rapid erosion of marls in synclines has left exposed steep dip slopes of limestone beds. These limestones themselves normally overlie a weak layer of marl, which cannot support the steeply dipping exposed limestone. As a result, the limestone has become detached and has collapsed in a great variety of patterns (figures 14.3 and 14.4), involving the underlying marl in the deformation. It is possible, but by no means necessary, that the synclinal valleys were eroding as the folding was still going on, resulting in an even greater rate of oversteepening of the limestone dip slopes.

**Figure 14.3**
Diagrams of gravity-collapse structures, representing stages in the development of indicated structures. (After Harrison and Falcon, 1934, as modified by Hills, 1963.)

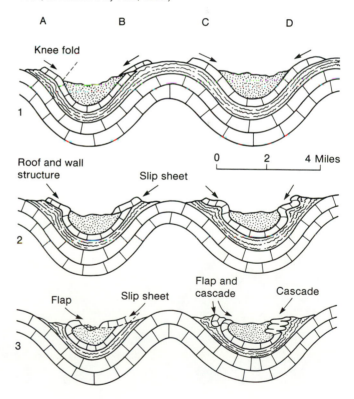

**Figure 14.4**
North flank of Kuh-e Gezeh (Fars Province, Iran) showing
gravitational slump structures in the Gachsaran Formation (Miocene)
limestones and marls. Such structures are common on many of the
Fars Province anticlinal flanks. Thin beds of gypsum often act as a
lubricating medium. Location: 54° 10'E, 27° 05'N. (Photograph by
H. McQuillan, published in Falcon 1969.)

**Figure 14.5**

Diagrammatic cross section, Baker Peaks, Arizona, showing rotation of Tertiary strata along listric normal faults. Note the relatively shallow depth at which the detachment fault at the sole and the autochthonous basement rocks presumably underlie the allochthonous rocks. Section below the dotted line (alluvial contact) is extrapolated. Scale: 1:18,500, with negligible vertical exaggeration. (From Pridmore and Craig, 1982.)

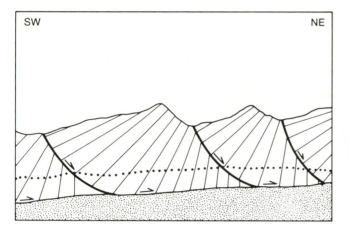

**Figure 14.6**

The south marginal fault of the Hunsrück plateau, West Germany, and its inferred downward continuation. The seismic velocity-depth curve is shown at left. Lowercase letters near top = Paleozoic formations: $M$ = Mohorovicic discontinuity (displaced); $C$ = Conrad discontinuity; horizontal hachuring = downward continuation of fault. Also shown are two first motion diagrams (arrows) with dates of the relevant earthquakes. (From Murawski, 1976.)

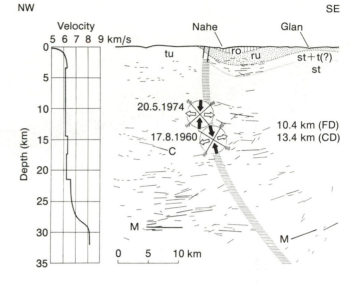

This last example comes very close to the ill-defined border between surficial and tectonic collapse structures. Strictly speaking, if the anticlines were still "alive" when collapse occurred, the collapse would have been tectonic, but if folding had already ceased, the collapse would have been purely surficial.

For evidence of large-scale phenomena of this type, we have to look for slope instability and for gliding under gravitational forces on a grand, tectonic scale.

## Listric Normal Faults and Growth Faults

Few faults maintain a uniform dip over a great extent. Many normal faults tend to flatten with depth; additionally, their strikes tend to be curved, giving the fault surface a spoon-shaped, concave-upward shape. Eduard Suess named such curved fault surfaces **listric** surfaces (e.g., figure 14.5).

The recognition of listric faults is of great importance in exploration: we need to know how to extrapolate known faults at depth. This brings us to a fundamental question: How do normal faults terminate at depth?

It is not difficult to understand that straight faults cannot terminate downward within a perfectly rigid crust. So straight faults may have to terminate in a yielding substratum. Most known normal faults that terminate in a yielding layer at depth are listric: that is, their dips flatten to some extent, at least until they reach the yielding layer. Or, they curve until they join a flat-lying boundary along which a rock layer has become detached (figure 14.5). We may suspect that many normal faults, if followed to sufficient depth, turn out to be listric. The question regarding any particular fault or set of faults is, how deep is the yielding or detachment layer? Examples are known (e.g., figures 14.6 and 14.40) in which a fault displaces the Mohorovičić discontinuity and may terminate at the lithosphere-asthenosphere boundary. In others, such as in figure 14.5, faults terminate along comparatively shallow detachment zones. The inference in each case is that faulting has been the result of extension in a brittle layer above a yielding layer or a detachment surface. The layer below may have participated in extension by yielding. In many known instances, the detachment surface coincides with the transition from the so-called brittle to the ductile regime in the crust, at a depth most commonly between 9 and 12 km.

## Figure 14.7

Gravity slide (slump structure) developed contemporaneously with rapid early Oligocene deposition of the Vicksburg Formation, Starr County, Texas. Note down-to-basin growth faults. (From Erxleben and Carnahan, 1983.)

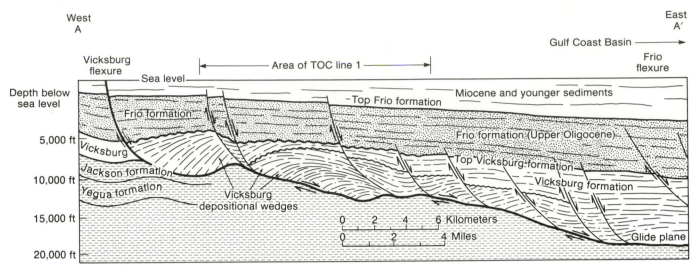

## Figure 14.8

Syndepositional listric normal fault (growth fault) assemblage in Tertiary Gulf Coast sediments, south Texas. Note basinward migration of rollover anticline crest with depth and differential compaction. (From Bruce, 1973.)

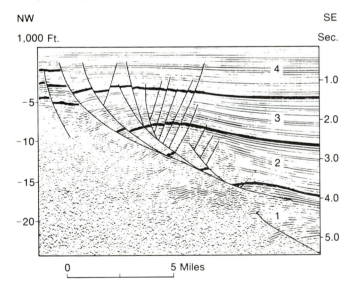

## Geometric Consequences of Listric Normal Faulting

Figure 14.5 shows that movement along a listric normal fault results in tilting the down-thrown block toward the fault, ideally keeping the angle between fault surface and bedding constant. In the deltaic environment (figure 14.7), and unless other faults intervene on the hanging wall side, this tilt effect dies out away from the fault, resulting in a structural form called a **"rollover."** Where the general dip of bedding is in the same direction as that of the fault (a common occurrence), the rollover becomes an anticline, and, in oil fields, a structural trap. We shall meet this situation again when we discuss growth faults.

Figure 14.8 shows that movement along listric faults is not possible unless the hanging wall side deforms to adapt to the listric shape. This can be either by yielding or by subsidiary faulting, as shown.

**Figure 14.9**
Antithetic faulting. (a) Antithetic normal faults; (b) Synthetic normal faults. (From Dennis and Kelley, 1980.)

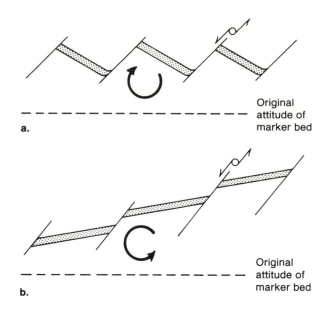

### Growth Faults

Growth faults are normal faults that displace an active surface of sedimentation. They are always listric. A consequence of this contemporaneity of faulting and sedimentation is that sedimentary units are thicker on the hanging wall side than on the footwall side; they expand in thickness or "grow" across the fault (figure 14.8). A common habitat of growth faults is in delta sedimentation, such as Tertiary and younger sedimentation in the Gulf of Mexico and in the Niger delta. Seismic reflection profiling seems to indicate that most growth faults eventually merge with bedding surfaces downdip; to compensate for the downward displacement of the hanging wall, they should show evidence of shortening downdip. Indeed, thrust faults have been found in the lower reaches of the Mississippi delta.

Growth faults, being listric, characteristically have rollovers in the hanging wall. They also change dip with lithology: since they are active during sedimentation, they are affected by compaction. Upon compaction, fault dips must flatten. Shales compact much more than do sands, so that eventually fault dips in sands become steeper than dips of the same fault in adjacent shales.

### Antithetic and Synthetic Faults

Normal faults may form in conjunction with crustal flexure—either updoming or subsidence. In such settings they tend to produce rotated blocks. Hans Cloos (1928) made a very useful distinction between faults that tend to *oppose* or attenuate the crustal flexure involved (**antithetic** faults) (figure 14.9a), and faults that tend to *enhance* the crustal flexure involved (**synthetic** faults) (figure 14.9b). In antithetic faulting, blocks are rotated in the sense opposite to that of the external shear that produced them. Here, they tend to "level" what would otherwise be pronounced tectonic uplift. In synthetic faults, the sense of rotation of the blocks is the same as that of the external shear, and the effect here is to enhance the flexure, usually subsidence. It is quite common for antithetic and synthetic faults to occur as conjugate systems in which one type is usually subordinate to the other (e.g., figures 14.12 and 14.19). In extended usage, "antithetic" and "synthetic" structures in the above sense may occur in association with any larger-scale structure whose development involves rotation.

**Figure 14.10**
Normal faulting above a salt dome. Reitbrook dome near Hamburg, Germany. (From Behrmann, 1949.)

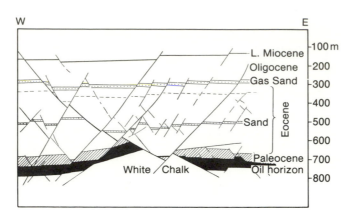

**Figure 14.11**
Model of Kettleman Hills structure (north dome), California, showing crestal fault pattern. The surface of this model represents the top of the oolite bed at the base of the Pliocene San Joaquin Formation. (Model courtesy of Martin van Couvering.)

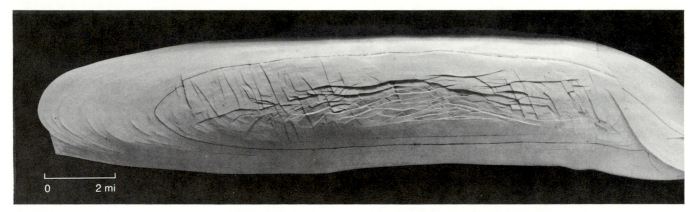

## Doming

Another extensional environment, and therefore a habitat of normal faulting, is over the tops of domes and diapirs and over some anticlines at shallow depth. Salt domes and igneous intrusions are characteristic examples (figure 14.10). The pattern of such faulting is difficult to evaluate in detail, since the shape of the rising body controls the orientation of consequent fractures. Only detailed mapping and subsurface exploration can reveal a three-dimensional pattern such as that shown in figure 14.11.

## Grabens and Rifts

Suess (1885) applied the German term *Graben* to several fault-trough structures, including the Red Sea rift and the upper Rhine Valley trough. In its original definition, which is still its true meaning, **graben** is a descriptive term referring to a linear topographic depression caused by subsidence along faults. Unfortunately, the boundary faults of graben zones are rarely accessible to direct observation. As a result, a number of theories on graben formation flourished until Hans Cloos brought together a considerable body of evidence for extension origin of the Rhine graben. He found that boundary faults dip toward the trough, mostly at angles between 45° and 80°. This, as we have seen, is characteristic for normal faults. Associated volcanic activity usually also indicates crustal extension. Observations on fault surfaces give clues to the movement through slickensides, pinnate joints, and similar features; in all observed cases, displacement has a large normal dip-slip component, a further indication of crustal extension. Perhaps most striking of all, model experiments with soft, wet clay result in subsidence and fracture patterns very similar to those derived from field observations on graben zones (figure 14.12). The experiments seem to show that subsidence need not extend to great depths, because of the compensating effect of a lessening dip with depth. We have already seen (p. 257) how listric fault surfaces enable normal faults to terminate at shallow depths.

**Figure 14.12**
Graben in clay experiment. Synthetic faulting and subsidiary antithetic faulting (*lower left*). Subsidence of the upper surface is attained without subsidence at the base. (From H. Cloos, 1929.)

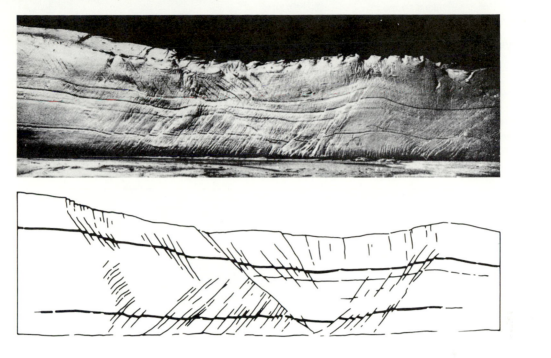

**Figure 14.13**
Cross section showing crustal structure under the Rhine graben, north of Karsruhe. P-velocity profile along B-B reveals two low-velocity layers. (From Müeller, et al., 1969.)

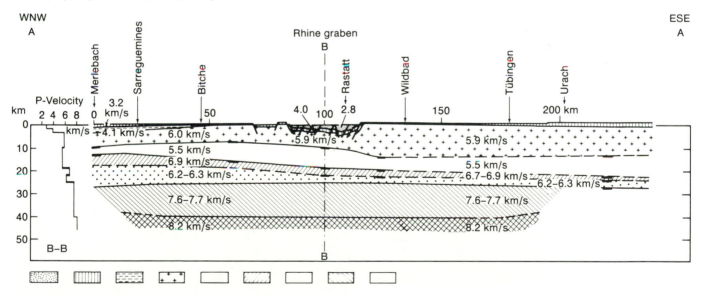

*Rifts* are long, linear zones along which the lithosphere, and with it the crust, undergo lateral extension and thinning. In the brittle upper regions of the crust, extension and thinning are accomplished by normal faulting, usually in the form of grabens. Rifting may eventually lead to continental separation.

Subsidence in rifts is the result of thinning of the lithosphere, and is reinforced by sedimentary load accumulating in the subsiding zone. Thinning is always associated with a certain amount of updoming, but it is not clear which is cause and which is effect. We know that some active rifts are underlain by a "pillow" (figure 14.13) of mantle material with abnormally low density

**Figure 14.14**
Crustal profile across Viking Graben. (From Ziegler, 1982.)

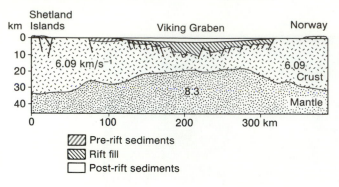

Pre-rift sediments
Rift fill
Post-rift sediments

**Figure 14.15**
Sketch illustrating how intracontinental basins commonly overlie rifts. The resemblance to a steer's head has led to this being known as the "longhorn condition" in Texas. (From Burke, 1979.)

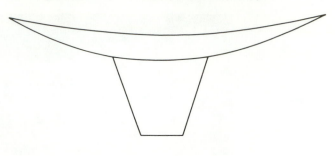

**Figure 14.16**
Map of the African rift valley from the Zambezi to the Ethiopian border. The Luanga-Mid-Zambezi rifts at the southern end are older features, now largely occupied by Mesozoic Karroo sediments.

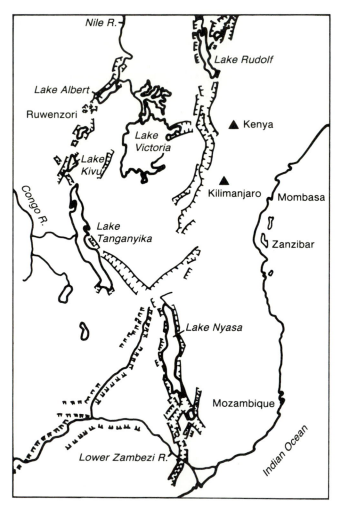

and low seismic velocities. The pillow vanishes once the graben becomes inactive; that is, when subsidence and extension cease (figure 14.14). The graben faults bottom well above the pillow, either by listric flattening, or because they enter a ductile layer, or both.

Many basins overlie fossil rifts, and this results in the characteristic "steer's head" cross section of some intracontinental basins (figure 14.15). Possible causes of this characteristic later basin subsidence are discussed by Burke (1979).

### The East African Rift System

The East African rift system (figures 14.16 and 14.17) is the type rift, for it is here that Gregory in 1894 coined the term "rift valley." The system consists of several individual grabens, each about 50 km wide. As an example, the Tanganyika graben is 700 km long and varies in width from 12 to 65 km. The net subsidence amounts to about 1,400 m. Volcanism with alkalic affinities is characteristic along much of the East African rift. Gravimetric surveys reveal a negative Bouguer anomaly along the subsided segments.

**Figure 14.18**

Sketch map of the Rhine upwarp, showing the rift valley of the upper Rhine between the Vosges mountains in France and the Black Forest in Germany, and its northerly bifurcations into the rifts of the Lower Rhine and Hesse. Volcanic areas in black.

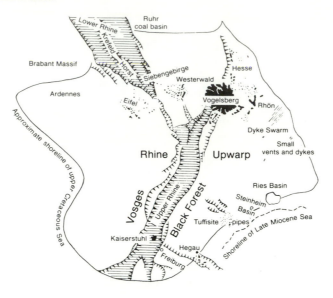

### The Rhine Graben

The Rhine graben is related to the Alpine collision of early Tertiary time. The upper Rhine Valley north of Basel is a graben about 300 km long and about 35 km wide (figure 14.18). Total relative subsidence is about 4,000 m, including 2,000–3,000 m of graben filling. The graben is rimmed by relatively raised blocks (*horsts*), which form the bordering mountain ranges; together they form a broad arch. The graben trends along this arch. The main faults dip toward the graben, forming several synthetic blocks (figure 14.19). Subsidiary antithetic faults dip outward, toward the rim. Fault dips average 60–65°, and, based on geophysical data, they extend to some 7 km in depth. There has also been volcanic activity along the graben, a further indication of crustal extension.

The Rhine graben is a zone of abnormally high geothermal gradient (8 m per degree Celsius) and has many thermal springs located near its borders. It is marked by a negative Bouguer gravity anomaly. It is still tectonically active: locally, relative subsidence averages 0.5 mm per year, enough to damage buildings and there is some seismic activity. Illies (1965) calculated that the graben has undergone lateral extension of 4–6 km since Eocene time.

Müeller et al (1969) found that the Rhine graben is underlain by a "pillow" of lighter mantle or possible crust-mantle mix (figure 14.13). This pillow is evidently responsible for arching of the crust along the graben and may, in fact, be the original trigger for graben formation.

### Marginal Rifts

Another environment of rift formation is along marginal orogenic belts (p. 23). These marginal rifts form in the late stages of, or follow, marginal or arc orogeny. Their trend is parallel to that of the associated orogen, and their subsidence is clearly related to contemporaneous magmatic activity, but their structure at depth is not well known. The bounding faults and associated secondary faults probably terminate downward in a shallow ductile zone due to a high geothermal gradient.

**Figure 14.19**
Block diagram of the Rhine graben north of Karlsruhe. (From Illies, 1965.)

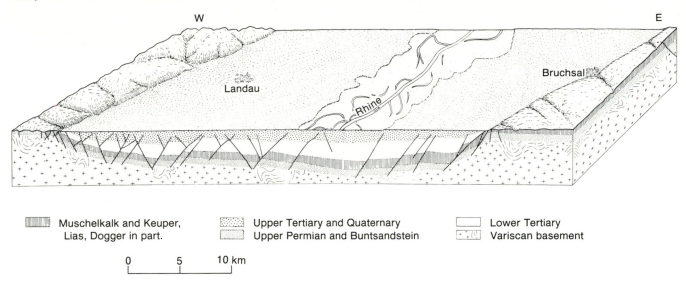

W

E

Landau

Rhine

Bruchsal

▨ Muschelkalk and Keuper,
Lias, Dogger in part.

▨ Upper Tertiary and Quaternary
▨ Upper Permian and Buntsandstein

☐ Lower Tertiary
▨ Variscan basement

0     5     10 km

**Figure 14.20**
Low-angle normal faulting, Marys River Valley, Nevada. (From Robison, 1983.)

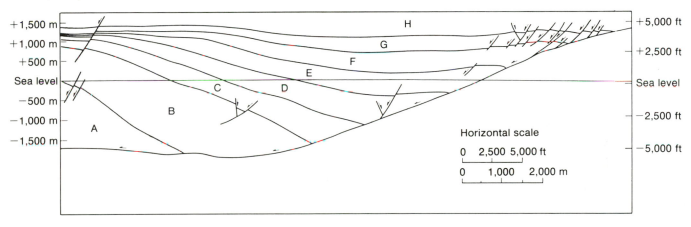

Horizontal scale

0   2,500  5,000 ft

0     1,000    2,000 m

The Central Valley of Chile and the Barisan "rift" zone in Sumatra are grabens that formed late in the history of an orogeny. They are all within their parent orogen, trend parallel to it, and are associated with copious volcanism. The volcanic character of this kind of graben is well illustrated by the Central Valley of Chile, which extends exactly the length of Andean volcanic deposits in Chile, and by the Semangko graben in the Barisan zone, which has a large caldera (Lake Toba) superimposed on it.

## Faulting in the Basin and Range Province of North America

The Basin and Range province of western North America has had a continuing history of block faulting and volcanism since Eocene time. The area eventually became divided into numerous emerging ranges and intervening broad valleys filled with detritus from the ranges, with subsidence along listric normal faults (figure 14.20).

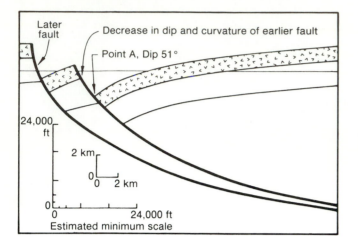

In the Yerington district of western Nevada, Proffett (1977) found listric normal faults with both high and low dips. The lower dip faults are older; they must have had steeper dips originally but were tilted by the younger faults. Tertiary and younger sedimentary and volcanic rocks, where present, show tilting of hanging wall rocks toward fault planes; this also strongly suggests that the faults concerned are listric (figure 14.21).

Over some of the Basin and Range province, normal fault systems with predominantly low dips dominate the near-surface structure. They tend to terminate against or merge along listric fault surfaces with a low-dip basal master fault, or **sole fault,** which is really a detachment surface (e.g., figure 14.5). Hence, these sole faults are **detachment faults;** the hanging wall rocks above such detachment faults are often called the "upper plate," and the footwall rocks below it are referred to as the "lower plate."

Some ranges in the Great Basin are cut by another type of low-angle normal faults much of whose displacement has been along bedding planes (figure 14.22). They resemble thrust faults by their low dips (see chapter 15), but, unlike most thrust faults, they always bring younger rocks to lie over older: where the faults follow bedding planes, they can be recognized because

**Figure 14.22**
Late Tertiary tectonic discordances of stratigraphically younger rocks overlying older rocks; characteristic detachment faults in Northern Virgin Mountains of Nevada and Arizona. (From Seager, 1970.)

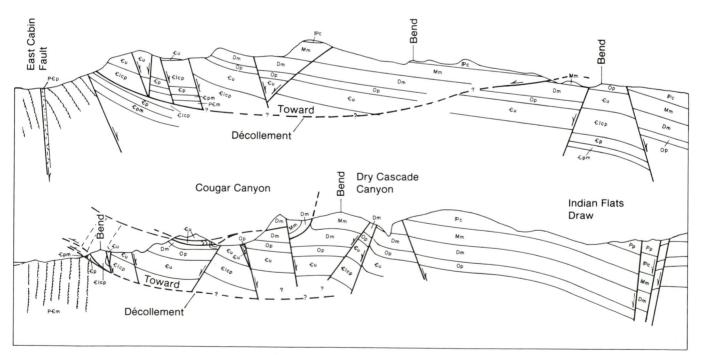

they cut out part of a stratigraphic succession that is known to exist in the area. Many of the fault surfaces are silicified, and some are mylonitized. These faults, too, are detachment faults.

While many of the faults in "upper plates" are listric (figure 14.5) and seem to merge with a basal detachment fault, others terminate quite abruptly against a detachment (figure 14.23). Clearly, some internal accommodation must occur within upper plate rocks; they cannot have moved as rigid blocks. Most of the accommodation is by antithetic faulting, some of it quite closely spaced. These fault systems are a response to extension in the crust. Below the detachment surface, extension may be ductile; above it, it occurs by normal faulting. In a number of places, the detachment surface is gently domed, revealing metamorphic basement in the so-called *metamorphic core complexes*.

## Passive Continental Margins

Rifting may lead to continental separation. Then new ocean floor forms along the rift, which is now in oceanic crust. The two flanks of the original rift, which had formed in continental crust, become continental margins. They are called *passive* or *trailing* continental margins. The original rift now becomes an oceanic rift.

The graben faults that formed in the course of the original rifting now form part of the passive continental margins: the margins inherit the crustal extension and thinning, and with it the listric faults that are part of it in the upper crust (figure 14.24). In addition, shelf and delta sedimentation accumulates and progrades over the original passive margins, associated, in many cases, with growth faulting. Thus, listric normal faults seem to be a characteristic structure of passive continental margins.

**Figure 14.23**
Development of stacked synthetic normal faults lacking curvature, from time 1 to time 3 in an extensional environment. As extension proceeds, new faults develop to provide progressive back rotation enabling the blocks or slices to adapt to the overall deformation. Many of the new faults are extremely closely spaced. This system of faults represents an approximation of the effect of a listric normal fault. (From Gross and Hillemeyer, 1982.)

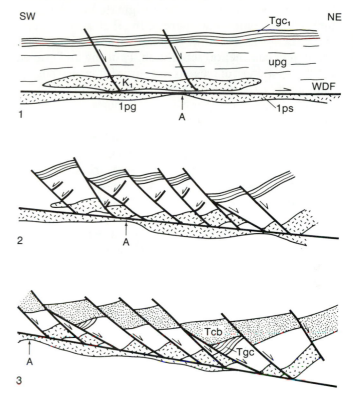

**Figure 14.24**
Schematic crustal section through the northern Bay of Biscay passive continental margin. Numbers indicate *P* wave velocity. (From Montadert et al., 1979.)

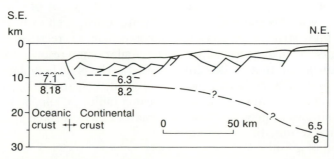

## Aulacogens

Aulacogens are a special type of rifts in continents; they strike into the continent from a concave reentrant in the continental margin (figure 1.21). The stratigraphic succession of their sediment fill correlates with that of the continental margin from which they enter the continent; this suggests that they originated at the same time as that continental margin. In other words, rifting along them was contemporaneous with the rifting that resulted in continental separation. Faulting and graben structures along aulacogens are very similar to those of "standard" rifts, which we have discussed, but their fill tends to be folded, suggesting a late phase of compression across the aulacogen.

The junction of an aulacogen with the related continental margin (possibly now an orogenic belt) has been interpreted as a former rift-rift-rift triple junction (Dewey and Burke 1974). Two of the rifts opened up to spread as an ocean basin, while the third or "failed" rift arm became an aulacogen.

Some rifts within continents form near sutures at the time of continent–continent collision: their fill is decidedly younger than that of the adjacent continental margin deposits; hence they are not aulacogens. Examples include the Rhine graben and the Baikal rift.

### Common Features of Rifts

We have discussed several different types of rifts on continents: the "type" rift in East Africa, marginal rifts, the Rhine Valley rift, and aulacogens. Each has a different tectonic setting, but their anatomies and evolution are very similar: they result from lithosphere extension and subsidence along faults, accompanied in many observed cases by a "pillow" of low-density mantle (figure 14.13), and doming.

## Horizontal Shear: Strike-Slip Faults

Most strike-slip faults have steep to vertical dips, and many extend for considerable distances along strike. Their straight traces are conspicuous on aerial photographs, where many strike-slip faults are revealed by zones of preferential erosion or textural contrasts due to contrasting rock terranes on opposite sides of the faults (figure 14.25). Until early in this century, most offsets of structures along a fault strike were considered "apparent" displacements or, in modern terminology, separations. The realization that strike-slip motion is real came slowly. The reason, no doubt, was that whereas dip-slip displacements can always be accommodated by unknown conditions at depth, it is far more difficult to account for the termination of strike-slip displacement at either end of a fault.

### Strike Slip in Oceanic Crust

Strike-slip faults in oceanic crust are active or former transform faults (see chapter 1). At the terminations of active transform faults, movement along them is "transformed" into other types of interplate movement: converging, diverging, or, occasionally, slip along another transform fault (figures 1.11 and 1.14). They are revealed both by ocean bottom topography and by displacement of ocean floor magnetic anomalies. Their strike is ideally (and nearly always *actually*) parallel to a small circle about the pole of rotation of the two plates that adjoin along the fault (figure 1.13). This means that, normally, neither compression nor tension is found across these transform faults. Active transform faults really form an integral part of plate divergence, and are most appropriately regarded as components of spreading systems. These active transform faults have inactive extensions along oceanic fracture zones, because the oceanic crust has moved beyond the active zone (figure 1.11).

### Strike Slip in Continental Crust

Most known fundamental strike-slip faults in continental crust are associated, directly or indirectly, with present or former transform motion. Transform motion along continental margins may be the result of oblique plate convergence, at a very acute angle. In such cases, relative motion between the plates tends to adjust to strike-slip motion. Since most continental margins do not have the outline of a smooth small circle about the pole of rotation of the two plates in contact, considerable adjustment must occur. The details need not concern us here. The result is often a fault zone consisting of several fault branches, and occasional jumping of active motion from one track to another. The fault trace at the surface may not be an accurate reflection of the true plate boundary at depth: the upper crust may be decoupled to some extent from the lower crust. Most important, because of the common deviation of the fault trace from an ideal small circle, many continental transform fault segments have components of either compression ("transpression") or tension ("transtension") across them, as we shall see (p. 273).

In some cases of oblique subduction under continental crust, such as in marginal orogeny, part of the component of relative motion parallel to the plate boundary may resolve itself into actual strike slip parallel to the plate boundary. The fault is then within the orogenic belt above the subducting slab, as in the Atacama fault in Chile, and in the Semangko fault in Sumatra (where strike-slip motion is along grabens).

**Figure 14.25**
Trace of the San Andreas fault, Tomales Bay, north of San Francisco, California. View is to the south. (Photograph by Robert E. Wallace and Parke D. Snavely, Jr., U.S. Geological Survey.)

In some collisions, small continental blocks may escape laterally between the colliding major continental blocks, resulting in strike-slip fault motion. Classic examples are provided by the prominent strike-slip faults in Tibet (figure 14.26) and by the Anatolian fault in Turkey (figure 14.27).

Most strike-slip faults of large, crustal dimensions also have large displacements. This has become known only fairly recently. Their classical representatives—the Great Glen fault in Scotland (figure 14.28) and the San Andreas fault in California (figure 14.29)—have been known for some time. It was only in 1946, however, that

## Figure 14.26

Schematic map of Cenozoic extrusion tectonics.and large faults in eastern Asia. White arrows: major block motions with respect to Siberia. Black arrows: directions of extrusion-related extensions. (From Tapponnier et al., 1982.)

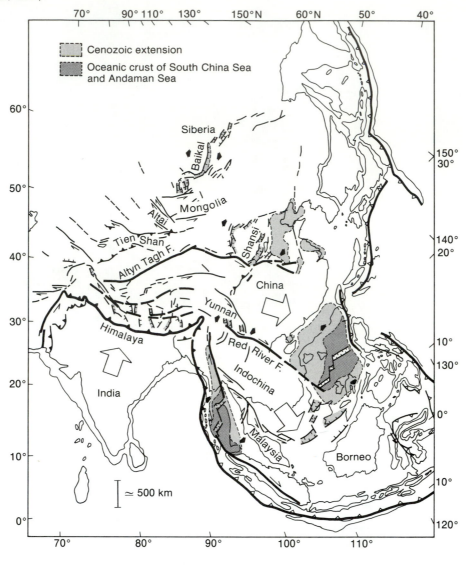

## Figure 14.27

Map of Anatolia and surrounding regions showing active plate boundaries or plate boundary zones and tectonic subdivisions. Heavy lines with half-arrows are strike-slip faults; lines with black triangles are thrust faults; lines with hachures are normal faults; lines with open triangles are subduction zones with triangles on the upper plate; simple solid lines are unspecified faults; dotted regions are depressions; broken lines with dots are boundaries of the tectonic divisions. G = Ganosdag; Ge = Gemlik Graben; I = Izmit/Sapanca Graben; S = island of Samothraki. Figures are elevations above sea level. Both Ganosdag and Samothraki may be regions of active thrusting in restraining bends along the North Anatolian transform fault. (From Sengör, 1979.)

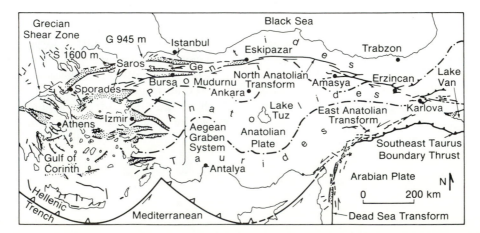

**Figure 14.28**

Great Glen fault. Unusual petrologic similarity between the Strontian and Foyers plutons (black) either side of the fault led Kennedy (1946) to interpret the distance between them along the fault trace as strike slip. Note the long straight trace (which is strongly reflected in the topography). This is characteristic of major strike-slip faults. (After Kennedy, 1946.)

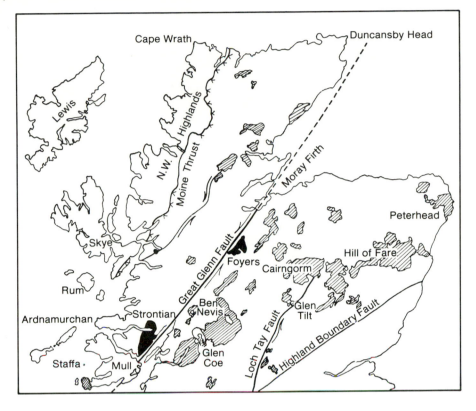

**Figure 14.29**

Faults of the southwestern United States. Abbreviations: S.F. = San Francisco; S.B. = Santa Barbara; L.A. = Los Angeles; S.D. = San Diego; Y = Yuma; B = Bakersfield; L.V. = Las Vegas; H.F. = Hayward fault; S.Y.F. = Santa Ynez fault; W.W.F. = White Wolf fault; S.G.F. = San Gabriel fault; S.J.F. = San Jacinto fault; W.F. = Whittier fault; L.V.F.Z. = Las Vegas fault zone; K.C.F. = Kern Canyon fault; S.N.F. = Sierra Nevada fault; P.V.F. = Panamint Valley fault; D.V.F. = Death Valley fault; F.C.F. = Furnace Creek fault; O.V.F.Z. = Owens Valley fault zone. (Source: Goddard Space Flight Center, January 1976.)

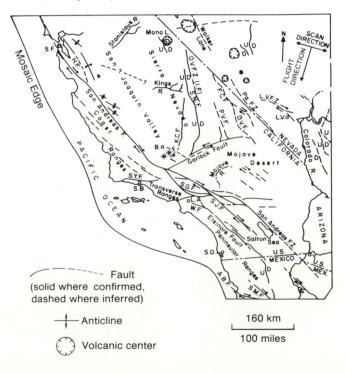

## Figure 14.30
Sketch map of idealized pull-apart basin. (From Crowell, 1974.)

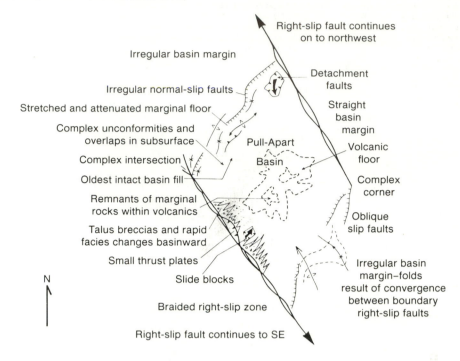

W. Q. Kennedy found evidence for some 150 km of strike slip along the Great Glen fault, in 1953 that Hill and Dibblee suggested some 500 km of strike slip for the San Andreas fault, and in 1955 that Wellman established a cumulative right slip of 480 km on the Alpine fault in New Zealand. The reason for the comparatively late discovery of these vast displacements is that with strike-slip displacements of that order, convincing matches between rocks and structures on each side of a fault are extremely difficult to find. This is particularly true where the most readily available correlating structures are layered sequences with low dips: a comparatively small dip-slip component will then yield an enormous strike separation, making such surfaces of questionable value in the determination of net slip.

There is another difficulty. Strike-slip faults reach the surface in zones of sheared and altered rock, up to hundreds and even thousands of meters wide; some consist of several subparallel zones. Such zones of weakness have particularly poor exposures. But probably the principal reason for the delay in recognizing great cumulative slip on strike-slip faults was the lack of appreciation that such displacements could occur; therefore, they were not sought on available regional geologic maps.

It seems clear that faults with such large displacements and such long, straight traces must extend to great depth. Geophysical evidence supports this. All currently active faults of this type mark active plate boundaries (e.g., the San Andreas fault in California; the Alpine fault in New Zealand), or they are related indirectly to active plate boundaries (e.g., strike-slip faults in Tibet). The Great Glen fault, essentially inactive since Permian time, still has seismic activity associated with it.

### Components of Movement Across Strike-Slip Faults

Where strike-slip faults do not follow the overall direction of crustal shearing—particularly where they deviate from the ideal small circle transform trend—there is either a component of extension or of convergence across them, depending on their trend. Where extension predominates, this may manifest itself in graben or basin structures ("*pull-apart basins*"), some quite small, some over 100 km in width, along the fault (figure 14.30). Extension may also induce volcanism, as in the Salton trough of southern California, and at times may create oceanic crust. Most of these extension structures are characterized by geothermal and gravity anomalies. Pull-apart basins that form in this manner tend

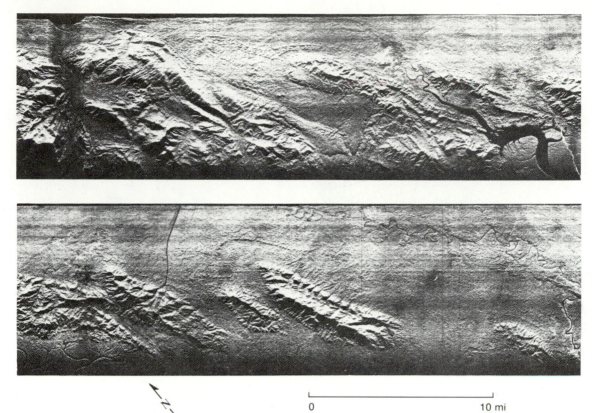

0                                        10 mi

to attract much sedimentation from adjacent fault scarps. If they contain favorable rocks and traps, they are prime targets for petroleum exploration. Harland (1971) has called extension across strike-slip faults **transtension.**

Convergence across strike-slip faults may result in folding of layered rocks along the fault or in reverse faulting. Harland (1971) has called such convergence **transpression.** Normally, folds so formed tend to be *en echelon,* with axes at an acute angle to the fault strike (figure 14.31), but they may also be nearly parallel to the fault (figure 14.32). Reverse faults form as subsidiary structures (figure 14.33) or by branching or deviation from the upper regions of strike-slip faults, as in the "*flower structure*" in figure 14.34. Figure 14.35 shows the relationships in diagrammatic form.

### Termination of Strike-Slip Faults on Continents

While oceanic transform faults terminate very clearly and abruptly along strike at limiting plate boundaries, the terminations along strike of continental strike-slip faults are less well understood and more complex. They tend to branch, distributing motion among several subsidiary faults, each taking up a fraction of the total displacement (figure 14.29).

### Structures Associated with Strike-Slip Faults

Clearly, strike-slip faulting cannot begin with a sudden, simultaneous break along its entire length: the energy required would be enormous. So, how can it happen?

## Figure 14.32
Folds with axes parallel to the trace of the San Andreas fault, southwest of Palmdale.

## Figure 14.33
Postulated structural evolution of the Painted Canyon fault. (a) Initial geometry of basement and overlying sedimentary strata. (b) Flexure of basement by movement along closely spaced fractures and shear planes within zone indicated by dashed lines. (c) Rupture of basement along Painted Canyon fault. (d) Secondary faulting in syncline, incipient buckling of beds in "book-cover structure." (e) Continued vertical displacement on Painted Canyon fault, rotation of secondary faults, and continued buckling of strata in "book-cover structure." (From Sylvester and Smith, 1976.)

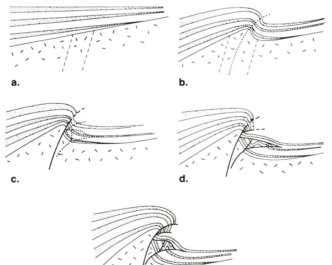

**Figure 14.34**
Seismic profile across a strike-slip fault in the Ardmore basin,
Oklahoma, showing flower structure. *Msp, Msy* = Mississippian
formations; *Ooc* = Ordovician. (From Harding and Lowell, 1979.)

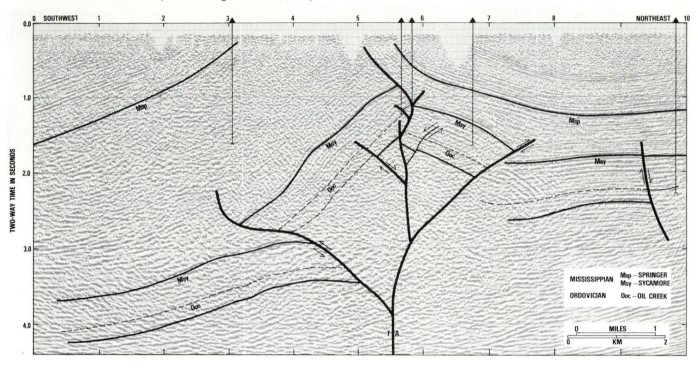

**Figure 14.35**
Schematic synoptic diagram showing minor structures associated
with a right-slip fault. *E* = extension; *C* = compression. (From
Sylvester and Smith, 1976; adapted from Harding, 1974.)

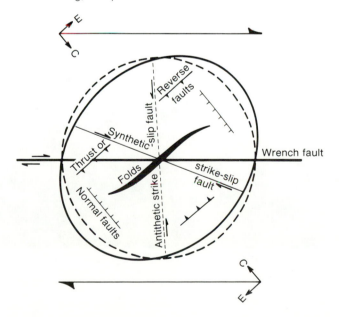

**Figure 14.36**
Clay model of left-slip fault (A–C = three stages, vertical views).
Note synthetic (Riedel) shears. (From Wilcox et al., 1973.)

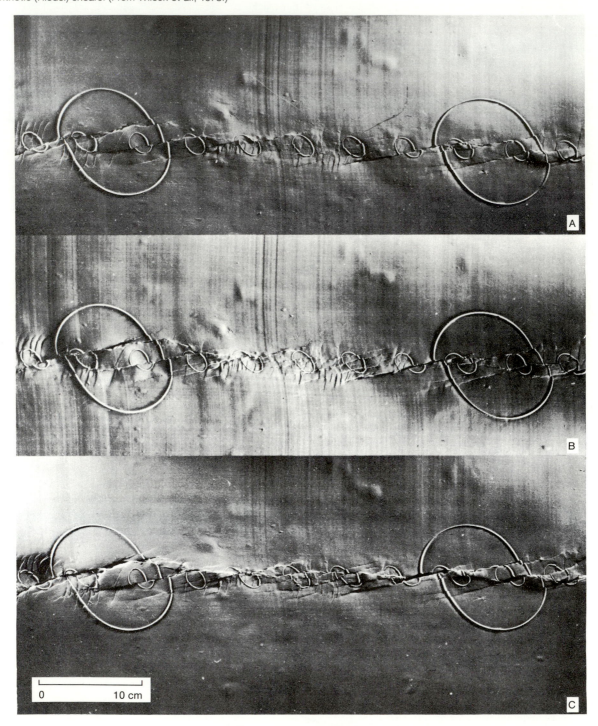

**Figure 14.37**
*En echelon* extension fissures, southern Iceland.

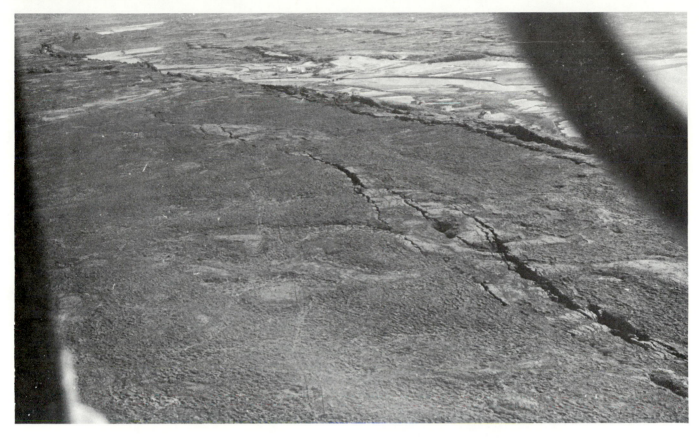

The clues come from clay-model experiments and from observations on new breaks that appear during earthquakes. Figure 14.36 shows what happens when a clay model is deformed to simulate strike-slip motion. Observations on faults newly activated during earthquakes and on other active faults (figure 14.37) show that this model is a good representation of what actually occurs in nature.

The first deformation is ductile shear along a narrow zone. Among the first structures to form are folds arranged *en echelon* along the trace of the incipient fault, in such a way that their axes are perpendicular to the shortening direction of the regional strain ellipsoid (figure 14.36). Initially, deformation in shear along the incipient fault is ductile, and strain is concentrated along a narrow zone. As shear proceeds, the strain ellipsoid within the shear zone rotates toward the fault trace, and eventually short fractures appear along the *synthetic* shear direction of the rotated ellipsoid; these fractures make a very small angle with the main shear, as shown. A little later, the conjugate shears may form. These will be *antithetic,* and at a large angle to the main shear (figure 14.36). As explained earlier (p. 259), the synthetic internal shears are in the same sense as the main (external) shear; the antithetic shears oppose it. These internal shear fractures have been called Riedel shears, after the German engineer who first described them under the direction of Hans Cloos. All such subsidiary structures tend to be rotated during the initial ductile shear along the main fault, often making analysis difficult in practice. Extension fractures may open in accordance with the Anderson relationship (figures 13.10 and 14.35). Once the main shear zone becomes a clean fault, ductile deformation practically ceases, and the fault inherits the family of subsidiary formed at its initiation.

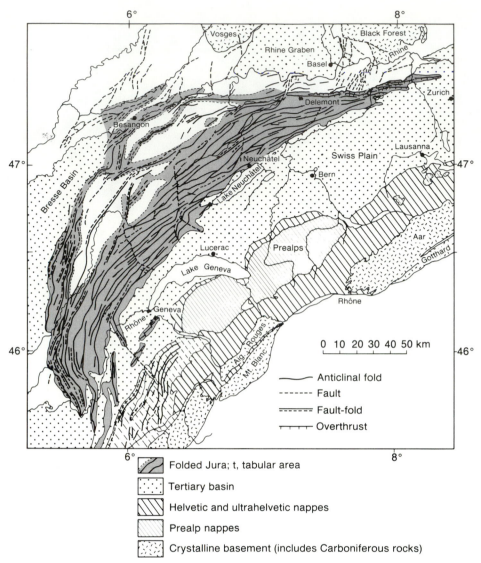

Folded Jura; t, tabular area

Tertiary basin

Helvetic and ultrahelvetic nappes

Prealp nappes

Crystalline basement (includes Carboniferous rocks)

## Tear Faults

Marr (1900) gave the name *tear fault* to strike-slip faults that cut detachment sheets or "lags" in the Lake District of England. More generally, they form when a detachment sheet or nappe (chapter 15) moves at unequal rates over a flat shear or detachment surface. In other words, they mark velocity discontinuities within the sheet. Such faults are confined to high tectonic levels in orogenic belts. Although many examples are known, the classic, and probably best-studied examples are in the folded Jura mountains of Switzerland and France.

The folded Jura is a 360-km-long arcuate mountain range straddling the border between Switzerland and France. It consists of a sheaf of parallel folds cut by a set of oblique strike-slip faults, as shown in figure 14.38. Folding is confined to Triassic and higher rocks, and detachment above a Triassic evaporite sequence is generally assumed. Thus, the strike-slip faults almost certainly do not extend below the base of the evaporite.

They are oblique to the trend of the folds they displace, and they form important topographic features. Displacement along them varies, but it is evidently modest compared with that of fundamental strike-slip faults. Some displace only one anticline. They can be observed to "vanish" in an adjoining syncline, "transforming" into small local thrusts. A number of displaced folds do not match well across the tear faults, indicating that folding on either side of the faults proceeded somewhat independently. Folding and faulting must have overlapped in time, and they are clearly linked genetically.

Other prominent examples of tear faults include occurrences in the central Rocky Mountains, such as those along the east flank of the Beartooth Mountains in Montana.

A look at the geological maps of deformed zones involving detachment reveals the widespread occurrence of tear faults. In fact, sets of tear faults in a deformed cover sequence would suggest detachment at depth.

**Figure 14.39**
Reverse faulting on borders of basement uplift. (a) Structure section through Boothia uplift and Cornwallis fold belt, Arctic Canada. (b) Cross-sectional view of dry sand scale-model experiment by Sanford (1959), after center portion was slowly raised. (From Kerr and Christie, 1965.)

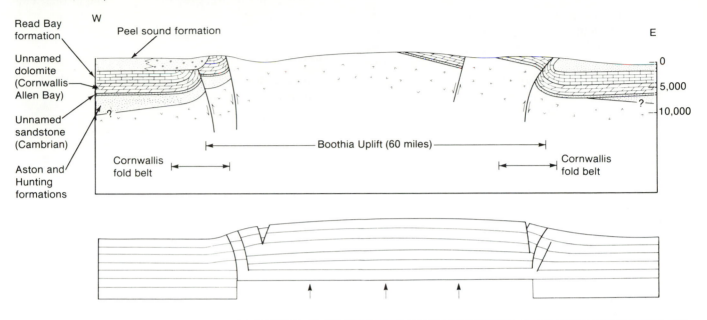

**Figure 14.40**
COCORP seismic profile across Wind River thrust, Wyoming. Sedimentary formations indicated by standard stratigraphic symbols. Vertical scale = approximately horizontal scale. (Reproduced courtesy Jon Brewer.)

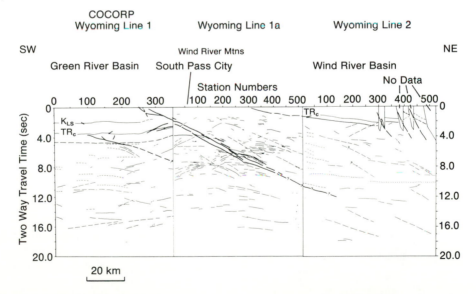

## Crustal Shortening: Reverse Faults

The dynamic setting of reverse faulting can be derived from figure 13.9, which for this case indicates horizontal shortening with upward extension. The movement picture is the same as that for folds, which form as a result of lateral shortening. Thus, most reverse faults are associated with fold belts. But horizontal shortening of any kind favors the formation of reverse faults.

In the central Rocky Mountains of the western United States, basement uplifts are bounded by dip-slip faults; some are normal, some reverse. Reverse faults generally indicate crustal shortening, but there is no agreement as to the immediate cause of compressive stress in this case. The upper part of the uplift may simply be spreading because of lack of lateral constraints, as in the Boothia uplift (figure 14.39), or the fault may continue straight into the lower crust, as in the Wind River thrust (figure 14.40). The border faults of the Boothia uplift in the Canadian Arctic were extrapolated to depth from theoretical and experimental considerations (Kerr and Christie 1965); the Wind River thrust was traced by seismic reflection (Smithson et al. 1978).

Many reverse faults are simply the brittle response to crustal shortening, where a ductile response would have resulted in folds. Both folding and reverse faulting may occur as a result of the same strain; in such cases, the reverse faults usually are initiated after the bulk of the folding has been completed.

Thrusts and overthrusts as defined on p. 222 will be discussed separately in chapter 15.

## Conclusion

Faults are most common near active and former plate boundaries: they reflect plate-boundary processes. Even intracontinental rifts such as aulacogens are, in a wider sense, related to processes characteristic of plate boundaries. But intraplate faulting also occurs, such as growth faulting in passive margin deltas, and minor faulting from intraplate isostatic adjustments. As a general rule—and it is only a general rule—thrust faults are characteristic of converging boundaries, normal faults and grabens of diverging boundaries and rifts, and strike-slip faults of transform boundaries. But an important class of strike-slip faults is related to plate convergence, particularly oblique convergence.

In a more restricted sense, faults, like all fractures, reflect the stress regime at the time of their formation. That may be a useful clue when searching for the structural control of mineral deposits, for vein deposits are controlled by the stress regime at the time of their emplacement.

## Review Questions

1. By means of clear, labeled sketches, explain the terms "antithetic faulting" and "synthetic faulting."
2. How do growth faults develop? Draw a representative cross section showing some characteristic features.
3. What fault patterns develop in extension? in compression?
4. Name and discuss some geological characteristics of graben structures.
5. Discuss hanging wall deformation in listric normal faults.
6. Compare the role of gravity in landslides and in listric normal faults.
7. How may dip-slip faults terminate at depth?
8. How may strike-slip faults terminate laterally?
9. Name and describe some regional structures that are the result of rifting.
10. In what tectonic settings do strike-slip faults occur?
11. Describe some near-surface structures commonly associated with strike-slip fault zones.
12. How do components of convergence and divergence affect strike-slip fault zones?

# Additional Reading

Aydin, A., and Nur, A. 1982. Evolution of pull-apart basins and their scale independence. *Tectonics* 1: 91–105.

Baker, B. H., Mohr, P. A., and Williams, L. A. J. 1972. Geology of the eastern rift system of Africa. *Geol. Soc. Am. Spec. Pap.* 136. 67 pp.

Bosworth, W., Lambiase, J., and Keisler, R. 1986. A new look at Gregory's rift: The structural style of continental rifting. *EOS Am. Geophys. Union Transact.* 67: 577, 582–583.

Brun, J. P., and Choukroune, P. 1983. Normal faulting, block tilting, and décollement in a stretched crust. *Tectonics* 2: 345–356.

de Charpal, O., Guennoc, P., Montadert, L., and Roberts, D. G. 1978. Rifting, crustal attenuation and subsidence in the Bay of Biscay. *Nature* 275: 706–710.

Freund, R. 1974. Kinematics of transform and transcurrent faults. *Tectonophysics* 21: 93–134.

Garfunkel, Z. 1981. Internal structure of the Dead Sea leaky transform (rift) in relation to plate kinematics. *Tectonophysics* 80: 81–108.

Gibbs, A. D. 1984. Structural evolution of extensional basin margins. *Geol. Soc. London J.* 141: 609–620.

Illies, J. H. 1981. Mechanism of graben formation. *Tectonophysics* 73: 249–266.

Kelley, V. C. 1979. Tectonics, Middle Rio Grande rift, New Mexico. Pages 57–70 in: R. E. Riecker, ed. *Rio Grande Rift: Tectonics and Magmatism.*

Mann, P., Hempton, M. R., Bradley, D. C., and Burke, K. 1983. Development of pull-apart basins. *J. Geol.* 91: 529–554.

Marshak, S., Geiser, P. A., Alvarez, W., and Engelder, T. 1982. Mesoscopic fault array of the northern Umbrian Apennine fold belt, Italy: Geometry of conjugate shear by pressure-solution slip. *Geol. Soc. Am. Bull.* 93: 1013–1022.

Mavko, G. M. 1981. Mechanics of motion on major faults. *Annu. Rev. Earth & Planet. Sci.* 9: 81–111.

Moody, J. D., and Hill, M. L. 1956. Wrench-fault tectonics. *Geol. Soc. Am. Bull.* 67: 1207–1246.

Nicholson, C., Seeber, L., Williams, P., and Sykes, L. R. 1986. Seismic evidence for conjugate slip and block rotation within the San Andreas fault system, southern California. *Tectonics* 5: 629–648.

Ramsay, J. G. 1980. Shear zone geometry: A review. *J. Struct. Geol.* 2: 83–89.

Reading, H. G. 1980. Characteristics and recognition of strike-slip fault systems. *Spec. Publ. Int. Assoc. Sediment.* 4: 7–26.

Scholz, C. H. 1977. Transform fault systems of California and New Zealand: Similarities in their tectonic and seismic styles. *J. Geol. Soc. London* 133: 215–229.

Segal, P., and Pollard, D. D. 1983. Nucleation and growth of strike-slip faults in granite. *J. Geophys. Res.* 88: 555–568.

Shelton, J. W. 1984. Listric normal faults: An illustrated summary. *AAPG Bull.* 68: 801–815.

Sibson, R. H. 1974. Frictional constraints on thrust, wrench and normal faults. *Nature* 249: 542–544.

Tchalenko, J. S., and Ambraseys, N. N. 1970. Structural analysis of the Dasht-e Bayaz (Iran) earthquake fractures. *Geol. Soc. Am. Bull.* 81: 41–59.

Wernicke, B., and Burchfiel, B. C. 1982. Modes of extensional tectonics. *J. Struct. Geol.* 4: 105–115.

Withjack, M. O., and Scheiner, C. 1982. Fault patterns associated with domes—an experimental and analytical study. *AAPG Bull.* 66: 302–316.

Zoback, M. L., and Zoback, M. D. 1980. Faulting patterns in north-central Nevada and strength of the crust. *J. Geophys. Res.* 85: 275–284.

# 15

# Thrusts and Nappes

Many well-known thrusts and nappes provide evidence of great displacements: tens of kilometers for some individual thrusts, and more than 200 km for entire thrust systems. In the nineteenth century, such displacements seemed so extraordinary that their recognition was held back for many years by simple disbelief. Arnold Escher (1807–1872) recognized the first overthrusts in the Swiss Alps in 1834. More were recognized in the following years. In the beginning, geologists would accept only displacements that made sense on a modest human scale, no more than a few kilometers. Field observations, however, gradually forced them to consider horizontal movements of thin sheets of rock of the order of 50, perhaps even 100 km.

The evolution of knowledge of Alpine structure is instructive. Briefly, Arnold Escher (1841) and Albert Heim (1878) had first interpreted a type structure in the Canton of Glarus, Switzerland, in such a way that a minimum of faulting and recumbent folding needed to be assumed (the famous "Glarus Double Fold"; figure 15.1). But in 1884, Marcel Bertrand showed that many difficulties in the structural interpretation could be removed by assuming large-scale overthrusting. Then, in 1893, Hans Schardt demonstrated that the Prealps of western Switzerland and Savoy, covering a 40-by-125 km area, were a group of "rootless" tectonic

sheets "floating" on younger rocks; these sheets must have traveled a considerable distance to reach their present site, a concept of very uncomfortable proportions to the geological thinking of the time. As a result, the new interpretation was severely attacked by many contemporary geologists. But Schardt's work rested on accurate mapping of the stratigraphy and a clear understanding of the geometry of the structure. His interpretation was not a matter for idle discussion. It was a matter of how soon his critics would check his field observations and his geometric constructions. This accomplished, they saw that the tremendous tangential displacement represented by the position of the Prealps was a direct and inescapable consequence of geometric and stratigraphic relations and could not be argued away. Schardt's great achievement rested on correct mapping (by himself and others, notably Maurice Lugeon) and on map interpretation. It carries one supremely important lesson: geological interpretations and syntheses are only as good as the geological maps on which they are based and on the geologists' correct understanding of those maps.

At about the same time, others, notably Lapworth (1883) in the Scottish Highlands and Törnebohm (1888) in the Caledonides of Norway, also discovered the necessity of postulating horizontal travel of thin sheets of rock by at least tens of kilometers.

**Figure 15.1**

"Glarus Double Fold" (*above*) and correct structural interpretation by thrusting (*below*) of this formation in the Swiss Alps. *m* = Tertiary Molasse; *n,f* = Tertiary and Cretaceous Flysch; *c* = Cretaceous, mainly limestones; *j* = Jurassic, mainly limestones; *t* = Triassic; *p* = Permian, mainly continental. (From Bailey, 1935.)

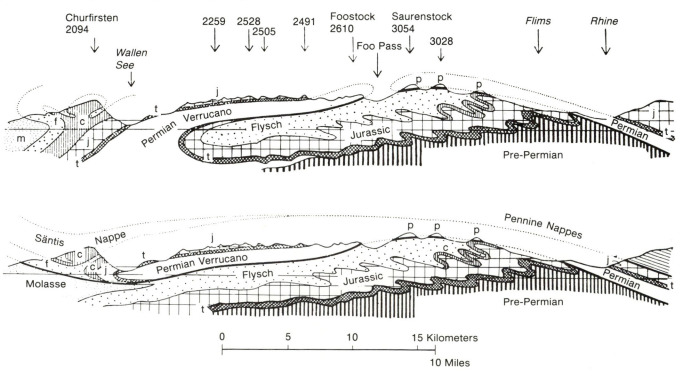

## Kinematics of Thrusting

A thrust commonly develops by detachment along a flat-lying surface, such as a bedding surface or a near-horizontal shear zone in crystalline basement rocks. It then has to break through upward toward a free surface, in order to allow the detached sheet to move. This simple mechanism generates several characteristic geometries, some rather complex.

The simplest form in cross section is the "sled-runner" thrust—a flat-lying surface terminating in a listric, concave-upward toe. After breaking through to the surface, the toe may erode as it emerges, or it may continue to travel along the ground surface, forming an **erosion thrust.** Or a thrust may at first follow a flat trajectory at depth, then break upward, follow a higher flat trajectory, break upward again, and so on (figures 15.2 and 15.3). Appropriately, flat portions of thrusts are called **"flats,"** steeper upward breaks are **ramps.** As

**Figure 15.2**

Relationship between frontal hanging wall (*HWR*) and footwall ramps (*FWR*). Beds are lettered a–d, in ascending order. (From Butler, 1982.)

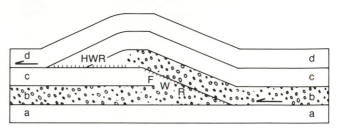

**Figure 15.3**

Sequential development of the Powell Valley anticline, a surficial rootless fold typical of thin-skinned deformation. (a) Subsurface splay thrusting in the early stages of the formation of the Pine Mountain décollement system results in the initial formation of the Powell Valley anticline. (b) Moderate northwest movement of the allochthonous sheet causes the fold to grow by shifting the northwest limb northwestward up the tectonic ramp and rotating cross-cut units onto the subhorizontal higher level décollement surface. This type of narrow-crested asymmetrical rootless anticline in which Cambrian or Ordovician rocks are exposed by erosion in the core is a common thin-skinned structure throughout parts of the Valley and Ridge and Appalachian plateaus. (c) Continued northwest movement enlarges the Powell Valley anticline by progressive duplication of beds above the higher level décollement. The end result is a broad, rootless anticline that has an undulating crestal region; on the surface, no single axial surface defines the fold.

Source: Harris, L. D. and R. C. Milici, "Characteristics of Thin-skinned Style of Deformation in the Southern Appalachians and Potential Carbon Traps" *U.S. Geological Survey Professional Paper 1018, 40,* 1977.

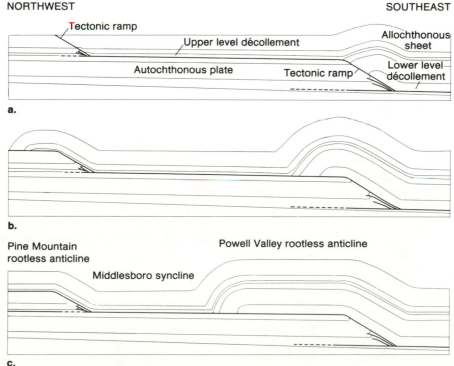

**Figure 15.4**
Burning Springs Anticline, West Virginia. An embryonic (blind) thrust fault. (From Woodward, 1959.)

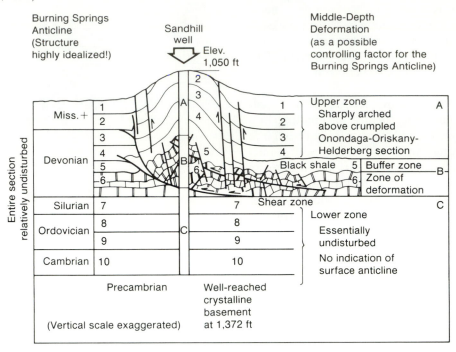

a thrust sheet climbs up a ramp and then bends to follow a flat, it folds into a gentle anticline. In some cases, usually in flat-lying stratigraphic successions, a thrust may not break through to the ground surface; then the deformation of such a **blind thrust** will be taken up by folding in the uppermost layers (figure 15.4).

The listric toe of a thrust may find its forward motion impeded; new upward breaks will then take up the motion, either in front of or behind the original break, which is abandoned. If new breaks develop successively (by *prograding*), new footwall material will be accreted to the thrust sheet, helping it to grow "piggyback" fashion, old slices riding on younger (figure 15.5). Where new breaks form successively *behind* the original break, it abandons its original toe and moves over it, climbing up a ramp (figure 15.3). Note that both listric thrusts and listric normal faults curve in the direction of easiest "escape."

Most thrust faults occur in networklike systems formed in a succession of new breaks and abandonment of old breaks (figure 15.5). A structure so formed by parallel, overlapping faults, rather like fish scales or overlapping roof tiles, is called an **imbricate structure** or *schuppen structure* (figure 15.6). Thrusts in such a system usually join asymptotically downward along a flat-lying **sole thrust** and, in some cases, asymptotically upward along a **roof thrust.** A structure enclosed entirely by a roof thrust and a sole thrust is called a **duplex** (figures 15.5 and 15.6). Each individual, entirely fault-bounded slice is a **horse.** Thus, Boyer and Elliott (1982) characterized a duplex as a "herd of horses." The mechanism appears to be as follows (illustrated in figures 15.5 and 15.7). Movement proceeds along the sole thrust of a duplex; its toe gets "stuck," and motion along the sole thrust then breaks through to the roof thrust behind the original toe thrust. As this process is repeated, an imbricate system will form.

**Figure 15.5**
Progressive collapse of footwall ramp builds up a duplex. This is a measured graphical experiment, assuming plane strain and kink folding, with angles and ratios of dimensions typical of natural examples. The roof thrust sheet undergoes a complex sequence of folding and unfolding, as shown by the black half-dots. (From Boyer and Elliott, 1979.)

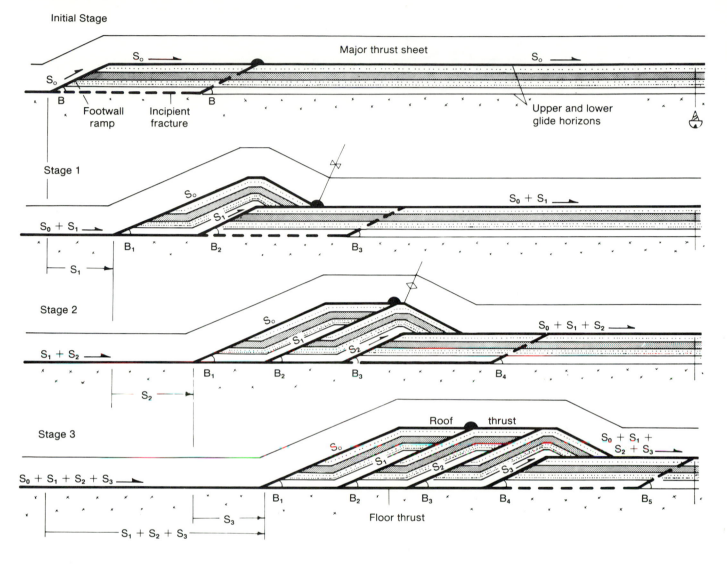

**Figure 15.6**
Cross section of hanging wall structure of the Lewis thrust sheet along the northwest side of Haig Brook, Clark Range, British Columbia. (From Fermor and Price, 1976.)

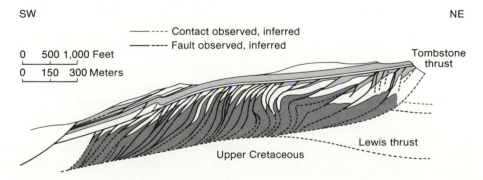

## Figure 15.7

Thrusts and folds formed by subglacial shear, Mon, Denmark. A competent permafrost layer overlies incompetent unfrozen ground. The glacier deforms the permafrost layer as shown: (a) proglacial thrusting. (b) Proglacial folding. (c) Stacking of thrust slices as a result of proglacial thrusting. Ice is shaded; unfrozen ground is vertically ruled. Note that here the driving force (for glacier flow) is gravity. (From Berthelsen, 1979.)

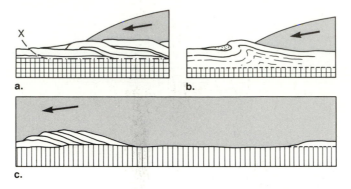

## Figure 15.8

Accommodation of differential thrust displacement. (From Dahlstrom, 1969.)

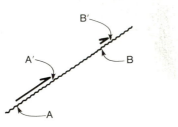

Apparent anomaly in amount of fault displacement
**a.**

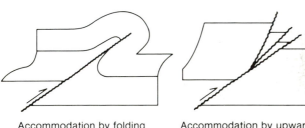

Accommodation by folding
**b.**

Accommodation by upward imbrication
**c.**

## Balanced Cross Sections

Since much subsurface information is incomplete, informed extrapolation becomes necessary. One of the most effective tools for this is the construction of *balanced cross sections* (chapter 7). This procedure assumes that all deformed layers have maintained constant thickness and constant bed length in cross section throughout. It assumes that all beds are parallel and that all folding is parallel. Also, all deformation is assumed to have occurred above a *detachment surface* and *within* the cross section, that is, by plane strain. Given these prerequisites, then, the original length of any layered succession shortened by folding or thrusting is clearly the cross-section area divided by the stratigraphic thickness of the sequence involved (figure 15.8). Hence, in a balanced cross section, folds and thrusts must be drawn so that original bed lengths are maintained, and the deformed cross section has the same area as the restored original cross section (figure 15.9).

A balanced cross section must accommodate available observations (surface, seismic, borehole) to these prerequisites. Balancing cross sections considerably restricts interpretation options and thus produces cross sections that have an increased likelihood of being correct. The procedure has been described by Dahlstrom (1969), and reviewed by Hossack (1979).

## Overall Shape

Cross sections and reconstructions show that most overthrust belts are *wedge shaped:* they thin in the direction of their motion (figure 15.10). The sole thrust emerges from deep crustal levels, while the surface boundary is an erosion surface that slopes toward the toe. Branches are added forming a complex thrust system.

**Figure 15.9**
Changes in thrust displacement in the Turner Valley structure, Canadian Rocky Mountains. The section is balanced by adding thrusts in the hanging wall. (Dahlstrom, 1969.)

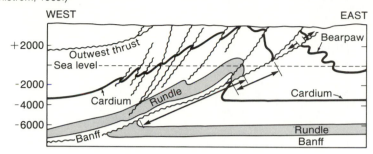

**Figure 15.10**
Simplified cross section through the Canadian Rocky Mountains. Pattern: basement. (After R. A. Price, 1981.)

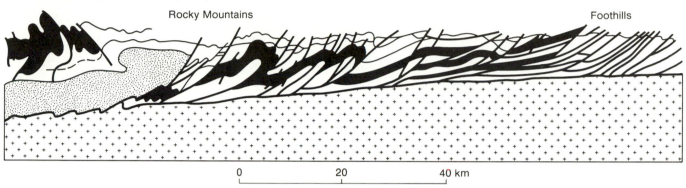

## Mechanism

How does a thrust sheet move, and what makes it move? By the beginning of the twentieth century, geologists, having recognized overthrusts, needed to find a mechanical explanation. The first step in that direction was taken by the Polish geologist Smoluchowski (1909). He showed that, assuming a coefficient of sliding friction of 0.15 (friction of iron on iron, an assumption that seemed reasonable) and a hypothetical across-strike width of a granite thrust sheet of about 165 km, the compressive stress necessary to move the sheet would be more than seven times the crushing strength of granite. The hypothetical figures Smoluchowski used are not wholly realistic, but they are well within an acceptable order of magnitude. Thus, in his words, "we may press the block with whatever force we like; we may eventually crush it, but we cannot succeed in moving it." According to Smoluchowski, this mechanical paradox could be resolved in two ways: (1) by sliding under the influence of gravity, down a slope of 1:6.5; or (2) if there is a thin pseudoviscous layer along the thrust plane or a pseudoviscous base below it, then "any force, however small, will succeed in moving the block. Its velocity may be small . . . , but in geology we have plenty of time; there is no hurry." Subsequent development of our understanding of thrust mechanisms has been almost equally slow.

## Conditions for Thrust Motion

We now turn to the first question at the beginning of this section: How does a thrust move? Three conditions must be met: (1) A sheet of rocks must become detached from its base; (2) the sheet or *nappe* must overcome resistance to forward travel, both along the detachment surface and at its toe; and (3) a driving force—usually closely linked to a crustal instability—must *move* the sheet. Let us look at each of these three conditions.

*Detachment (décollement)* or, initially, *potential* detachment is a prime precondition of thrust faulting. This can occur along structurally predestined surfaces, such as bedding surfaces or other flat-lying mechanical heterogeneities. The lithosphere-asthenosphere boundary probably facilitates lithospheric subduction; but at higher levels any thermally softened basement might serve. Thermal softening may favor backarc areas (see Armstrong and Dick 1974). Detachment will also occur in conjunction with some of the mechanisms discussed.

*Reduction of frictional resistance* to forward travel is the central problem of the "mechanical paradox of large overthrusts" (Hubbert and Rubey 1959). Shear stresses of no more than 10 to 20 bars (1–2 MPa) should be sufficient to move a thrust sheet. Several possible

**Figure 15.11**
Classical section at the Lochseite, near Schwanden, Glarus, Switzerland. 1, Eocene Flysch; 2, Lochseiten limestone (limestone mylonite); 3, fault gouge; 4, green Verrucano conglomerates with strongly flattened clasts; 5, purple Verrucano conglomerates with less deformed clasts. (From Trümpy, 1980.)

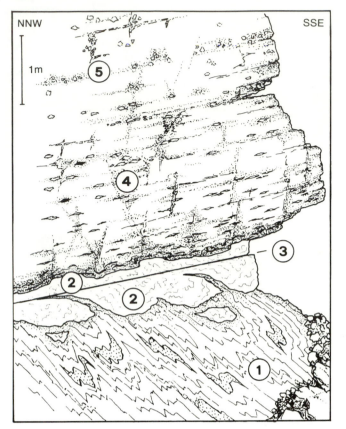

mechanisms have been proposed, supported by theoretical, experimental, and field evidence. Smoluchowski's first suggestion of a slope greater than 1:6.5 to overcome normal sliding friction has never been seriously considered, not even by Smoluchowski himself, because such a slope would not fit any known examples of true overthrusts. But his second suggestion has never been contested, and it appears to apply in many known examples. Kehle (1970) gave us a good modern account of this phenomenon: a weak, pseudoviscous layer along the thrust surface ensures that resistance to forward motion is not through friction, but through viscous resistance to flow, which even exceedingly small stresses can overcome.

It is not altogether clear which of the intimate mechanisms responsible for steady-state creep, as discussed in chapter 6, acts along yielding thrust surfaces. It may not be the same mechanism in all cases. Certainly, mylonites are common fault rocks along thrust surfaces; but so far it has not been possible to replicate experimentally characteristic deformation structures found in some thrust-mylonite fabrics. All we can say for certain is that in many cases the mylonite fabric suggests steady-state creep, followed by actual loss of cohesion. This latter, however, may have occurred only during the last stage of thrust motion as in figure 15.11. On the other hand, the mylonite cannot have been present at the very beginning of motion. In other words, some thrusts have moved along yield horizons that were not yielding when motion was initiated, but only became yielding as a result of motion. This is really part of a more general problem: How do shear zones become individualized? We know that, in the transition field between brittle fracture and pervasive ductile shear, deformation occurs along discrete ductile shear zones; but it is not clear how they individualize in a particular location. The fact is, they do. An even more puzzling problem is presented by thrusts such as major portions of the Keystone thrust of Nevada and California, where no preferred yield horizon is known over large areas (Burchfiel et al. 1982).

Another mechanism for overcoming basal friction along thrust surfaces may be at work here: the action of abnormally high pore pressure below the thrust surface. This is the subject of an important pair of papers (Hubbert and Rubey 1959; Rubey and Hubbert 1959). Applying the Coulomb criterion (chapter 13) to a sheet of rocks 1 km thick, Hubbert and Rubey showed that the greatest length of such a sheet that could be "thrust" by external forces in a horizontal direction, using $7 \times 10^8$ dynes/cm$^2$ as a representative value for the crushing strength of rock, was 8 km. Since thrust sheets of the order of 30-km length and greater are known, simple horizontal pushing or thrusting is a mechanical impossibility. This had been intuitively recognized by pioneer investigators such as Bertrand, Schardt, and Cloos. The only readily available body force is gravity. Again, Hubbert and Rubey showed that, assuming a ready-made break, a similar rock sheet would need a basal slope angle of 30° before it would overcome friction and slide under the influence of gravity.

The above figures assume dry rock; but rocks contain pore fluids such as water and, possibly, oil. In chapter 13, we saw that pore pressure could theoretically reach the value of the lithostatic pressure and would then "float" the rock mass. Drilling experience in oil fields where rocks are compacting, or where tectonic stresses have reduced pore space, show values of such "abnormal" pressure exceeding 90% of that of the lithostatic pressure—close to the flotation threshold. This mechanism might reduce friction to the point where large-scale sliding of rock masses becomes a mechanical possibility.

According to equation 13.1

$$S = p + \sigma_e \qquad \text{(15.1)}$$

where $\sigma_e$ is the effective normal stress, $S$ is the total normal stress through any plane in fluid saturated rock, and $p$ is the pore pressure of the fluid in the rock. The critical shear stress, $\tau_c$, required to slide the mass, is given by

$$\tau_c = \tau_o + \sigma_e \tan \pi \qquad \text{(15.2)}$$

where $\pi$ is the angle of sliding friction, and $\tau_o$ is the shear strength at $\tau_c = 0$. From equations 15.1 and 15.2 we obtain

$$\tau_c = \tau_o + (S - p) \tan \pi \qquad \text{(15.3)}$$

Thus, as $p$ approaches, $S$, $\tau_c$ approaches $\tau_o$, which is independent of $\pi$. Clearly, increasing $p$ does not reduce the angle (coefficient) of friction but, rather, the normal stress on the sliding surface, and *hence* the friction (expressed as $\tau_c$ above).

## Field Evidence

Rubey and Hubbert applied their hypothesis to one area in particular, the overthrust belt of western Wyoming and adjacent states. This belt, which is several hundred kilometers long and about 100 km wide, lies in a region of long-continued sedimentation, ranging from Lower Paleozoic through Cretaceous and early Tertiary time. It consists of a series of large detachment faults, each continuing for tens of kilometers along the outcrop. Horizontal displacement, according to Rubey, is conservatively estimated at 18–25 km for each fault. Thickening and, in some beds, coarsening toward the west indicate a contemporary rising zone in that direction. Rapid deposition caused abnormal pressures in compacting sediments, reducing friction along critical bedding planes. Detachment and gliding may have followed any disturbance of equilibrium once pore pressure had reached abnormally high values, but it may have been delayed for as long as high pore pressure was maintained.

Data from the Louisiana Gulf Coast show that abnormal pore fluid pressures had been maintained for at least 40 million years (since Lower Oligocene time), at a fluid-pressure-to-lithostatic-pressure ratio of 0.87. In the presence of tectonic movements, such pressures might be set up in comparatively old rocks.

Good independent field evidence exists for the validity of this mechanism. In southern Nevada, the Lower Paleozoic Goodsprings Dolomite overlies Jurassic Aztec sandstone along the Keystone Thrust (figure 15.12). Along this contact there is repeated evidence of abnormally high pore pressure, as illustrated in figure 15.13. Fluidized material from the Aztec has intruded the Goodsprings under pressure along numerous fractures and fissures, some of them exceedingly fine and narrow.

Where mylonites or other fault rocks that reduce shear strength are absent, only abnormal pore pressure seems capable of reducing sliding resistance to forward motion. There is, however, another mechanism worth considering for this role: vertical acceleration during earthquakes. A finite difference in vertical acceleration between hanging wall and footwall, caused by differences in response time, may lessen the load on the thrust plane and hence of friction, as in a shaking conveyor belt. Similarly, seismic stresses may cause a transient increase in pore pressures below the thrust surface, also reducing friction. Since seismic activity must be quite common at the time of thrusting, there may be a role for it in lessening friction.

## Rates of Thrusting

The toes of many thrust sheets or nappes rest on sediments derived from them, indicating that they have moved from areas of high relief toward the lower foreland. Where it is possible to date both nappe emplacement and the accumulation of nappe debris, we can compute rates of nappe movement: this turns out to be of the order of a few millimeters per year. However, nappe movement probably occurs episodically, rather than continuously, so that computed rates are likely to be averages, and maximum rates may be multiples of average rates.

**Figure 15.13**
Clastic dike of Jurassic Aztec sandstone material intruding the overlying lower Paleozoic Goodsprings Dolomite near Las Vegas, Nevada. The dike material consists of wholly disintegrated Aztec sandstone, with inclusions of Goodsprings Dolomite. This is evidence of abnormally high fluid pressure under the Goodsprings-Aztec contact, which is a fault contact, the Keystone Thrust.

## Initiation and Driving Mechanisms of Thrusting

Thrust sheets become detached and move because, for some reason, they have become unstable in their original location. Two principal kinds of instability have been identified as possible generators and driving forces of thrusting: horizontally acting compression; and topographic relief exceeding a critical limit—that is, gravitational instability. These instabilities may not only initiate thrusts, they may also continue to drive them until stability is restored.

In horizontal compression, the driving force is applied laterally, generating boundary stresses that are transmitted along the thrust sheet. In relief instability, the driving force is a body force—gravity—which acts on every particle of the moving sheet. Let us look at each in turn.

### Horizontal Compression

Plate motions may provide localized compressive stresses in three different settings.

1. At a time of plate reorganization, usually following collisions of fragments of continental lithosphere, one or more pairs of converging plates become sutured along continental margins, effectively stopping convergence between them. This motion must be taken up elsewhere in order to conserve the zero vector sum of plate motions. As a result, horizontal compressive stress builds up and eventually results in failure along one or more zones of weakness: a fracture (thrust fault) forms a new break in oceanic lithosphere, creating a new converging plate boundary along which one newly individualized plate overrides the other, at a rate that conserves the global zero vector sum of plate motions. Motion along the new thrust fault continues as long as the newly created pattern of plates and plate motions continues to require convergence at this new boundary.
2. Convergence between two plates commonly ends soon after a continent riding on the subducting plate collides with a continent on the overriding plate. On a small scale, the same thing happens when a backarc or other small ocean basin closes. In such cases, the leading edge of one continent may override the other. Subsequently, thrust geometry may evolve in complex patterns such as previously described.
3. A form of subduction may occur *within* continents, along the inner edge (toward the continent) of orogenic belts. Its mechanism is not entirely clear, but detailed mapping and resulting balanced cross sections, notably in the Canadian Rocky Mountains, have shown that basement has been shortened underneath the sedimentary cover; seismic profiling suggests subduction of foreland basement under the adjacent orogenic belt (figure 15.10).

To distinguish subduction types 1 and 3, each has been given a different name: oceanic, true interplate subduction is called *lithospheric* or *B*-subduction ("B" for Benioff, because it is associated with Benioff zones). Continental, intraplate subduction is called *crustal* or *A*-subduction ("A" for A. Amstutz, who first described this type of subduction). The part of the craton adjoining an orogenic belt is also called the *foreland* of that orogenic belt. Hence, thrust belts associated with *A*-subduction, in which thrusts ride out over the foreland, as in the Canadian Rocky Mountains and in the southern Appalachians, are also called *foreland thrust belts*.

A fourth mechanism has been proposed in the form of a hypothesis: the horizontal push could sometimes be provided by magmatic pressure. This is actually an old idea, first advanced by Hutton in 1788. A. G. Smith (1981) showed that it might be possible for the fluid pressure generated by magma emplacement to exert a lateral force sufficient to generate the detachment of wedge-shaped rock bodies typical of overthrust belts, provided, of course, a suitable detachment and glide horizon is available. Price and Johnson (1982) calculated that, in the setting of the Sierra Nevada and the contemporary Sevier thrust belt of California, Nevada, and Utah, the lateral pressure of repeated magma injections in the Sierra batholith could have driven thrusts that crop out several hundreds of kilometers to the west.

### The Role of Gravity

After large-scale overthrusts had been recognized in the Alps and elsewhere, a number of geologists, notably Schardt (1898), proposed that thrusting may take place by gliding under the influence of gravity. We shall see that gravity may, indeed, have a role in the emplacement of thrust sheets. However, the need to validate gravity as a driving force of overthrusts is no longer as compelling as it was before the general acceptance of plate tectonics.

**Figure 15.14**

Three types of mass movement that may result from gravitational instability: (a) gravity sliding entirely downslope; (b) gravity sliding partly downslope, the remainder upslope; and (c) gravity spreading. (From de Jong and Scholten, 1973.)

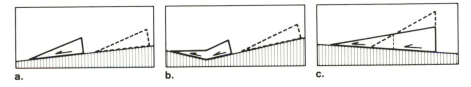

a.        b.        c.

Relief created by tectonic movements is normally graded by surface agents, erosion, and sedimentation, faster than an unstable slope can form. Under certain conditions, however, relief instability may develop in orogenic belts. Equilibrium, then, tends to be restored by two main adjustment mechanisms: (1) sliding along low dip surfaces (downslope gliding), and (2) lateral spreading of excess relief mass (gravity spreading), generally upslope along a basal surface (figure 15.14). Transitional conditions between these two processes are also possible. We shall examine the validity of these mechanisms and then proceed to see to what extent known structures fit this picture.

*Downslope Gliding*

A necessary prerequisite for downslope gliding is the creation of tectonic elevations, either by plutonic activity or by compression. This style of gliding may have operated as a secondary, local mechanism in orogenic deformation.

If friction along a glide plane is sufficiently reduced, either by abnormal pore pressure or by means of a pseudoviscous gliding horizon, even a small slope (7° or less) is sufficient to initiate downslope gliding. It occurs, for instance, when blocks of rock overlying a tilted or tilting glide horizon in a submarine deforming basin break off and slide toward the center of the basin, and come to rest as **exotic blocks**; in such an environment individual stratiform rafts of rock are known as **olistoliths.**

Some "structureless" shales have lost internal cohesion in the process of gravity sliding. The best-known case is that of the Argille Scagliose of the northern Apennines in Italy. Formerly continuous sedimentary layers are in a jumbled mass of disconnected slabs and blocks (figure 15.15); these overlie, along a tectonic contact, autochthonous rocks of similar (mainly Oligocene) age. The source, as deduced from move-

**Figure 15.15**

Discordant contact (heavy line) between black shale formation and Macigno sandy siltstone. Remnants of white limestone beds are enveloped in black shale, which shows extreme structural confusion. Underlying Macigno, shown by dot-and-dash pattern, is less disturbed. Contact is a surface of differential movement *within* the allochthon. (View is northward at nearly vertical exposure 12 m high, near the crest of the Apennines.) (From Page, 1963.)

**Figure 15.16**
Scrope's (1825) conception of the relationship between mountain building and igneous activity. Note that folding is represented as a secondary process resulting from the upward movement of granite and the downward pull of gravity.

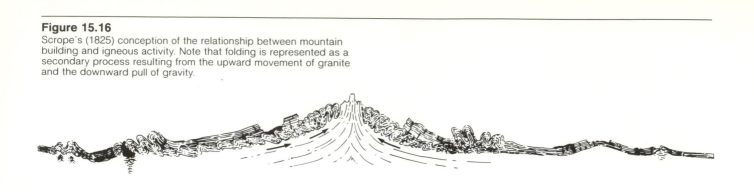

**Figure 15.17**
The Säntis (Sä) and Glarus (Gl) thrusts, Helvetic Alps of eastern Switzerland. The Säntis thrust is a classical décollement structure. Note duplex. (Cross section by O. A. Pfiffner, in Funk et al., 1983.)

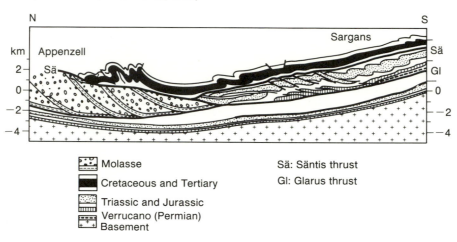

Molasse

Cretaceous and Tertiary

Triassic and Jurassic

Verrucano (Permian)
Basement

Sä: Säntis thrust

Gl: Glarus thrust

from movement traces such as disconnected folds, appears to be to the southwest, the present site of the Mediterranean sea. Nevertheless, transport in a northeasterly direction seems to be well established (Merla 1951; Page 1963).

*Gravity Spreading*

Movement under the influence of gravity originally was assumed to occur downslope (figure 15.16). This seems to agree with some observations (e.g., figure 15.17), but most thrust surfaces seen today actually slope *up* in the direction of presumed motion. If gravity was to be the driving force, this posed a problem.

A classical experiment by Bucher (1956) pointed the way to a solution: since the motion of many thrust sheets resembles that of pseudoviscous masses rather than rigid plates, "uphill" motion is possible (figure 15.18). This was illustrated by Price and Mountjoy

(1970) and, in some detail, by Elliott (1976). Elliott was able to show analytically that, in gravity spreading, it is not the basal slope along which movement takes place that governs motion; rather, it is the slope of the ground surface above (figure 15.19) that drives the nappe forward. This makes sense intuitively if we see the nappe as a viscous "hill" that spreads under its own weight. Thus, relief that is not flattened by erosion at a sufficient rate will tend to flatten by spreading under the influence of gravity, because rocks behave pseudoviscously over extended time spans.

If a **glide horizon,** or weak layer exists at the base of a wedge of rocks (figure 15.20), gravity will tend to flatten the wedge and drive it forward. Ramberg (1977) found that the internal fabric of nappes in the Caledonian belt of Scandinavia may document such forward extension: there is schistosity parallel to the thrust surface and extension lineation parallel to thrust motion.

**Figure 15.18**
Structure produced by solid flow in layers of clear stitching wax interbedded with layers of grease, resulting from lateral spreading of a block of clear stitching wax. (From Price, 1972, after Bucher 1956.)

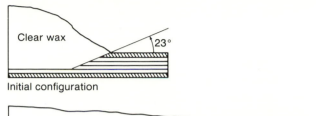

**Figure 15.19**
Portion of a thrust sheet, with surface and basal slopes $\alpha,\beta$ measured from the horizontal. Rectangular Cartesian coordinates have origin O resting on the base of the thrust; $X_1$ parallels the surface and lies at an angle $(\alpha + \beta)$ to the base. The thickness $H$ of the sheet is measured along the $X_2$ coordinate. (From Elliott, 1976.)

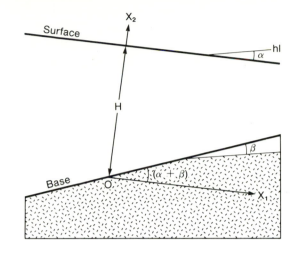

**Figure 15.20**
Half profile of symmetric spreading rectangular bodies (a) with free slip along the base; and (b) with coherence along the base. Dashed lines indicate initial cross section. Full curves outline shape at an early stage of spreading. (From Ramberg, 1977.)

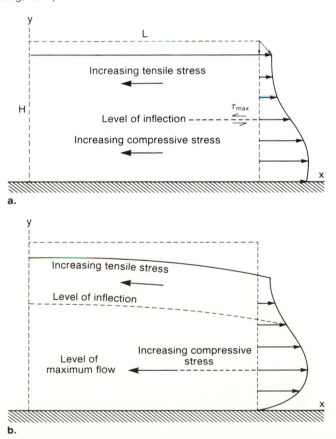

**Figure 15.21**
Reef dolomite overlying ductile shale, Dolomite Mountains, Alto Adige (South Tyrol), northern Italy. (From Engelen, 1963.)

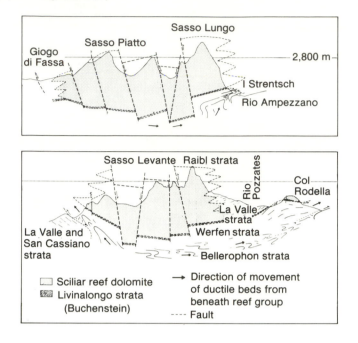

**Figure 15.22**
Two structural cross sections through the Naukluft range, Namibia. The folded sedimentary rocks fill a basin formed by the basement (black). Note thin fault rock (dolomite) at base of main thrust. (From Korn and Martin, 1959.)

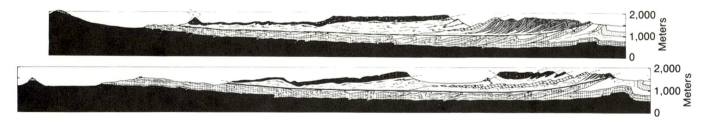

## Other Examples

Gravity spreading may occur without benefit of a glide horizon. For instance, tectonic uplifts may develop "overhangs" along border faults. The Boothia uplift in the Canadian Arctic is an excellent example (figure 14.39).

Relief instability may also result from unusually rapid erosion. In the Dolomite mountains of the southern Alps, scarps developed in the course of Pleistocene glacial erosion have begun spreading laterally to an extent that has not been entirely checked by current erosion (figure 15.21). This spreading has been facilitated by layers of ductile shales underlying massive blocks of dolomite; the shales have spread laterally in the same way that the valley bulges in England (figure 6.28) have bulged upward, under unequal load distribution resulting from erosion.

The strain pattern in the Naukluft mountains of Namibia suggests at least some movement under the influence of gravity. Figure 15.22 is a geological panorama constructed from actual field observation in canyons, and requires little interpretation. Several great nappes have moved into a subsiding tectonic basin.

**Figure 15.23**
Coast cliff section on the north side of Frederick E. Hyde Fjord, Greenland. Profile height is approximately 500 m. Note ductile deformation style. (From Pedersen, 1980.)

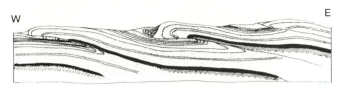

W                                 E

Frederick E. Hyde Fjord

**Figure 15.24**
Schematic structural profile across Japan Trench and inner trench slope off Hachinoe, northeast Japan, showing basal reflection plane (acoustic basement, BM), thrust faults, accretionary prism, and some seismic velocities and earthquake foci. (From Shiki and Misawa, 1982.)

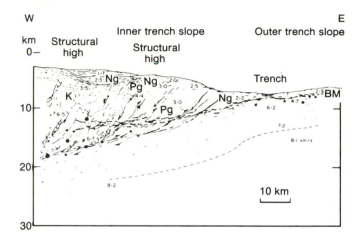

Movement traces are most conspicuous along the steepest part of the basin. The frontal portions of the nappes are deflected upward against the far (south) side of the basin, as is common in gravity spreading. The strain pattern is clearly one of southward flow. Initiation of thrusting may have been by compression, but gravity must have played a role in thrust motion here.

Even thrust sheets that show no evidence of strain in their fabric do not move as rigid "plates" but, rather, as floppy sheets. This is well illustrated in figure 15.23, which was drawn from nature, and thus shows true geometry. This evident weakness in deformation underlines the old stipulation that, unless basal friction is considerably reduced or eliminated, sheets of rock, if pushed, will crumple but will not move forward.

In a theoretical analysis, Chapple (1978) found that a "push from behind" is necessary to move many known thrust sheets, as in the southern Appalachians and in the Canadian Rocky Mountains, and that their overall strain pattern suggests shortening of the nappe pile in the direction of motion. It seems that the strain pattern of nappes is a good indicator to help differentiate between the "push from behind" and the gravity-spreading mechanisms in individual cases. Both may well act on the same sheet.

## Occurrence and Setting of Thrusts

We can recognize three broadly defined plate tectonic settings of thrusts: (1) the thrust systems along subduction zones; (2) thrust belts between a magmatic arc and its cratonic foreland; and (3) thrust belts associated with collision zones (sutures). The last may be associated with either of the other two. Thus, overthrusts are confined to areas adjoining converging plate boundaries. Some thrusting occurs along transform boundaries with a degree of convergence (p. 274), but true "thin-skinned" overthrusts along which thin flat-lying sheets of rock have been displaced for many kilometers are characteristic of converging plate boundaries.

## Examples of Overthrust Belts and Their Setting
### Thrusts Along Subduction Zones
Subduction itself, of course, is the prime example of thrusting. The ocean floor fractures or yields along a zone of weakness, and this break transects the whole of the lithosphere. The fracture forms under horizontal compression, but fracturing alone will not necessarily relieve the instability that caused it: convergence of the two newly outlined plates will continue for as long as the global plate-motion pattern requires it. So one plate (always an oceanic one) will dive under the other into the asthenosphere, while the other plate, usually carrying a continent, overrides it in a wedge-shaped thrust sheet or complex. There has been some dispute as to whether this is underthrusting or overthrusting. In reality, the motion is relative: no one plate is more "active" than the other.

The actual fault surface cuts through the trench fill sediments and tends to *prograde;* that is, it periodically creates a new break ahead of an older one, and in so doing adds footwall material to the hanging wall (figure 15.24). Thus, subduction thrust systems consist of imbricate listric faults that enclose what is commonly called a **subduction complex** or **accretionary wedge.** The wedge form is characteristic. It is an expression of original arcward thickening of the sediments, as well as foreshortening during deformation. However, along some trenches, subduction proceeds without fault prograding; that is, without accretion of trench fill to the overriding plate. In other words, along some subduction zones *all* the sediment is carried below the thrust by the downgoing slab.

**Figure 15.25**
Many regular folds of the central Appalachian Valley and Ridge Province are underlain by duplex involving thick, competent Cambrian and Ordovician carbonates. Small displacements on subsidiary faults that cut through large ramps produce periodically folded roof thrusts. Sole is comparatively smooth in Lower Cambrian Wainsboro shales. (From Boyer and Elliott, 1982.)

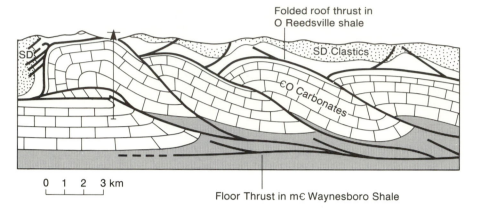

Folded roof thrust in
O Reedsville shale

SD Clastics

€O Carbonates

SD

0  1  2  3 km

Floor Thrust in m€ Waynesboro Shale

## The Canadian Rocky Mountains

Foreland thrusting in the Canadian Rocky Mountains is the listric, "sled-runner" style (figure 15.10). The sole faults extend westward subsurface for more than 200 km, a fact that was revealed only when seismic data for that region became available. Reconstructing the pre-thrusting cross section (called a palinspastic cross section) results in a greater total length of key cover strata parallel to the cross section than for the basement surface; at least 200 km of basement cross section has been lost, presumably "swallowed up" or subducted into the lower crust. The Canadian Rockies are typical for a thrust belt between a magmatic arc and a cratonic foreland. The main overriding thrust in the southern part of the Canadian Rockies is the Lewis thrust, which has a displacement of at least 35 km and carries a stratigraphic succession about 6 km thick (before erosion), including basement.

## The Southern Appalachians

Cross sections through the Southern Appalachians of Kentucky, Tennessee, and the Carolinas display a geometry not unlike that of the Canadian Rocky Mountains thrust belt: sled-runner thrusts merging downward into sole faults (figure 15.25). Again, seismic exploration has revealed a great eastward extent of the sole fault, and a thin thrust wedge overlapping the foreland to the west (figure 15.26). This thrust belt is associated with the collision that closed the Iapetus ocean at the end of Paleozoic time. Here again, there is an overriding basement thrust, the Blue Ridge thrust. The total present thickness of its main allochthon is from 6 to 15 km, and total displacement is estimated to have been at least 260 km (Cook et al. 1979). This basement thrust appears to have generated the belt of foreland thrusts and folds ahead of it. However, in the northern Appalachians of Vermont and Quebec, a similar basement thrust, the Champlain thrust, has generated no comparable foreland fold or thrust belt. Much of the displacement information comes from deep seismic reflection profiling.

## The Western Alps

Thrusting in the western Alps (figure 15.27) is more complicated than that shown in the first two examples. But some elements are similar: shortening of basement, and thin thrust sheets or *nappes* that overlap the foreland. Basement shortening across the western Alps is of the order of 200 km; individual nappes have moved as much as 40 km. Here, too, nappes carrying crystalline basement over younger sedimentary rocks, and of the order of 6 km thick, appear to have been the prime movers. While crustal shortening must have been the chief cause of décollement and thrusting, gravity appears to have had a role in shaping the architecture of today's mountain range: this follows from the internal fabric of some of the nappes, as well as from their external form (figures 8.29, 15.17, and 15.27).

**Figure 15.26**
Generalized cross section of the southeastern United States and
present offshore Atlantic shelf, based on an interpretation of the
COCORP traverse and on offshore cross sections. Note vertical
scales. (From Cook et al., 1979.)

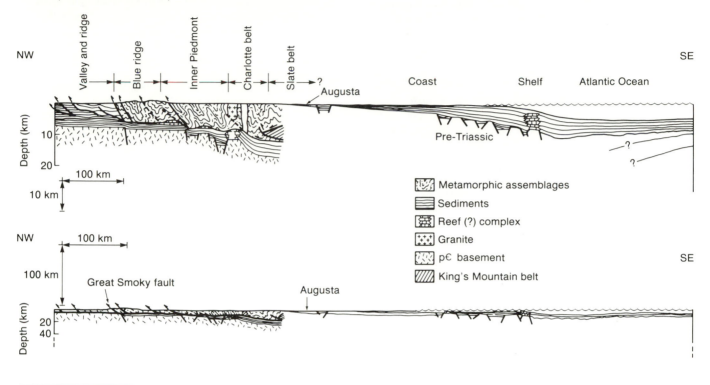

**Figure 15.27**
Finite strains in the western Helvetic nappes, Valais, Switzerland.
(From Ramsay and Huber, 1983.)

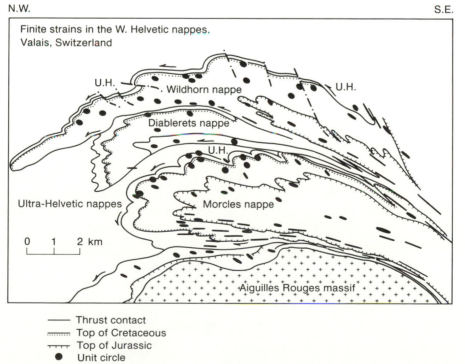

## Conclusion

Thrust faults evidently occur under very special conditions; but these conditions cannot be uncommon in orogenic belts. They include availability of a detachment surface, a mechanism for reducing sliding resistance, and a driving force. The instability that generates the driving force may be crustal shortening, high relief (gravitational instability), or emplacement of plutons. Some of these instabilities may have acted in conjunction; all three occur in orogenic belts. We know much more about thrust faults than we did only a few years ago. But we do not yet have a complete explanation.

## Review Questions

1. This textbook puts thrusts in a chapter by themselves. Explain why this seems a good (or bad) idea.
2. How would you distinguish between a thrust and an unconformity in the field?
3. Describe how a thrust develops along "flats" and "ramps."
4. Draw a labeled cross section of a duplex and explain how it develops.
5. What is the "mechanical paradox" of thrusting?
6. What three conditions must be met for motion along an overthrust?
7. How may basal friction along overthrusts be reduced?
8. In what regional tectonic settings may overthrust belts occur?
9. How may gravity act as a driving force for overthrusts? Distinguish between gravity sliding and gravity spreading.
10. Discuss strain within thrust sheets.

## Additional Reading

Bally, A. W. 1981. Thoughts on the tectonics of folded belts. *Geol. Soc. London Spec. Pub.* 9: 13–32.

Coward, M. P. 1983. Thrust tectonics, thin skinned or thick skinned, and the continuation of thrusts to deep in the crust. *J. Struct. Geol.* 5: 113–123.

Elliott, D. 1976. The energy balance and deformation mechanisms of thrust sheets. *R. Soc. London Philos. Trans.* A283: 289–312.

Gilotti, J. A. and Kumpulainen, R. 1986. Strain softening induced ductile flow in the Sarv thrust sheet, Scandinavian Caledonides. *J. Struct. Geol.* 8: 441–455.

Gretener, P. E. 1977. On the character of thrust faults with particular reference to the basal tongues. *Bull. Can. Pet. Geol.* 25: 110–122.

Hatcher, R. D., Jr., and Williams, R. T. 1986. Mechanical model for single thrust sheets Part I: Taxonomy of crystalline thrust sheets and their relationships to the mechanical behavior of orogenic belts. *Geol. Soc. Am. Bull.* 97: 975–985.

Johnson, M. R. W. 1981. The erosion factor in the emplacement of the Keystone thrust sheet (South East Nevada) across a land surface. *Geol. Mag.* 5: 501–507.

Leonov, M. G. 1978. Olistostromes and their origin. *Geotectonics* 12: 333–342.

Longwell, C. R. 1949. Structure of the northern Muddy Mountain area, Nevada. *Geol. Soc. Am. Bull.* 60: 923–968.

Moores, E. M., Scott, R. B., and Lumsden, W. W. 1968. Tertiary tectonics of the White Pine—Grant Range region, east-central Nevada, and some regional implications. *Geol. Soc. Am. Bull.* 79: 1703–1726.

Pierce, W. G. 1963. Reef Creek detachment fault, northwestern Wyoming. *Geol. Soc. Am. Bull.* 74: 1225–1236.

Price, R. A. 1986. The southeastern Canadian Cordillera: thrust faulting, tectonic wedging, and delamination of the lithosphere. *J. Struct. Geol.* 8: 239–254.

Sanderson, D. J. 1982. Models of strain variation in nappes and thrust sheets: A review. *Tectonophysics* 88: 201–223.

Schmid, S. M. 1975. The Glarus overthrust: Field evidence and mechanical model. *Eclogae Geol. Helv.* 68(2): 247–280.

Suppe, J. 1983. Geometry and kinematics of fault-bend folding. *Am. J. Sci.* 283: 684–721.

Tocher, D. 1975. On crustal plates (Presidential Address). *Seismol. Soc. Am. Bull.* 65: 1495–1500.

Vann, I. R., Graham, R. H., and Hayward, A. B. 1986. The structure of mountain fronts. *J. Struct. Geol.* 8: 215–227.

Voight, B. 1976. Mechanics of thrust faults and décollement. *Benchmark Papers in Geology* vol. 32. Stroudsburg: Hutchinson & Ross, Inc. 471 pp.

Weber, K., and Ahrendt, H. 1983. Mechanisms of nappe emplacement at the southern margin of the Damara Orogen (Namibia). *Tectonophysics* 92: 253–274.

Wiltschko, D. V. 1981. Thrust sheet deformation at a ramp: Summary and extensions of an earlier model. *Geol. Soc. London Spec. Publ.* 9: 55–63.

# Part Four

# Structures of Igneous Rocks

# 16

# Tectonics of Igneous Rocks

C rustal structure and igneous activity are intimately linked. Structure guides and controls igneous activity, and igneous activity, in turn, influences structure. Igneous rocks are themselves structural elements of their setting.

The tectonic evolution of igneous rocks usually passes through three stages: the hearth or "factory," in which the mineral assemblage develops; the movement of this assemblage through its external frame; and its final emplacement and consolidation as a rock. We shall consider two main settings: cratonic and orogenic. We shall not be concerned with oceanic igneous activity beyond its mention in chapter 1.

## Cratonic Setting

Igneous activity in cratons is of two main kinds: volcanic and plutonic. It is normally confined to extensional environments.

### Flood Basalts (Plateau Basalts)

Extensive accumulations of tholeiitic basalts in flows, sills, and associated dikes exist in several continental areas. Examples include the Deccan traps in India, the Stormberg volcanics in southern Africa, the Columbia River basalts, and many more. Flood basalts have two main settings: emplacement under purely hydrostatic conditions, and emplacement under excess pressure, due to dissolved gases.

The Paraná basin in Brazil contains what is perhaps the most extensive succession of sills and flows on the continents, and it provides the best example of what is probably hydrostatic emplacement. The known (Lower Jurassic) occurrence is 1,800 km long by 800 km wide and has been estimated to contain about 1 million cubic kilometers of mafic rocks (figure 16.1). Exploratory drilling for petroleum has revealed its structure. Consistent shallow water marine facies of sedimentary rocks intercalated with basalt flows indicates that basin subsidence kept pace with sedimentation and lava accumulation. The lavas were very fluid, as evidenced by the thick vesicular tops and the large extent of single flows (several hundred kilometers for the uppermost flow).

Sills are intercalated in all the older rocks of the basin. They may reach up to 200 m in thickness and up to 100 km in extent, although average dimensions are much more modest. They have not assimilated any country rock. According to Bischoff (1966), feeder-dike thicknesses average 10–20 m but may reach 100 m. They are spaced at average intervals of 5–8 km. These dikes are in four distinct sets, and they have fed at least 13 different generations of flows. Hence, Bischoff estimates that the feeder interval per eruptive period must have been on the order of 20 km.

Sill intrusion is possible only if the intruding magma "floats" the overlying rocks. This requires a pressure head on the magma greater than the lithostatic load on the country rock at the level of the sill; the magma column balances the rock column, which it floats in hydrostatic equilibrium.

Effusive activity of heavy basaltic magma (specific gravity about 2.9) over comparatively light country rock (specific gravity about 2.5) needs an explanation. Where the rising magma is under pressure, probably because of exsolving gases, this will no doubt give it "lift." In the Paraná basin and in similar Gondwana basins containing extensive volumes of flows and sills, the extrusion of lava appears to have been quite passive. Two explanations for a purely hydrostatic rise of magma have been suggested. According to Vollbrecht (1964), basaltic magma under these basins results from selective melting of the mantle below, at a source depth of about 50 km. Beyond this depth, a column of basalt weighs less than an equivalent-volume column of rock, which includes not only the light surface rocks, but also the denser mantle material. Hence, the lighter (overall) lava will extrude. Bischoff (1966) proposed a source at a depth of 35 km in the lower crust and included the basin margin in the hydrostatic calculations. If the margins are sufficiently high, they might counterbalance a column of lava that reached the surface in the basin and spilled over into its lower elevations.

The Paraná situation is common in the Gondwana continents, and it is particularly interesting to note that a refitting of South America and Africa would bring the similar African area of Lower Jurassic volcanic activity, the Kaoko volcanic area, into a remarkably good fit with the Paraná basin. The areas fit, and so do the feeder-dike alignments and spacings. The space needed for the feeder dikes is good evidence that the areas concerned were in extension at the time of magmatic activity. Geotherms must have been high. The time, Lower Jurassic, is noted for widespread volcanic activity and

**Figure 16.1**
Map of Paraná basin, southern Brazil. (From Bischoff, 1966.)

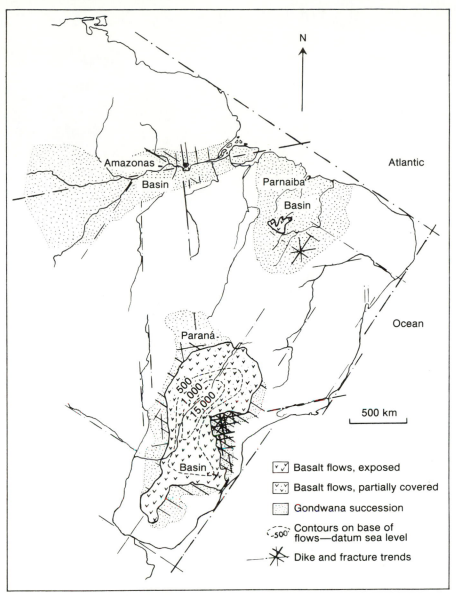

N

Amazonas
Basin

Parnaiba
Basin

Atlantic

Paraná·

Ocean

500
1,000
15,000

Basin

500 km

☑ Basalt flows, exposed

☑ Basalt flows, partially covered

⬚ Gondwana succession

⌒-500⌒ Contours on base of flows—datum sea level

⟶✳ Dike and fracture trends

is the probable time for the initiation of the separation of Africa and South America. The timing also corresponds to the appearance of similar flows and sills in Madagascar, India, Australia (Tasmania), and eastern Antarctica. This was not long after the same type of activity in the eastern United States and in western Europe. These vast, worldwide eruptions closely match the initiation of spreading cycles in the Atlantic and Indian oceans. All of this can hardly be coincidence.

The second setting for flood-basalt emplacement is under excess pressure from dissolved gases. Most Tertiary volcanic plateaus require lava to rise to greater elevations than in the subsiding Mesozoic basins. Rising magma requires excess gas pressure to bring it to the surface; it may, however, be counterbalanced by a heavier column of rock, as in oceanic crust. This problem needs further study. Examples include the Columbia River plateau and the Djebel es Soda area in Libya.

**Figure 16.2**
Contact of Lower Carboniferous sandstone (*low area at left*) and
slightly younger tuffisite pipe (*right*), Dunbar, Scotland. Note
characteristic (but not universal) inward dip of layered tuff. The
sandstone also dips slightly inward. (West is at left.)

## Volcanic Vents

Central volcanoes cap volcanic pipes or vents, along
which the products of volcanism reach the surface.
These vents tap reservoirs of magma at varying depths.
All are the result of high-pressure explosive eruption
and are structurally controlled, usually along fracture
intersections. Vents of some extinct volcanoes have become exposed by erosion.

It seems clear that the first eruption, which
"drilled" the vent, must have been explosive, with very
little, if any, lava involved. This is followed by lava
eruption; and at the end of volcanic activity, lava consolidates in the vent, forming a volcanic pipe. Differential erosion tends to shape pipe outcrops into
conspicuous volcanic necks.

Some volcanic vents are filled with originally solid
fragments (**tephra**) rather than lava. The composition
of the filling varies considerably. It is derived both from
the wall rocks and from depth. Sizes vary from ash size
to large blocks and boulders. The proportion of igneous
to wall rock material varies widely.

Vent filling of this type is called tuffisite, to distinguish it from tuff, which it resembles but which settles
at the surface. The pipes, then, are tuffisite pipes. Near
the top, the vent filling may be stratified in the form of
a vent agglomerate. The bedding in some such vent ag-
glomerates has a characteristic inward dip near the rim
(figure 16.2). Similarly, low-dip bedding in the country
rock tends to dip toward the pipe, denoting inward sag
at the end of volcanic activity.

**Figure 16.3**
Section through a tuffisite pipe, Urach area, southern Germany. White, massive blocks of Jurassic limestone detached from the walls and displaced by subsidence in rising streams of fluidized tuff. Gray tuffisite consists mainly of microlapilli of olivine-melilite basalt and small fragments and dust of Jurassic rocks. (From H. Cloos, 1941.)

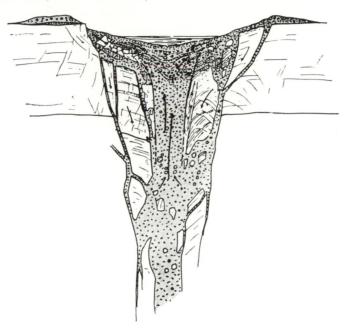

Tuffisite not only fills the pipes, but also penetrates fractures in the form of apophyses. Rock fragments are suspended in it as if "floating" (figure 16.3). In fact, tuffisite material appears to have behaved very much like a coherent fluid at the time of emplacement. Suspensions of solid particulate matter in fluids are known as fluidized systems (Reynolds 1954). In tuffisites, the system consisted of both juvenile and wall rock clasts suspended in a stream of volcanic gases. Such systems are powerful drilling agents, and they can widen vents to several hundred meters in diameter. Clasts may become rounded, as in stream erosion, and quartz grains may become rounded like sand, as in the well-known Tsumeb pipe in Namibia.

Examples of tuffisite pipes include the system of pipes in the Urach area of southern Germany, the Firth of Forth area in Scotland, and the kimberlite pipes in different parts of the world (figure 16.4). The most famous kimberlite pipes are those of southern and eastern Africa. A groundmass of mainly serpentine and calcite contains inclusions of ecologite and of wall rock material derived from rocks now present and from rocks formerly above the present surface, eroded long ago. Most important, kimberlite contains diamond, a mineral that requires at least 40 kilobars to form. This pressure corresponds to a depth of about 125 km. In other words, kimberlite pipes tap a region well within the upper mantle, and probably within the asthenosphere.

**Figure 16.4**
Kimberlite pipe in South Africa, showing transition at depth to dike. Country rock omitted. Height of pipe is 5–7 km. (From H. Cloos, 1954; *Gespräch mit der Erde,* C. R. Piper & Co. Verlag, Munich, 1947.)

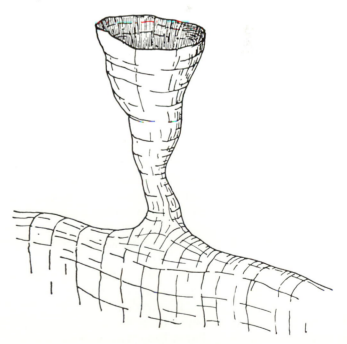

**Figure 16.5**
Two genetic types of caldera: (a) outward dipping, and (b) inward dipping.
Source: Smith, R. L., R. A. Bailey, and C. S. Ross, "Structural Evolution of the Valles Caldera, New Mexico, and its Emplacement of Ring Dikes" in *U.S. Geological Survey Professional Paper 424D*, pp. 145–149, 1961.

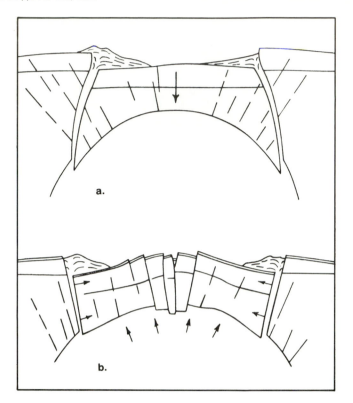

## Calderas

Physiographic depressions called **calderas** (from the Spanish word for "cauldrons") surround many central volcanoes. They range in scale from a local widening of the crater to circular depressions several kilometers in diameter.

Historical eruptions, especially the 1888 eruptions of Bandai-San, Japan, and of Krakatau, Indonesia, have shown how some calderas may have formed. At Bandai-San, the cause was a series of very violent explosions. No solid or liquid products appeared. The resulting caldera was about 2.5 km in diameter and 400 m deep. At Krakatau, an explosive eruption of great masses of ashes and pumice preceded collapse of the central area, resulting in a caldera. These are calderas of the *explosive* type.

*Collapse* calderas result principally from movement and consequent withdrawal of magma from below the caldera. Volumes of extruded material, where checked, have been found insufficient to be the sole cause of collapse. Collapse has often followed the extrusion of ash flows and pumice, as in the Krakatau eruption of 1888. Collapse following quiet effusion of basic lava is typified by the Kilauea caldera in Hawaii. Collapse calderas result from subsidence of roughly cylindrical plugs, along cylindrical or ring-shaped fractures (figure 16.5).

Classical areas for the study of calderas include the Devonian and Tertiary volcanic districts of Scotland, where the subsurface anatomy of the calderas has been revealed by erosion. The fractures that accompanied subsidence have nearly all been filled by magma, and the whole now forms complex systems of igneous intrusions, mainly dikes of circular and radial outcrop patterns known as *ring complexes* (figure 16.6).

### Ring Complexes

In many parts of the world, dikes that are ring-shaped in plan surround a volcanic center. Good examples exist of such ring complexes in western Scotland, New Hampshire, Norway, Namibia, Nigeria, and Australia (Victoria).

There are two types of circular-outcrop dikes: ring dikes and cone sheets. **Ring dikes** are concentric dikes, dipping at a small angle from the vertical either toward or away from a common center, and normally hundreds of meters thick. Most are of granitic or alkalic silicic

**Figure 16.6**
Anatomy of a ring dike and caldera complex. Surface structure
includes caldera, central cone, crater, vent, radial dikes, cone
sheets, and sills. Deep structure includes cone sheets and ring
dikes (the latter partly continuing the caldera ring dike above). The
dips of the ring dikes are conjectural. (Based on data by J. E.
Richey; from H. Cloos, 1936.)

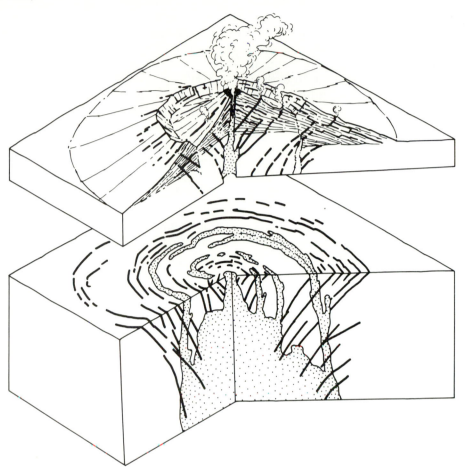

rock. Ideally, they would form a complete ring; more
commonly, they are crescent-shaped in outcrop. Several generations of ring dikes may congregate around
one center. They appear to form as the result of the
subsidence of a central plug of country rock.

The other type of circular-outcrop dikes are known
as **cone sheets.** These are thin, inward-dipping sheets,
usually of basaltic rock, forming an inverted cone pattern in three dimensions. They are associated with some
ring dikes and are normally a few meters or less in
thickness. Individual sheets are not continuous around
one center. The hypothetical apex of the cone formed
by the cone sheets is probably the focus of upward volcanic pressure, which induced the cone fractures. In any
one episode, the cone sheets became emplaced before
the ring dikes. Thus, the sequence of events must be

upward pressure, cone fracturing, and emplacement of
cone sheets, followed by subsidence of a central plug
along cylindrical fractures, accompanied by the emplacement of ring dikes. The ring-dike magma is always more silicic than the cone-sheet magma. This
sequence of events may be repeated a number of times
around one center, the succession of dikes leaving
"screens" of country rock between them. The center
may also shift a few miles in the course of the volcanic
history of the region. The western Highlands of Scotland offer excellent examples of ring-dike complexes
from which such a history may be read. In most cases,
radial dikes and parallel peripheral dikes (a little removed from the center) accompany the ring complex.
A complete complex may be from 300 m to 30 km in
diameter.

Such ring complexes represent the "roots" of many volcanoes. Not all the features mentioned above are necessarily present in each case, and their relative degree of development varies widely. Cone sheets are rare outside Scotland, for example, and only vestigial cone sheets exist around the Brandberg in Namibia (Africa), one of the largest known ring structures.

There appear to be two types of magmatic evolution in fracture-controlled ring complexes and other subvolcanic structures in a cratonic environment: alkalic mafic (or ultramafic), differentiating to carbonatite; and calc-alkalic to weakly alkalic, differentiating to rhyolite (emplaced as ash flows).

## Alkalic Igneous Rocks Along Continental Fracture Zones

In contrast to the areally extensive plateau lava setting, continental igneous rocks may also be emplaced along narrow linear zones, or lineaments that are, or are inferred to be, fracture zones. Examples include the Cameroon Mountains in West Africa (figure 11.12) and the Brighton line in New England extending into the Monteregian Hills (figure 11.13). The tuffisite and kimberlite pipes belong to this setting (p. 309). But many igneous bodies along fracture zones are subvolcanic; that is, the magma that rose along them did not reach the surface. Rather, it formed great plutons, many of them ring complexes, at a high crustal level.

There are few continental rifts without evidence of alkalic mafic volcanic activity. Good current examples include most continental rift systems.

## Mafic Dike Swarms

Dike swarms are extensive sets of parallel dikes. Where they occur, they are evidence for an extensional regime at the time of emplacement. The most common dike rocks concerned are diabase or dolerite; they may be eroded feeder dikes of plateau lavas. The dikes are useful tectonic markers, not only because they document crustal extension, but because they also help in tracing plutonic history. Furthermore, by knowing something of their original emplacement pattern, one may use their present pattern to document strain (figure 16.7).

**Figure 16.7**
Part of the Kangamiut dyke swarm in West Greenland. Note the two generations of dikes and the more highly deformed zone to the northwest. (After Escher, Escher, and Watterson, 1975.)

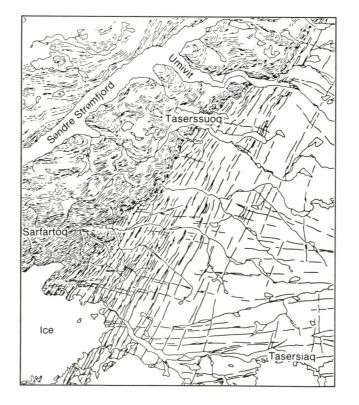

## Summary

To summarize, we may classify the more familiar volcanic and subvolcanic rocks of the stable areas of the continental crust into three categories, according to their tectonic setting:

1. Alkalic volcanic rocks and plutons along rifts and graben zones, such as the east African rift, the Baikal rift, and the Rhine graben.
2. Alkalic plutons along fracture zones that trend into continental margins, such as the Cameroon Mountains, the Brighton line, and the Waterberg-Brandberg line.
3. (a) Flood basalts, also usually near continental margins, of the Thulean, Columbia River, and Deccan type; (b) dike swarms, sills, and other small bodies that may have been associated with flood basalts, now eroded away from a higher level.

## Explosion Structures

A number of puzzling structures exist whose outcrops are ring-shaped but not associated with any known direct evidence of volcanic activity; many may not, in fact, be of volcanic origin. The classic ones are the large Ries structure of Nördlingen (figure 16.8) and the smaller neighboring Steinheim basin, both in southern Germany. The Steinheim basin (figure 16.9) was first described by Branco and Fraas (1905) as a "cryptovolcanic" structure implying "hidden" volcanism.

**Figure 16.8**
Relief model of Ries crater, southern Germany. (Photograph by Schönherr of model in municipal museum, Nördlingen, West Germany.)

**Figure 16.9**
Cross section of Steinheim explosion structure, southern Germany. (From Groschopf and Reiff, 1969.)

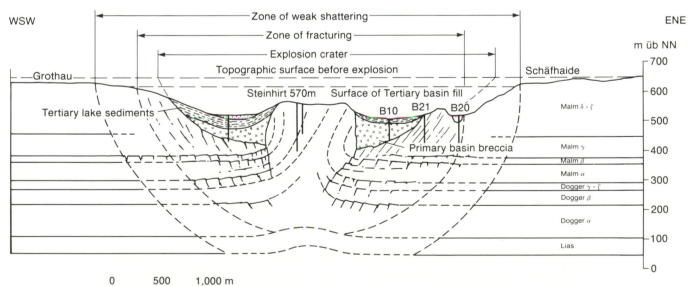

**Figure 16.10**
Crooked Creek polygonal explosion structure, central Missouri.
(From Amstutz, 1960. After Hendricks, 1947.)

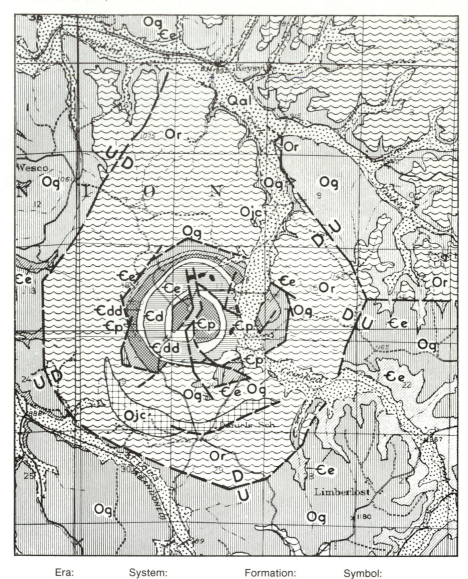

| Era: | System: | Formation: | Symbol: |
|------|---------|-----------|---------|
| Cenozoic | Quarternary | Alluvium | Qal |
| Paleozoic | Lower Ordovician | Jefferson City | Ojc |
| | | Roubidoux | Or |
| | | Gasconade | Og |
| | Upper Cambrian | Eminence | Є𝖾 |
| | | Potosi | Єp |
| | | Derby-Doerun | Єdd |
| | | Davis | Єd |
| | | Bonne terre | Єb |
| | | (black spots) | |

Other, similar structures have since been found. They resemble some lunar craters in outline and dimension and may well be related to them. They have near-circular or polygonal raised rims, which are the result either of faulting or of upturning of surface rocks (figure 16.10), from anywhere between a few hundred meters to many kilometers in diameter. Most, but not all, also have a raised, disordered, or brecciated central portion; and some have small, conical striated structures, called shatter cones, in associated rocks (Dietz 1959). These shatter cones probably result from shock, that is, a transient high-pressure event. Laboratory studies reveal other unusual mineral modifications, all of which require pressures well above those obtained in the upper crust. In many cases the modifications are incomplete, suggesting a transient event, possibly shock from the impact of a meteorite.

Coesite and stishovite, high-pressure forms of silica, and other indications for transient high pressures, have been found associated both with confirmed meteorite craters and with "cryptovolcanic" structures. Stishovite requires a minimum pressure of 99 kilobars at 100° C, and pressures of this order are unknown in volcanic near-surface processes. Other criteria of high-pressure shock origin include sets of closely spaced planes in quartz crystals, especially planes parallel to (0001) and to (1013). These planar features are known in quartz that has been subjected to nuclear explosions, where they are indicative of shocks of at least 50–100 kilobars. Again, such pressures are unknown in near-surface geological processes. Isotropic quartz and feldspar, in some instances, indicate pressures of several hundred kilobars. On the other hand, if a ring structure is tectonically controlled (by some singular structural trend or, better, by the intersection of trends in the earth's crust), then a meteorite origin seems unlikely, since meteorites do not select specific structural trends for impact (Bucher 1963). It is not clear, however, what internal event could generate such shocks.

## Orogenic Setting

### Characteristic Igneous Rocks

The characteristic plutonic rocks of this environment are "granitic" plutons, ranging in composition from diorite to granite. As a matter of convenience, and following a widely accepted convention, we shall henceforth refer to these rocks simply as *granites*.

Evidence from petrology shows that granitic magma cannot become differentiated on a large scale from basaltic material, nor can it be derived directly from such a source by partial melting. The source of the bulk of the granitic plutonic rocks must therefore be in the sialic crust. Many of these granitic rocks are derived from the upper crust by transformation of pre-existing rocks, through a process known as **palingenesis.** Palingenesis operates on rocks in situ but, once transformed, the new granitic material may migrate and become emplaced at different tectonic levels in a number of ways.

We must bear in mind the different tectonic and physicochemical controls that operate during palingenesis and during the subsequent movement and emplacement of granitic material. First, there is external pressure due to load, and internal pressure due to dissolved fluid phases. Next, granitic material at depth, in a framework of metamorphic and, possibly, mafic igneous rocks, is lighter than its surroundings. In fact, the transformation of deep rocks into granite almost always involves a volume increase. The dominant process is feldspathization of micaceous minerals, and this occurs with a volume increase of 5% or more and a concomitant decrease in density. Finally, granitic material at depth is more mobile than the surrounding rocks.

Because all these factors create inherent instability, granite is predestined to move. Wherever it moves, it must make way for itself, taking up space formerly occupied by other rocks. To find out how this happens, we must study remaining indications of strain and of movement in the fabric, but we must also bear in mind the probability of chemical reactions at the highly unstable interface between the granite and its frame.

Working from concepts developed by Wegmann (1935), Read (1948, 1949), and Buddington (1959), geologists have come to recognize characteristic styles of granite emplacement according to tectonic levels. Overall classification distinguishes emplacement in **infrastructure, transition level,** and **superstructure.**

### *Infrastructural Emplacement*

Granitic material is generated down in the highly mobile infrastructure. The process, **anatexis,** is structurally recorded in rocks consisting of a mixture of interfingering country rock and granitic material; this mixed assemblage is **migmatite.** Country rock layers

**Figure 16.11**
Marginal relations of an ascending plutonic diapir (diagrammatic cross section). 1–Preexisting crystalline schists and mafic dikes; 2–gneisses with the older structural trends; 3–migmatites, showing draining of mobilized fluids (ichors) toward the diapir; 4–more or less massive granites. The contact is diffuse in the lower levels but becomes more clear-cut toward the upper levels. (From Wegmann, 1965.)

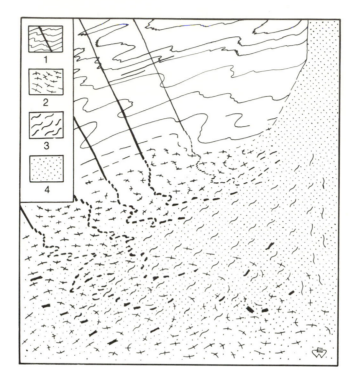

**Figure 16.12**

Diagrammatic illustrations of an infrastructural cycle. (a) Granite body cut by two sets of mafic dikes. (b) Dikes are folded and sheared, with resulting boudinage, gneisses and augen gneisses. (c) Following remobilization (anatexis) the granite is at the same time "younger" and "older" than the inclusions. (d) A new cycle begins with another generation of dike intrusions. (From Wegmann, 1965.)

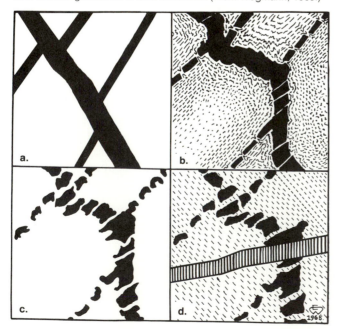

terminate at the granite contact and may be traced inside the granite as "ghost" layers, sometimes even into inclusions that are perfectly aligned with the corresponding layer in the country rock. Well-defined structures gradually disappear, become nebulous ("nebulites"), and grade into seemingly homogeneous granite. While anatexis is the *breakdown* of preexisting rock to form migmatite and granitic material, palingenesis more specifically refers to the *genesis* of new granitic material in the environment.

It is the evidently fluid style of the site of palingenesis (and of the whole infrastructure) that catches the eye: more or less irregular flow surfaces displacing preexisting structural surfaces of all kinds; small, intricately folded ptygmatic veins of granite and quartz; and other traces of a highly fluid environment (figure 16.11). It is no wonder that some of the granitic material behaves diapirically toward less mobile enclosing rocks. Some segments of the infrastructure may show traces of a cyclic history. Markers such as mafic dikes may reveal more than one infrastructural cycle. In figure 16.12a, massive granite is cut by two sets of mafic dikes. In figure 16.12b, deformation of granite and dikes is at an intermediate stage. The dikes are sheared and buckled. Some have acquired boudinage structure; their matrix is partially changed. The granites have become

**Figure 16.13**
Migmatite body in the upper zone of the infrastructure, Caledonian orogen in northeast Greenland. Recumbent fold faces southwest; cliff height is about 800 m. Such tongue and mushroom-shaped bodies are characteristic in this region for this tectonic level. (From Haller, 1956.)

gneissic granites, or augen gneisses; the minerals are partly fragmented, partly bent, the whole forming large and small folds. After remobilization and recrystallization (figure 16.12), we find a paradoxical situation, first noted by Sederholm, in which the granite is simultaneously older and younger than its inclusions. After a flow phase (weak, in this example), the dikes and their matrix recrystallize and become massive. Finally, the whole is once more in a condition in which it is cut by shear and slip surfaces; a new set of dikes traverses it and marks the beginning of a new cycle.

The infrastructure is also the source of gneiss domes. Such domes may rise from their original site and form structures at any scale, large or small (figure 16.13). They originate in remobilized basement and dome upward into country rock, which they deform and displace (figure 16.14). Foliation within them normally is parallel to their walls.

**Figure 16.14**
The evolution of gneiss domes. (From Eskola, 1949.)

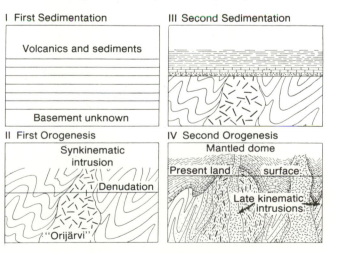

**Figure 16.15**

Duncan Hill pluton, Cascade Mountains, Washington. Projected cross section obtained by projection along b-axes, down a 10° southeasterly plunge (compare with figure 6.2b). (After an unpublished drawing by C. A. Hopson, 1970.)

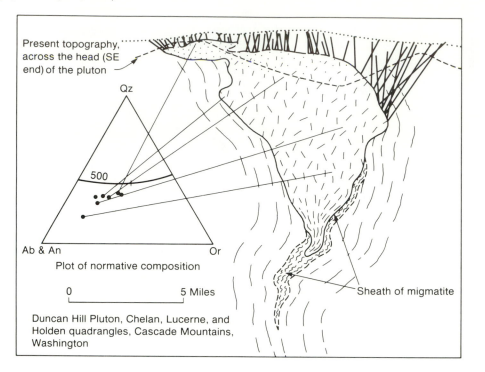

The granites of the infrastructure, called autochthonous and para-autochthonous by Read (1948), have not moved far from their original site. Emplacement is plainly by replacement and infiltration of country rock, or, in domal structures, by displacement of overlying metamorphic rocks. The latter must not be regarded as rigid blocks barring upward rise of underlying material. Rather, they yield and flow out plastically to make room for the rising domes.

In summary, granitic material, whether crystallized, partially crystallized, or liquid, is born in extremely mobile surroundings in the infrastructure, and has a tendency to rise owing to a greater mobility and lower density than that of the enclosing rocks. Tectonic movements during orogeny (called synorogenic) aid this process and are part of it.

### Transition-Level Emplacement

Between the infrastructure and the superstructure is the transition level, the site of regional metamorphism. Two processes of granite emplacement are tectonically important at this level: displacement of country rock consequent upon diapiric rise, with or without subsidiary granitization and fissure filling along the borders (figure 16.15); and replacement of country rock (figure 16.16).

The transition level is the level in which most intrusive granites are emplaced. Again, intrusive relationships between granite and enclosing rocks may imply greater mobility for the granite, but not necessarily a truly liquid state. The granitic material will either rise straight upward through relatively ductile rocks, or it will seek out paths of least resistance, which may be sill-like and even horizontal. Being both hot and very reactive during much of its rise, it will tend to react with country rock; this sometimes results in diffuse and

**Figure 16.16**
Hypothetical history, in four stages, of the Loch Doon pluton in southern Scotland. Stage 1 is the oldest, stage 4 the final (present) stage. The top face of the blocks corresponds to the present surface. (From Oertel, 1955.)

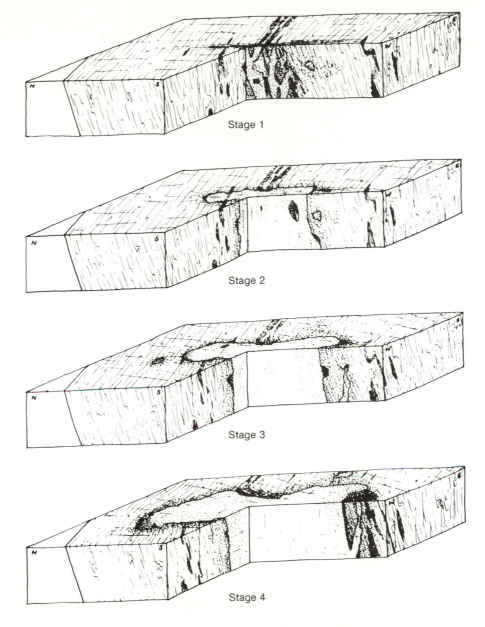

Stage 1

Stage 2

Stage 3

Stage 4

migmatitic borders. Some granites have sharp contacts along part of their periphery, and migmatitic ones along others; they may be cross-cutting in one part and conformable in another. Conformable contacts tend to be sharper, since fluid migration is easier *along* layering (figure 16.17). Contact metamorphic aureoles, where present, tend to be wide and have some of the characteristics of regional metamorphism. Chilled margins are absent, since the frame is generally at an elevated temperature at the time of emplacement.

Replacement of country rock may occur by a process called zone melting. In zone melting, the granite reacts exothermically with overlying country rock. As long as this process occurs in an environment in which little heat is conducted away, the heat of reaction furnishes ready energy for a continued chain reaction upward, and results in a gradual upward enlargement of the granite body. The end relations are those of a replacement body with possible "ghost" remnants of preexisting rocks (figure 16.18).

Granites of the infrastructure and of the transition level may, for one reason or another, become partially or completely molten. In such cases, the resulting material migrates as a true liquid along fissures and becomes emplaced like a classical intrusive granite. Often the magma reaches the superstructure and may even reach the surface as lava or ignimbrite.

### Superstructural Emplacement

Igneous activity in the superstructure includes the emplacement of "allochthonous" plutons (Read 1948) and "epizonal" plutons (Buddington 1959). Superstructural plutons (figure 16.19) have characteristic sharp contacts. Most have rounded outlines in plan. In fact, structurally, the superstructural environment of orogenic igneous activity resembles that of stable cratons in many respects. In orogenic zones, superstructural igneous emplacement appears to be late in the development of the orogen, but this is probably because any record of earlier high-level activity has been destroyed.

**Figure 16.18**
Xenoliths and "ghosts" in various stages of assimilation. Rogart granodiorite, Sutherland, Scotland.

**Figure 16.19**
Schematic relation of the Vrådal pluton to its country rocks. A rim syncline in the country rocks is indicated by the folded amphibolite (black) and granite gneiss. The foliation of the gneiss is most pronounced near the pluton and is weak or indistinct from 2–3 km from the contact. (From Sylvester, 1964.)

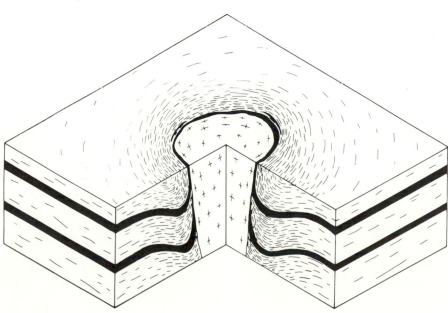

**Figure 16.20**
Shiveluch, Kamchatka fiery cloud, in 1947. The ash flow is at the
bottom of the cloud. (Courtesy of Geological Faculty, Moscow
[USSR] State University.)

## Ash Flows

Ash particles and droplets of lava may become suspended in streams of ascending and evolving volcanic
gases, and the gas stream with its suspension then becomes a fluidized system. It can erupt and flow like a
lava stream of high mobility or can intrude in fissures
as dikes, sills, or pipes. The best-known records of such
**ash flows** are the ignimbrites.

By definition, **ignimbrites** are deposits formed by
"fiery clouds" or *nuées ardentes;* these have been observed, on a small scale, in several eruptions. Notable
examples include the cloud emitted by the Montagne
Pelée ("pealed mountain") on Martinique in 1902, and
the cloud observed at Shiveluch, Kamchatka, in 1947
(figure 16.20). In each case, the fiery dust in the air
above was far more spectacular than the true ash flow
rushing along the bottom. Ash flows have been observed to move at speeds of 100 km per hour and more.
They are true density currents, consisting of hot gases
holding droplets of lava and hot ash particles in suspension. Some, perhaps most, of the gas may escape
from the droplets during flow. **Ash flows** behave like liquids and come to rest like liquids, filling out topographic depressions and keeping a level top surface. This
behavior serves to distinguish them from normal **ash
falls,** which settle directly from the air and drape the
topographic surface. Mafic ash flows are rare, because
of the ready release of gases from the more mobile mafic
lavas. Most ignimbrites are rhyolitic or dacitic. Most
sheetlike rhyolites on geologic maps are ignimbrites, and
most ignimbrite vents are in calderas.

These rocks are characteristically unsorted. Grain
size is variable but is mainly of ash size (less than 3
mm). The presence of glass shards and pumice fragments is diagnostic. Cooling produces columnar
jointing, as in lavas. Ignimbrites are not internally
bedded. But load deformation may impart a wavy foliated ("eutaxitic") texture, the result of flattening and
draping of hot glassy fragments and of pumice over
phenocrysts (figure 16.21). Some ignimbrites or zones
of ignimbrite units become indurated by crystallization, others do so by a process called welding. The latter
are colloquially known as **welded tuffs.**

The original horizontality of ignimbrite tops is more
reliable than that of water-laid strata. It is reflected in
a foliation due to overburden. Hence, apart from deflections due to differential compaction, the attitude of
ignimbrite tops is a very valuable reference datum for
post-ignimbrite deformation.

Among ignimbrite occurrences, the Basin and
Range province of the western United States deserves
special mention. It is unusually wide (700 km). Much
of the volcanism has been rhyolitic, and much of it has
been in the form of ignimbrite eruptions; considerable
volumes, now largely eroded, must have been produced
in mid-Tertiary time. Making use of the premise of
originally horizontal ignimbrite tops, Mackin (1960)
was able to establish that post-ignimbrite movement in
the Basin and Range province was predominantly vertical.

## Tectonic Framework of Orogenic Igneous Activity

Igneous activity in orogenic belts is associated with
converging plate boundaries. Figure 16.22 shows the
characteristic orogenic environments distinguished by
Pitcher (1979) for batholiths. **Alpinotype** structures,
which originate in and near the trench, usually include
sutures between continents or continental fragments.
They have undergone considerable shortening. Igneous
activity in such an environment is minimal. The most
characteristic igneous rocks are ophiolites (p. 18), whose
emplacement in that environment is purely tectonic
(ophiolites were originally formed at oceanic spreading
centers).

## Figure 16.21
(a) Miocene ignimbrites showing well-developed columns, Nye County, Nevada. (b) Glassy welded tuff, Rio Blanco, near Guadalajara, Mexico. Horizontal alignment of shards indicates marked compaction. (From Ross and Smith, 1961.)

a.

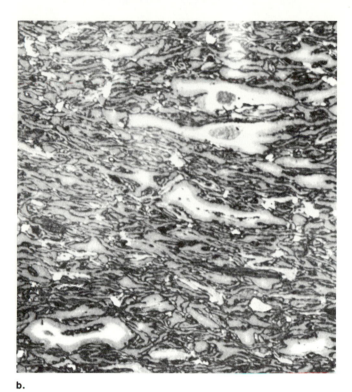

b.

## Figure 16.22
Schematic representation of batholith emplacement in relation to orogenic environment. (From Pitcher, 1979.)

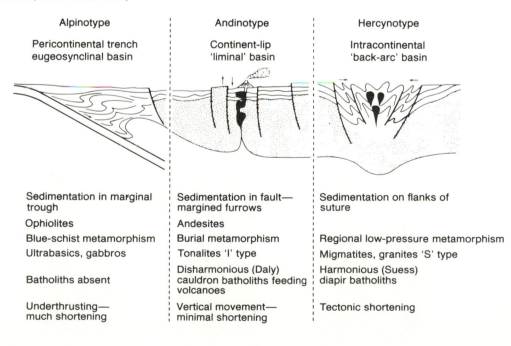

| Alpinotype | Andinotype | Hercynotype |
|---|---|---|
| Pericontinental trench eugeosynclinal basin | Continent-lip 'liminal' basin | Intracontinental 'back-arc' basin |
| Sedimentation in marginal trough | Sedimentation in fault—margined furrows | Sedimentation on flanks of suture |
| Ophiolites | Andesites | |
| Blue-schist metamorphism | Burial metamorphism | Regional low-pressure metamorphism |
| Ultrabasics, gabbros | Tonalites 'I' type | Migmatites, granites 'S' type |
| Batholiths absent | Disharmonious (Daly) cauldron batholiths feeding volcanoes | Harmonious (Suess) diapir batholiths |
| Underthrusting—much shortening | Vertical movement—minimal shortening | Tectonic shortening |

**Figure 16.23**

Cross section through the Sierra Nevada batholith. Upper cross section through Donner Pass; lower four cross sections through Yosemite National Park. 1-Calaveras formation; 2-amphibolite; 3-metamorphic rocks of eastern flank and septum; 4-Mariposa formation; 4-metadiorite; 6-granodiorite; 7-fine-grained granite; 8-volcanics overlying pluton at Castle Peak; 9-plutonic rock and flow structure; 10-inclusions; 11-schlieren; 12-fan fractures; and 13-pegmatites. (From E. Cloos, 1936.)

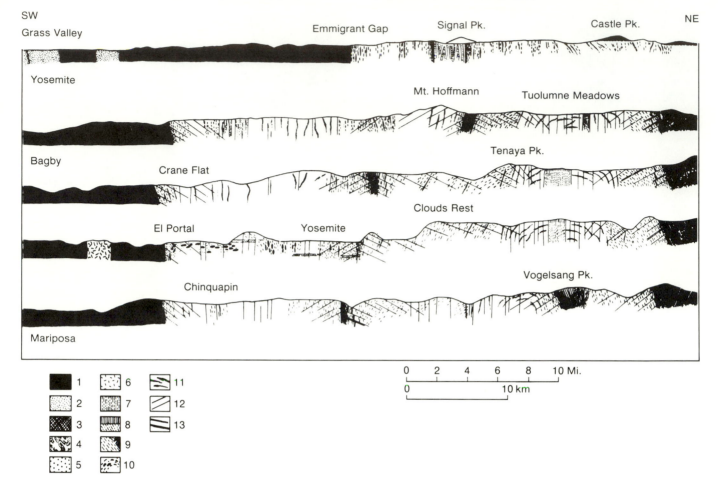

Igneous activity in the oceanic **island arc** environment is almost wholly volcanic. Volcanism resembles that of the andinotype environment but tends to be more mafic. Plutons are rare, but they do occur (as on Fiji).

**Andinotype** igneous activity occurs along marginal orogens. A chain of volcanoes—some active, some extinct and of batholiths, mainly Mesozoic in age—extends for almost the whole length of the circum-Pacific belt and along the Indonesian arc. Volcanism is mainly andesitic, but it is also dacitic and basaltic. The large batholiths of the Cordilleran system, such as the Sierra Nevada, appear to be composed of many closely adjacent domal intrusive bodies, as determined from internal structure (figure 16.23). Interestingly, similar patterns of dome-shaped plutons are characteristic of some Archean terranes (figure 16.24). Andinotype plutons are mainly tonalitic and granodioritic, but there is a whole range of magma compositions.

B. W. Chappell and A. J. R. White (1974) distinguished between two types of granitic rocks: I-type and S-type. Each has a characteristic range of compositions

## Figure 16.24
Structure of the Zimbabwe basement, interpreted as "gregarious batholiths." Although there is much interpretation in this map, the essential structural style is fairly well-documented. (From Macgregor, 1951.)

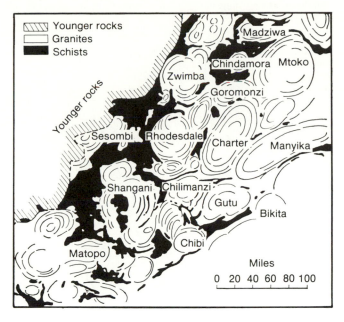

and of isotope ratios. I-type granites are believed to be mantle-derived, by purely igneous processes, whereas S-type granites appear to result from recycling of sedimentary rocks. The I-type is mainly characteristic of the andinotype environment, whereas the S-type occurs in both the andean and the hercynotype environments.

**Hercynotype** igneous activity in Pitcher's model is associated with the backarc environment (figure 16.22). Here, large volumes of S-type and a few I-type granites were emplaced in thick piles of regionally metamorphosed rocks while the region was under *compression*. The type example is the Hercynian orogenic belt of Europe. Continental backarc regions in *extension* tend to be settings for ignimbrite eruptions, as in the Basin and Range province of the western United States and in the North Island of New Zealand.

## The Fabric of Plutonic Rocks

Hans Cloos (1885–1951) initiated the study of the fabric of igneous rocks *(Granittektonik)* as an indicator of pre- and paraconsolidation movement. His work has been made accessible to English readers by Balk (1937). The fabric elements concerned fall into three classes: primary flow structures, potential parting planes, and primary fractures. We shall briefly consider the first two; the third was discussed in chapter 13.

### Primary Flow Structures

Both linear and planar penetrative fabric elements may reflect strain in plutonic rocks. Common linear elements include alignments of elongated crystals, for example feldspar phenocrysts and amphiboles. Even in the absence of visible indicators, nonequant grains may be oriented. This is seldom obvious, and must be checked

**Figure 16.25**

Flow-line markers in granite. (a) Lineation not in an *s*-plane. Plunge at about 20° SW. Trend can be measured on faces *GEFH* and *EDFB*; plunge can be measured on face *BAIH* only. (b) Flow lines and their relation to flow planes; drawn after granite blocks from Milford, Massachusetts. *TR* is the trend of the flow lines, which plunge at about 30° SW, as shown on face *BCD*. Flow planes strike about WNW, dipping 35° SSW. (From Balk, 1937.)

**Figure 16.26**

Schematic orthographic projection of a block of porphyritic quartz monzonite demonstrating the planar fabric as defined by the parallelism of the shaded faces (010) of microcline megacrysts. The fabric was determined by a study of the orientation of Carlsbad twin planes and of the cleavage planes. (From Sylvester, 1964.)

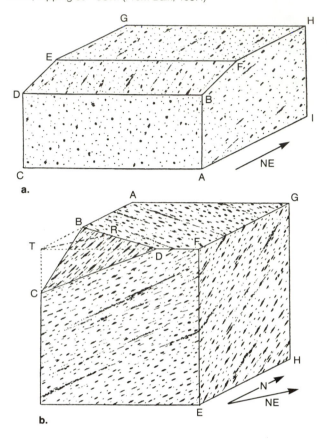

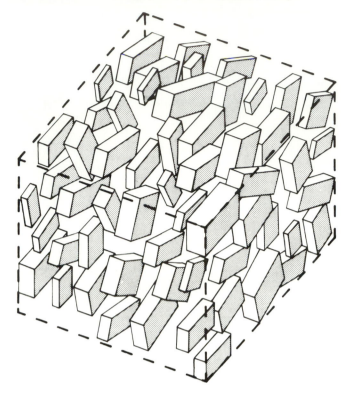

by means of many careful attitude readings. Lineations, in most instances, lie within planar fabric elements. Where they do not, they indicate constrictive flow (figure 16.25).

Petrofabric analysis may reveal lineations not evident to the unaided eye. Apparently, randomly oriented micas may yield a good girdle (see figure 9.3) and therefore a good lineation.

Planar elements include planar alignments (foliation) of platy minerals, such as micas and feldspar phenocrysts. Planar fabric, however, may not always be mesoscopically apparent as foliation surfaces (as in figure 16.26), for example, where planar fabric became apparent through microscopic study of crystallographic orientation.

Some lineations and foliations in igneous rocks are revealed only by numerous individual field measurements. The Drachenfels ("Dragon Rock"), in the Rhineland, Germany, is a well-known example. The foliation pattern (figure 16.27), established from detailed observations, serves as a guide to the dome's outer form and its inner flow pattern, and shows that it never reached the surface. Figure 16.28 shows a reconstruction of this dome by Hans and Ernst Cloos.

**Figure 16.27**
Fabric map of Drachenfels volcano showing foliation (flow plane)
attitudes. (From Cloos and Cloos, 1927.)

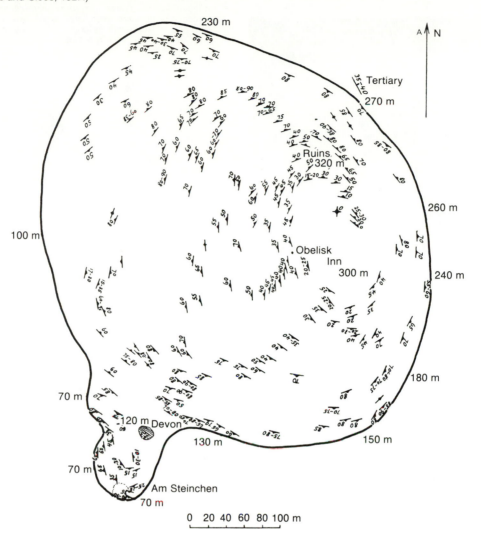

**Figure 16.28**
Cross section at bearing of 120° constructed from map shown in
figure 16.27, bringing out the internal structure.

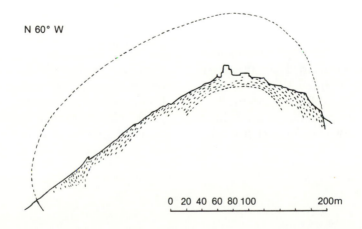

Figure 16.29
Strained schlieren, Southern California batholith, San Diego County, California.

Figure 16.30
Sketch map of the Ardara pluton, Ireland, showing strain pattern. Numbers refer to the dips of internal foliations, and ellipses represent the finite strain recorded by xenoliths. Filled ellipses represent deformation associated with intrusion of the Ardara pluton. Open ellipses represent deformation associated with intrusion of Main Donegal Granite. (From Holder, 1979.)

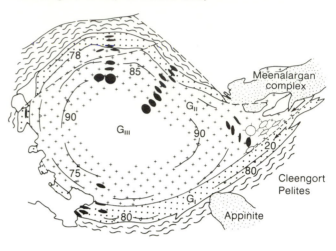

Many igneous bodies contain segregations of the minerals composing it; the segregations are called **schlieren.** Along and near the borders of diapiric plutons, these tend to be flattened in a foliation parallel to the contact, and to have their long dimension aligned along a lineation (figure 16.29). Foreign inclusions or xenoliths may be similarly aligned.

Schlieren and inclusions, being three-dimensional, can be used to evaluate strain. Holder (1979) mapped strain in the Ardara pluton in Ireland. From the strain pattern, he derived an emplacement history (figure 16.30), following a concept of J. G. Ramsay. In the fluid interior, strain is quasi-isotropic. Hence, there is little or no measurable strain. As new magma intrudes, older

material is pushed outward and, upon cooling, consolidates to a plastic crystalline state. In this state, schlieren register strain as illustrated in figure 16.30. The strain pattern reflects increasing inflation by repeated addition of magma.

## Potential Parting Planes

Bodies of granitic rock contain potential parting planes that are closely associated with the flow fabric. They are also used extensively in quarrying. The **rift** is along steep planes through the flow direction. It is usually the easiest direction for splitting. The **hardway** is steep and at right angles to the rift. As its name suggests, it is the hardest-splitting direction. The **"bedding"** (grain) lies flat and usually (but not necessarily) is parallel to lineation where this also is flat-lying, commonly near the top of a pluton. The names given to these parting planes vary locally.

## Conclusion

Clearly, tectonics and igneous activity are intimately linked. Igneous bodies, liquid or solid, move through a framework of surrounding rock, which itself participates in the movement and transmits it. Hutton rightly assigned igneous activity a dominant tectonic role.

## Review Questions

1. What forces are capable of raising basaltic magma through the crust to the earth's surface?
2. Distinguish between two types of calderas.
3. Describe the structure of ring complexes and explain their origin.
4. How may some igneous rock bodies indicate a crustal extension direction at the time of their emplacement?
5. What evidence in rocks indicates exceptionally high transient pressures such as would be associated with explosion structures?
6. What main processes generate granitic magma in the earth's crust?
7. In what tectonic settings may granite be emplaced in the earth's crust?
8. Distinguish between ash falls and ash flows.
9. Distinguish between different orogenic environments of igneous activity.
10. a) Describe fabric elements of plutonic rocks.
    b) What inferences does the fabric of plutonic rocks allow concerning their emplacement history?

## Additional Reading

Chapman, C. A. 1967. Magmatic central complexes and tectonic evolution of certain orogenic belts. Pages 41–51 in: J. P. Schaer, ed. *Etages Tectoniques*. Neuchâtel: La Baconnière.

Cobbing, E. J. 1982. The segmented coastal batholith of Peru: Its relationship to volcanicity and metallogenesis. *Earth Sci. Rev.* 18: 241–251.

Dennis, J. G. 1971. Ries structure, southern Germany, a review. *J. Geophys. Res.* 76: 5394–5406.

Gretener, P. E. 1969. On the mechanics of the intrusion of sills. *Can. J. Earth Sci.* 6: 1415–1419.

McLellan, E. 1984. Deformational behavior of migmatites and problems of structural analysis in migmatite terrains. *Geol. Mag.* 121: 339–345.

Marre, J. 1986. *The Structural Analysis of Granitic Rocks*. Amsterdam: Elsevier. 196 pp.

Marsh, B. D. 1982. On the mechanics of igneous diapirism, stoping, and zone melting. *Am. J. Sci.* 282: 808–855.

Mitra, Shankar. 1986. Duplex structures and imbricate thrust systems: geometry, structural position, and hydrocarbon potential. *AAPG Bull.* 70: 1087–1112.

Miyashiro, A. 1975. Volcanic rock series and tectonic setting. *Ann. Rev. Earth Planet. Sci.* 3: 251–269.

Pitcher, W. S. 1983. Granite type and tectonic environment. Pages 19–40 in: K. Hsu, *Mountain Building Processes*. London: Academic Press.

Soula, J. C. 1982. Characteristics and mode of emplacement of gneiss domes and plutonic domes in central-eastern Pyrenees. *J. Struct. Geol.* 4(3): 313–342.

Sylvester, A. G., Oertel, G. Nelson, C. A., and Christie, J. M. 1978. Papoose Flat pluton: A granitic blister in the Inyo Mountains, California. *Geol. Soc. Am. Bull.* 89: 1205–1219.

Talbot, C. J. 1971. Thermal convection below the solidus in a mantled gneiss dome, Fungwi Reserve, Rhodesia. *Geol. Soc. London J.* 127: 377–410.

Wyllie, P. J. 1981. Plate tectonics and magma genesis. *Geol. Rundsch.* 70: 128–153.

# Appendix A

# Laboratory Exercises

The methods presented here are designed both to solve practical problems in structural geology and to give the student facility in visualizing geological structures in three dimensions. No effort has been made to solve difficult and unusual space relations. Instead, the discussion is limited to basic problems that commonly confront the structural geologist in the field and in the preparation of reports and maps.

## Orthographic Projection

A geologic structure has a definite, fixed position and attitude in the earth's crust. Once these have been determined, the observer should never imagine the structure to be moved or turned. If he wishes to view it in another direction, he must look directly along that direction. Consistent use of this simple rule of "direct viewing" will help the viewer to avoid many pitfalls.

To illustrate certain essential definitions, imagine a sedimentary formation to be dipping 30° due south. From this dipping formation, a segment is cut out and placed inside a hollow transparent block, in its original attitude. Figure A.1 is a sketch of the result, which illustrates the following definitions.

*Orthographic projection* uses parallel lines of sight at right angles to an image plane. It is a "right angle" type of projection.

The *image plane* is the plane on which a view is projected. It is always between the observer and the geologic structure, and it is always perpendicular to the lines of sight for that view. The sides of the large block are image planes.

A *line of sight* is a straight line from the observer to the object. All lines of sight are parallel for each view. The arrow-tipped lines are lines of sight, and an infinite number could be drawn.

The *plan view* is the basic view for structural geologists. It corresponds to the earth's surface and to the familiar areal geologic map. Compass directions and bearings are shown in the plan view.

*Elevation views* are the familiar geologic cross sections in which the image plane is vertical and the lines of sight are horizontal. The south and east sections in figure A.1 are elevation views.

A *folding line* is the intersection of two image planes. The edges of the large block are folding lines.

*Projection lines* are straight lines at right angles to the folding line; by projection, they connect identical points in two adjacent views. Two projection lines are shown by dotted lines in figure A.1.

**Figure A.1**
Pictorial sketch of a segment from a bed dipping 30° due south.

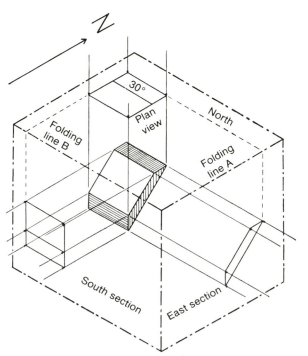

## Purpose of Folding Lines

In graphic solutions all views and measurements are made in one plane, which is the plane of the drawing paper. Therefore, all the image planes (sides of the large block in figure A.1) must be brought into a common plane, the plane of the paper on which the drawing is made. This is done by rotating the image planes around the folding lines.

Since the plan view is the basic view for structural geologists, it may be considered fixed. All other image planes are then rotated about their folding lines into the plane of the plan view, which becomes the plane of the drawing paper. This has been done in figure A.2, and comparison with figure A.1 will show how it has been accomplished.

Note the following:

1. There is only one plan view.
2. Bearings and compass directions are shown in the plan view.
3. Any number of elevation views or cross sections may be made from the plan view.
4. Any number of inclined views may be made from the elevation views.
5. *The true dip of 30° is seen only in that elevation or cross section that is at right angles to the strike or bearing of the formation*—in this case, east or west sections.

## Notation

To avoid confusion, especially in the solution of problems that require several views, follow a simple system of numbering the views, folding lines, and points.

Since the plan view is basic, it is given the number 1. All other views are given numbers according to the order in which they are drawn. The numbers of the views are placed on each side of the folding line, which is marked *FL*. As shown in figure A.2, the folding line lying between the plan view and the northwest section carries the numbers 1 and 4. The folding line lying between the south section and the inclined view carries the numbers 3 and 5. This indicates that the south section was drawn third, and the inclined section was drawn fifth. The additional notation *A* and *B* on two of the folding lines is merely for convenience in comparing figures A.1 and A.2.

Points in each view may be lettered, and corresponding points in all views retain the same letters. Each letter has a subscript, the subscript number being the same as the view number.

At the outset, number and letter all views, folding lines, and points. Later, with increased experience, much of this notation may safely be eliminated. Use additional notation sufficient to make the drawing clear and explicit.

**Figure A.2**
Orthographic projection of a segment from a bed dipping 30° due south.

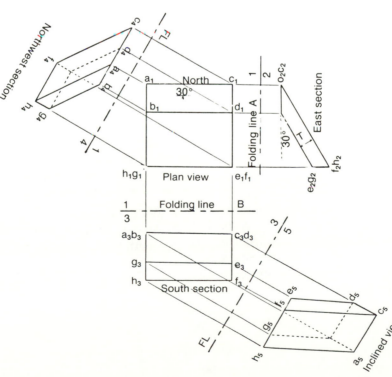

## Measurements from Folding Lines

Views are *related* to each other if (a) they have a common folding line; and (b) two views of any point lie on the same projection line, which is at right angles to the folding line common to the two views.

In figure A.2, views 1 and 2, 1 and 4, 1 and 3, and 3 and 5 are related. Notice also that views 2, 3, and 4, although not directly related to each other, are related to a common view, view 1. In the same way, views 5 and 1 are related to a common view, view 3.

The correct measurements to take in order to obtain new views are controlled by perpendicular distances from folding lines. Stated simply, the rule is: *In all views that are related to a common view, the object, or any point on the object, is the same distance away from the folding line.*

For example, all cross sections or elevation views are related to a common view, the plan view. Hence, any point on an object will appear the same distance below the folding line in all elevation views. Thus $e_2$, $e_3$, and $e_4$ are the same distance below the folding line. Moreover, the distance $e_2f_2$, $e_3f_3$, and $e_4f_4$ is identical in all three elevation views, as it should be, since it represents the apparent thickness of the formation that would be penetrated by a vertical drill hole. Incidentally, this should not be confused with the true thickness of the formation, which is shown by the letter $T$ and is measured perpendicular to the bedding.

View 5 is not related to view 1, and point $c_5$ is much farther below the folding line than is $c_2$, $c_3$, or $c_4$, which are related to view 1. View 5 and view 1 are related through the common view, view 3. Therefore, $c_5$ should be the same distance from the folding line as $c_1$; and the distances to all points from the folding line in view 5 should check the distances from the folding line in view 1 to all corresponding points.

If a new inclined view 6 is constructed from view 5, then all measurements below the folding line in view 6 would be taken from view 3 to the corresponding points. If this is difficult to visualize, perhaps it may help if the common view is temporarily considered to be a plan view. For instance, view 3, which is common to views 1 and 5, may temporarily be called a plan view; then views 1 and 5 would be temporary elevations or cross sections; and any object would appear the same distance below the folding line in these two views. Also, in the same way, view 5 may be considered a temporary plan view, and a new view 6 and 3 will be temporary elevation views or cross sections, with all corresponding points the same distance below the folding line.

One reason for lettering and numbering points is to aid in making correct measurements from folding lines. Also, careful notation is very helpful in visualizing the object. For example, the top of the bed is apparent in the plan view as $a_1b_1c_1d_1$ and in the cross sections as $a_2b_2c_2d_2$, $a_3b_3c_3d_3$, and $a_4b_4c_4d_4$, even without the notations. But in view 5, $a_5b_5c_5d_5$ might seem to be the bottom of the bed if it were not lettered. This seeming contradiction will disappear if view 3 is folded into its former vertical position, and view 5 is then turned an additional 90° around its folding line. The position of $a_5b_5c_5d_5$ would now be directly below $a_1b_1c_1d_1$ of the plan view. This emphasizes again the identical character of the measurements in views 1 and 5 and their common relation to view 3.

## Accuracy of Graphic Solutions

The accuracy with which problems in structural geology are solved graphically is largely dependent on the scale of the drawing and on the drafting precision. All construction lines should be very fine and accurate, with the object lines somewhat more pronounced. All angles, as well as measurements from folding lines, must be laid off precisely.

In the illustrations, the folding lines are shown as heavy, dashed lines so as to better distinguish the views. Actual scale drawings will be more accurate if the folding lines are shown as fine, solid lines.

The scale of the drawing must be large enough to be read to the accuracy desired. For instance, a scale of 1:50,000 can be used to within an accuracy of about 25 m; a scale of 1:1,000 would permit graphic solutions to within about 25 cm of the correct mathematical answer. As a test of precision, solve problem 1 graphically and check your solution against the mathematical answers given in problem 2.

## Practice Problems

### Problem 1

A red shale with a thickness of 500 feet strikes N 40 W and dips 30° to the south. A well is drilled 1 mile due south of the upper contact of this shale. Assume the ground to be level. Using a scale of 1,000 feet to 1 inch, determine the following:

1. Width of outcrop
2. Amount of shale penetrated by the drill
3. Depth to top of shale in the drill hole

A small-scale orthographic projection is shown in figure A.3.

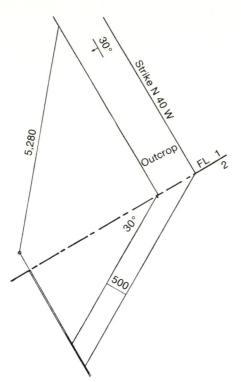

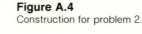

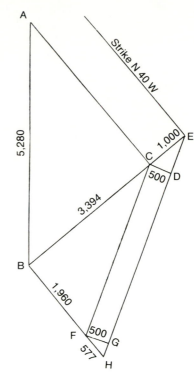

## Problem 2

Using the data from problem 1, derive precise answers by plane trigonometry. Those whose trigonometry is somewhat hazy may use the setup of figure A.4. The simple equations are as follows:

$BC = \sin 40° \times 5,280 = 3,394$ feet

$BF = \tan 30° \times 3,394 = 1,960$ feet depth to shale in drill hole

$FH = 500/\sin 60° = 577$ feet of shale penetrated by drill

$CE = 500/\sin 30° = 1,000$ feet width of outcrop

## Problem 3

Figure A.5 shows the plan view and a front-elevation view of a bisphenoid. Complete two additional elevation views and two inclined sections in the positions shown by the folding lines. Number all points, and show all construction lines. Note that lines connecting $a_5b_5$, $c_5b_5$, $d_5b_5$, $c_6b_6$, and $c_3d_3$ should be dashed, because they represent concealed edges of the bisphenoid.

**Figure A.5**
Plan view (top) and front elevation (bottom) of a bisphenoid.

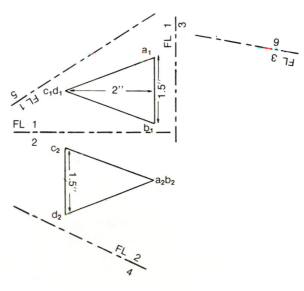

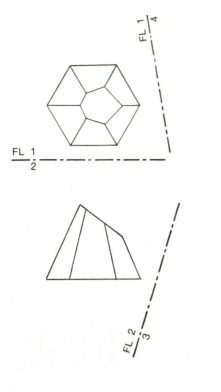

## Problem 4

Figure A.6 shows the plan view and a front-elevation view of a truncated hexagonal pyramid. Complete an additional elevation view and one inclined view in the positions shown by the folding lines. All concealed edges of the truncated hexagonal pyramid should be indicated by dashed lines.

## True and Apparent Dip Problems

The *apparent dip* (chapter 3) and its relation to the true dip are important in many practical problems of structural geology. True and apparent dip problems may be solved by graphical methods; trigonometric formulas; and tables and various forms of alignment diagrams prepared from trigonometric formulas. The following true and apparent dip problems will be solved by each of these methods. A few additional problems are given at the end of this section for practice.

## Problem 5

A bed dips 30° to the S 60 W. What is the apparent dip of the bed in a cross section along a north-south line?

## Problem 6

In a rock cut along a bend of a railroad track, two apparent dips are 20° to the S 40 W and 35° to the S 10 W. Find the direction and amount of true dip.

## Problem 7

A thin, easily distinguished key bed dips 1.5 feet in 320 feet to the S 15 E, and 0.8 feet in 200 feet to the S 80 E. Find the direction of true dip and also the amount of true dip, in feet per mile.

### Orthographic Solution

In an orthographic solution, remember the following:

1. Bearings are shown only in the plan view.
2. The true dip is in a section at right angles to the strike.
3. The strike is a level line on a geological surface connecting points of the same elevation.

### *Examples*

**To Determine Apparent Dips from a Known True Dip (problem 5 above).** In figure A.7, the bearings of the true dip and strike are shown in view 1, together with the direction of the desired cross section. The amount of true dip is shown in view 2, a section taken in the direction of the dip. Depth $x$ to the bed in view 2 is any unit depth to the bed, taken along the projection of the strike. This same depth $x$ is laid off along the projection of the strike in view 3, which is proper, since the strike line connects points of the same elevation on the bed. By so doing, the apparent dip of 16° is obtained in the required north-south section. Any number of apparent dips in other sections may be determined in the same way.

**To Determine the True Dip from Two Apparent Dips (problem 6 above).** Bearings of the two apparent dips, from an initial point $a_1$, are shown in view 1 of figure A.8. The amounts of the apparent dips are shown in views 2 and 3, the sections being taken in the directions of these apparent dips. Depth $x$ in views 2 and 3 is identical and is any unit distance. Projecting the position of depth $x$ into the plan view gives two points of the same elevation on the bed and determines the bearing of the strike line.

View 4 is a section at right angles to the strike. The common depth $x$ is laid off in view 4, along the projection of the strike line, and this determines the true dip of 40°. The bearing of the true dip, S 24 E, is found in the plan view by drawing a line to the initial point $a_1$, that is perpendicular to the strike.

**Figure A.7**
Apparent dip from a known true dip, by orthographic projection.

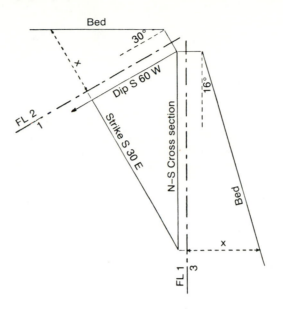

**Figure A.8**
Strike and dip from two angular components, by orthographic projection.

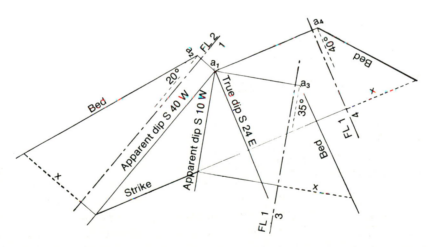

**Figure A.9**
Strike and dip from two angular components, by orthographic
projection, compressed construction.

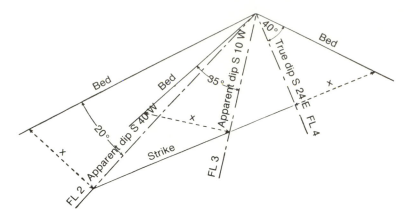

To conserve space, the folding lines may be drawn directly on the bearing lines of the dips. Also the sections may be rotated directly into the plan view, instead of being projected before rotation, so as to clear the plan view. This has been done in figure A.9. The disadvantage is that construction lines are more readily confused. Of course, the sections may be rotated on either side of their folding line.

**To Determine the True Dip from Two Apparent Dips of Small Magnitude (problem 7 above).** The orthographic-projection method is useful in solving true and apparent dip problems in the field, because only a scale and a protractor are needed. The method can be adapted for dips of small magnitude.

Since the answer to problem 7 is required in feet per mile, it will be more convenient to change the data to that form before starting the solution. Thus, the two apparent dips become 24.75 feet per mile S 15 E and 21.1 feet per mile S 80 E. Since 1° is more than 90 feet per mile, the apparent dips are obviously too small to plot as angles with an ordinary protractor.

Remember that dips represent rates of slope—a fixed ratio of horizontal distance to vertical distance for any particular angle of dip—and thus do not necessarily have to be measured by angles. The vertical distances may be exaggerated, without invalidating the solution, provided that all vertical measurements are exaggerated by the same amount. Of course, this must be taken into account in the final answer, and no angle measurements of dip slopes should be made. In the following solution, the vertical measurements have all been exaggerated 100 times.

In figure A.10, the directions of the two apparent dips are shown along folding lines 1 and 2. A horizontal distance *ab* of 1 mile is laid off on folding lines 1 and 2, using any scale desired. *In all graphical solutions, once the scale is decided upon, it should not be changed.*

Along folding lines 1 and 2, apparent dip sections exaggerated 100 times are constructed at the mile point *b*. The identical unit depth *x* to the bed is then laid off in the two apparent dip sections; it determines the position and bearing of the strike line in the plan view. The direction of the true dip S 41 E is found in the plan view by drawing a perpendicular from the strike to the initial point *a*.

This dip direction is then used as a folding line, namely folding line 3; from it, along the line of strike, is laid off the unit depth *x*. To save space, this section along folding line 3 is directly rotated into the plan view. The depth to the bed, at the 1-mile point *b* in this true dip section, is 27.4 feet × 100 on the exaggerated scale. The actual true dip is 27.4 feet per mile. Repeat this problem, using meters and kilometers instead of feet and miles.

## Vectors

Simple and accurate graphical solutions of true and apparent dip problems may be obtained by using vectors. In this method, the length of a plotted vector represents the rate of dip (gradient), and the bearing of the vector is the dip direction. Note carefully that the *rate* of dip, shown by the length of the vector, is not proportional to the *angle* of dip. For instance, the rate of dip for 45° is unity, and the rate of dip for 90° is infinity, not twice unity, as the increase of angle would seem to indicate. Therefore, angles should not be plotted graphically as vectors, because they do not obey the laws of vector addition and resolution.

In this application of vector analysis, the scalar field is the horizontal plane or plan view; hence, the bearings of the vectors are true. The gradient is the rate of change of elevation of a point on an inclined surface or line, with horizontal distance. This gradient may be expressed as the tangent of the dip, if the measurement of the dip is in degrees. If the dip is expressed in feet per mile or some similar gradient, one should solve by vectors directly without any conversion.

**Figure A.10**
Strike and dip from two angular components of small magnitude, by orthographic projection.

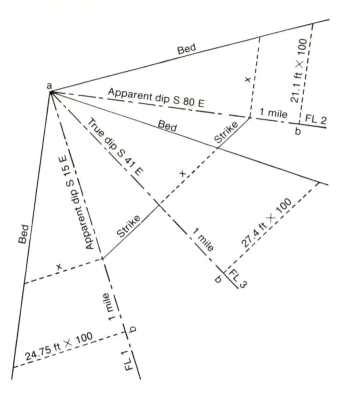

**Figure A.11**
Relation of true dip to apparent dips, by tangent vectors.

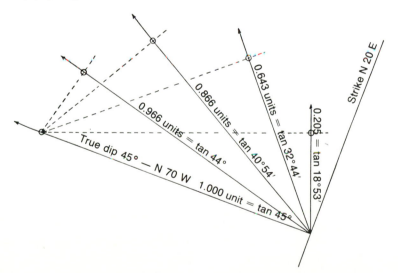

## Tangent Vector Method[1]

Figure A.11 illustrates how rates of apparent dip, expressed by tangents of the angle of dip, are related graphically by vectors to a true dip. The true dip of 45° to the N 70 W is plotted as a vector with a value of one unit, which is the tangent of 45°. It is evident that the

vectors of the apparent dips in the directions N 55 W, N 40 W, N 20 W, and due north have gradient values determined by their perpendiculars.

In the following constructions, note that *the perpendiculars are always drawn to the apparent dip vectors and that the true dip direction always has the longest vector.* We shall solve problems 5 to 7 as examples to demonstrate the usefulness of the tangent vector method.

---

[1]Cotangent vectors may also be used.

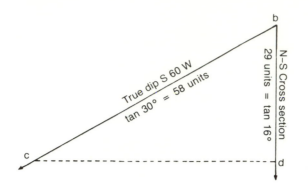

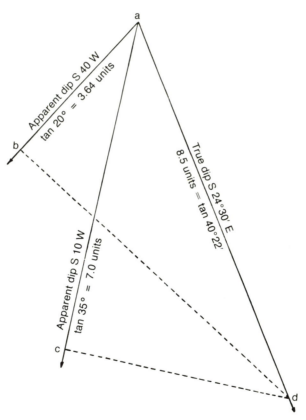

*Examples*

**Apparent Dips from a Known True Dip (problem 5 above).** In figure A.12, the tangent of the true dip is plotted as a vector *bc,* a line is drawn from point *c* perpendicular to the north-south cross section, and the value of the apparent dip vector *db* is scaled.

**True Dip from Two Angular Components (problem 6 above).** In figure A.13, the tangents of the two apparent dips are plotted as vectors *ab* and *ac*. The intersection of perpendiculars from points *b* and *c* determines the bearing and value of the true dip vector *ad.*

**Figure A.14**
True dip from two angular components of small magnitude, by
tangent vectors.

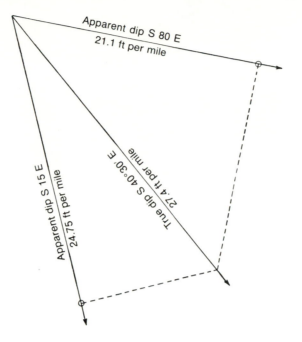

**Figure A.15**
Tangent conversion on the oblong protractor.

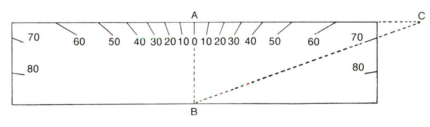

**True Dip from Two Angular Components of Small
Magnitude (problem 7 above).** Since the answer is required in feet per mile, it will be convenient to change
the dip data into that form, namely, 24.75 feet per mile
S 15 E and 21.1 feet per mile S 80 E. The rate of dip
now being in the form of the tangent, the data are
plotted as feet per mile. In figure A.14, the true dip is
scaled directly as 27.4 feet per mile to the S 40° 30'E.

*Use of Oblong Protractor*
If the width of an oblong protractor, *AB* in figure A.15,
is taken as a unit, tangents may be scaled directly by
applying this unit distance, from point *A*, along the top
edge of the protractor. For example, 30° would scale
0.58 part of the unit distance (tangent of 30° is 0.58);
70° would scale 2.75 units, or the distance *AC*. By using
the ordinary 6-inch oblong protractor, the accuracy in
converting angles to tangents is usually adequate for
most true and apparent dip problems.

**Figure A.16**
Average true dip by tangent vectors. Readings to the N 30 E and to
due north should be discarded. The average true dip is N 20° 30' E.

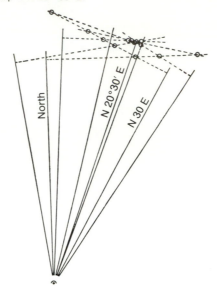

**Figure A.17**
True and apparent dips by tangent vectors.

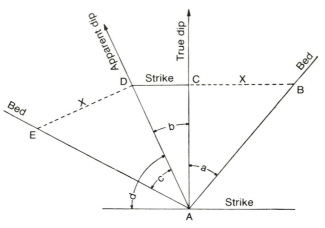

## *Use of Tangent Vectors to Determine an Average True Dip*

In areas of even but poorly defined bedding, the true dip, if based on only one or two readings, is seldom dependable. Increasing the number of readings and taking an average is often not helpful because some of the dips may be apparent rather than true dips. Also, some of the data may be erroneous and therefore should be discarded and not averaged.

If all the dips are plotted as tangent vectors, the perpendiculars will give intersections lying within a relatively small area for all trustworthy readings, and the average true dip may be determined. Erroneous readings can be recognized because their perpendiculars will give intersections that are inconsistent with the other data. This is illustrated in figure A.16.

## Formulas, Tables, and Diagrams

Even before graphical methods became established, trigonometric formulas were devised for the solution of true and apparent dip problems. These formulas are the basis of alignment diagrams and networks. They may be derived quite simply.

In figure A.17, the directions and angles of true and apparent dip are shown. Then:

$$BC = AC \tan a$$

$$AD = \frac{AC}{\cos b}$$

$$DE = BC = AC \tan a$$

$$\tan c = \frac{DE}{AD} = \frac{AC \tan a}{AC \cos b}$$

$$\tan c = \tan a \cos b$$

or

$$\tan c = \tan a \sin d \quad (\text{since } \cos b = \sin d)$$

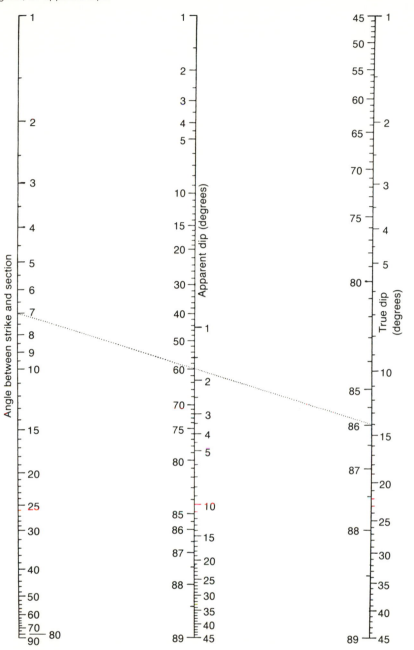

where *a* is the angle of true dip; *b* is the horizontal (compass) angle between the dip and apparent dip directions; *c* is the angle of apparent dip; and *d* is the horizontal (compass) angle between the apparent dip and the strike directions.

Several true and apparent dip tables, based on these formulas, have been published. Probably the earliest was by Jukes (1859). Somewhat later, Oldham (1884) offered a network type of diagram. Since then many varieties of networks and diagrams have been used. Perhaps the most practical is an alignment type of diagram or *nomogram,* such as is commonly used today in engineering and physics. Its construction is easy.

Figure A.18 is a nomogram for true and apparent dips, depending on the horizontal (compass) angle between the section and the strike. The true dips are laid off as tangents of their angles on logarithmic paper with a 10-inch base. To save space, the scale is doubled back on itself at the 45° mark. The apparent dips are laid off on logarithmic paper with a 5-inch base, and the scale, from 1° to 45°, is repeated so as to take care of the doubled-back feature of the true dip. The horizontal (compass) angle between the strike and the section is plotted as the sine, on logarithmic paper with a 10-inch base. The scales are placed at equal, convenient distances apart. Of course, the nomogram may be

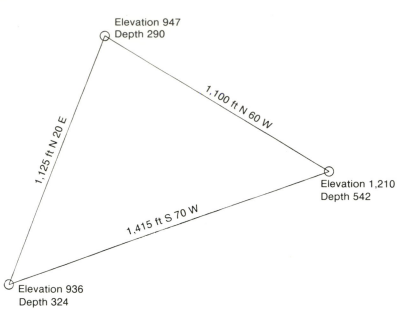

Elevation 947
Depth 290

1,100 ft N 60 W

1,125 ft N 20 E

Elevation 1,210
Depth 542

1,415 ft S 70 W

Elevation 936
Depth 324

set up just as readily for the angle between the section and the direction of dip, if cosines instead of sines are used.

Nomogram solutions are obtained by applying a straightedge from the known angle of true dip to the horizontal (compass) angle between the section desired and the strike. The angle of apparent dip is read from the intersection of the straightedge. For example, in figure A.18, with a true dip of 86° and a horizontal (compass) angle between the strike and section of 7°, the apparent dip along the desired section will be 60°; or if the true dip is 14°, the apparent dip will be 1° 45'.

Frequently, in the construction of an accurate cross section, a large number of true dips must be converted to apparent dips. In that event, the nomogram is very useful because it is speedy; it also reduces the chance of errors, as setting a straightedge and reading an intercept are comparatively simple operations, and speedier than solving by pocket calculator.

## Practice Problems

### Problem 8
A bed of coal dips 30° to the N 45 W. At what distance due north, on level ground, will the top of the bed be 300 feet deep?

### Problem 9
A series of formations are followed along their strike of N 45 E. In a north-south railroad cut, the apparent dip is 30° to the north. What is the true dip? Solve by tangent vectors and by orthographic projection.

### Problem 10
In an area of gently dipping beds and no relief, two apparent dips were determined on thin sandstone beds— 120 feet per mile due east and 60 feet in 3,000 feet to the S 10 E. Find the direction of true dip and the amount of true dip in feet per mile. Determine by orthographic projection and by tangent vectors.

### Problem 11
The drill holes, shown in figure A.19, found the top of a key bed or "marker" horizon at the depths shown. What is the true dip of the key bed? (This familiar type of three-point problem may be solved by using two apparent dips.)

### Problem 12
In an area of poorly defined but even bedding, the true dip could not be determined accurately from one or two readings. Therefore, several apparent dips were taken, specifically 21° 30' N 52 W, 25° N 36 W, 13° N 60 E, 29° N 23 W, 23° N 45 E, 28° N 17 E, and 24° 30' N 31 E. Determine the average true dip from only the trustworthy data, and point out which readings should be discarded.

### Problem 13
A geologic map, modified from the Tyrone Quadrangle, Pennsylvania, is shown in figure A.20. The area is approximately 7 miles north-south by 4 miles east-west. Using the nomogram in figure A.18, construct cross sections along lines A-A' and B-B'.

**Figure A.20**
Modified geologic map from the Tyrone Quadrangle, Pennsylvania. This is a base map for problem 13. (Copyright 1949 by Charles Merrick Nevin.)

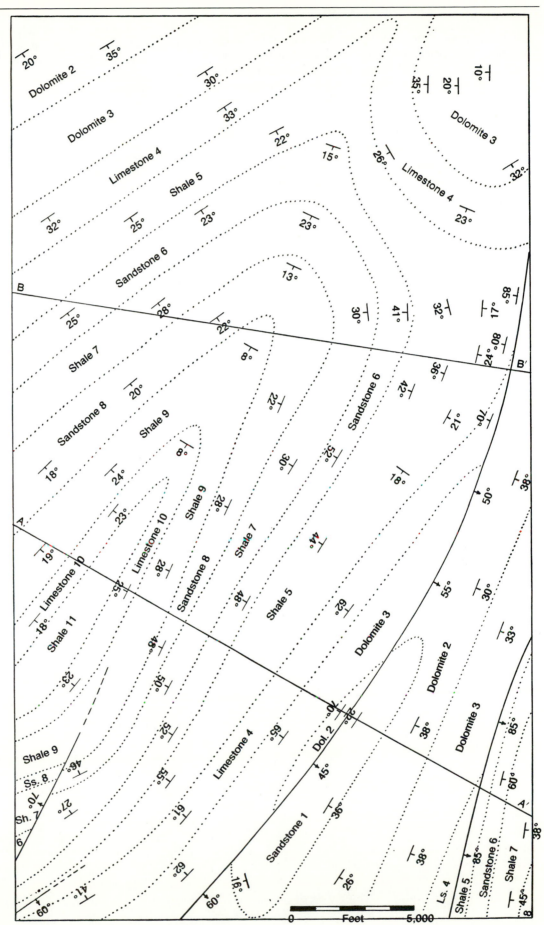

## Thickness and Outcrop Width

Since thickness is measured perpendicular to bedding, for *horizontal* bedding the difference in elevation between the top and the bottom will give the true thickness. If the bedding is *vertical,* a horizontal distance at the surface, perpendicular to the strike, will give true thickness.

By measuring the width of outcrop of a *dipping* bed (horizontal distance between the top and the bottom), in a direction perpendicular to the strike, its thickness may be calculated from the following relation:

Thickness = Width of outcrop × sine of angle of dip

Note that the ground should be approximately level, so that the width of outcrop is a true horizontal distance; that the measurement must be perpendicular to the strike; and that the bed must remain constant in attitude.

If the above conditions are satisfied, use a nomogram such as that in figure A.21. For example, with a dip of 26° and widths of outcrops of 3.2 feet, 32 feet, 320 feet, or 3,200 feet, the corresponding thicknesses would be 1.4 feet, 14 feet, and 140 feet. With a dip of 10° and widths of outcrops of 2.3 feet, 23 feet, 230 feet, and 2,300 feet, the corresponding thicknesses would be 0.4 foot, 4 feet, 40 feet, and 400 feet. With a dip of 1° and the same widths of outcrops, the corresponding thicknesses would be 0.04 foot, 0.4 foot, 4 feet, and 40 feet.

The thickness nomogram is also useful in estimating dips and thicknesses on geological maps, provided that the influence of topography on the width of outcrop is not large—that is, if the dips are fairly steep or if the topographic relief is relatively small.

The surface of the ground covered by the outcrop of a thick formation is seldom level. Sometimes the ground slopes with the dip of the formation, sometimes in an opposite direction or in both directions. This affects the width of outcrop and, unless allowed for, will give an erroneous thickness. Frequently a traverse cannot be made perpendicular to the strike of the formation.

In addition, the attitude of beds is seldom constant over a wide area. Each change in attitude may affect the width of outcrop. Moreover, the assumption of an average dip and strike for the entire width of outcrop may be inaccurate, as there is no simple way to evaluate the effect of considerable variations in dip and strike.

One of the most practical thickness formulas is the general equation by Secrist, which can be used in a programmable calculator:

$$t = \pm d \sin \alpha \sin \beta \pm h \cos \alpha \qquad \text{(A.1)}$$

where

$t$ = thickness in feet

$d$ = horizontal distance in feet

$\alpha$ = dip of the bed in degrees

$\beta$ = angle between traverse and strike of the bed

$h$ = difference in elevation in feet

$d$ is positive if traverse is in direction of dip

$h$ is positive if the traverse is uphill

The disadvantage of using an equation is that errors may creep into the computation. A graphic solution is usually quicker, and any mistakes come to light more readily. However, the graphic solution is valid only if the dip and strike are constant or if changes in the dip and strike can be evaluated with a fair degree of certainty. The following problems illustrate these points.

**Figure A.21**
Nomogram relating angle of dip to width of outcrop and thickness units.

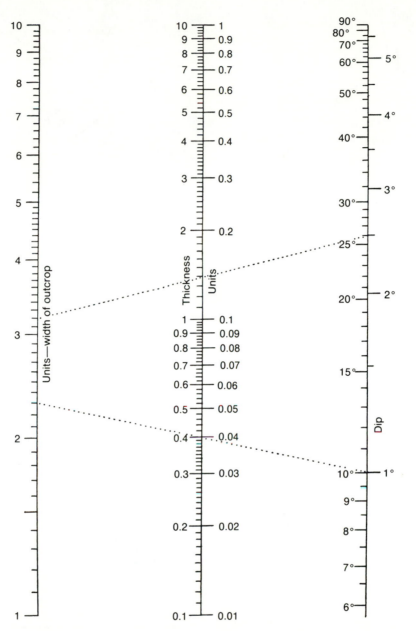

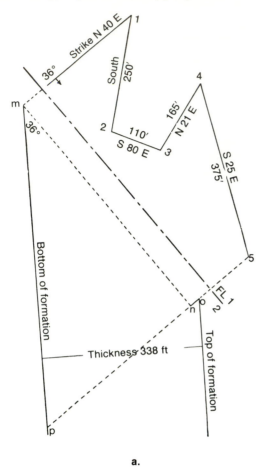

a.

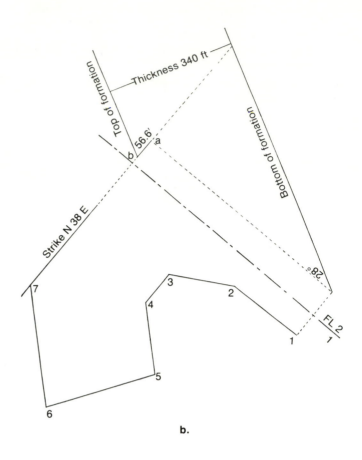

b.

## Problem 14

A formation outcrops along a ravine and dips 36° to the S 50° E (figure A.22). The following data were secured by a traverse across the outcrop, between well-defined lower and upper contacts. Determine the thickness graphically, and check by means of equation A.1.

| Station | Distance (in feet) | Traverse Bearing | Strike | Dip | Elevation Difference |
|---------|---------|----------|--------|-----|-----------|
| 1–2 | 250 | South | N 40 E | 36° SE | −27.4 |
| 2–3 | 110 | S 80 E | N 40 E | 36° SE | −18.1 |
| 3–4 | 165 | N 21 E | N 40 E | 36° SE | +25.4 |
| 4–5 | 375 | S 25 E | N 40 E | 36° SE | +41.3 |

Recasting into terms of equation A.1, we get

| Station | d | $\alpha$ | $\beta$ | h | t |
|---------|-----|-----|-----|------|------|
| 1–2 | +250 | 36° | 40° | −27.4 | +72 |
| 2–3 | +110 | 36° | 60° | −18.1 | +41 |
| 3–4 | −165 | 36° | 19° | +25.4 | −11 |
| 4–5 | +375 | 36° | 65° | +41.3 | +233 |

Thickness = 346 − 11 = 335 feet.

Since the dip and strike are constant, problem 14 may be solved graphically, as shown in figure A.22. The vertical distance *on* represents the difference in elevation of 21.2 feet, between the ends of the traverse. Note that changes in topography and in the direction of the traverse are easily accommodated in a graphic solution. The calculated thickness of 335 feet compares well with the graphic measurement of 338 feet.

The next problem illustrates a common situation: the dip and strike change somewhat, but not enough to affect the estimate of an average dip and strike for the entire formation.

## Problem 15

The following data were secured in a traverse to determine the thickness of a formation:

| Station | Distance (in feet) | Traverse Bearing | Strike | Dip | Elevation Difference |
|---|---|---|---|---|---|
| 1–2 | 205 | N 54 W | N 48 E | 25° NW | +18.2 |
| 2–3 | 171 | N 82 W | N 42 E | 26° NW | +11.3 |
| 3–4 | 97 | S 37 W | N 38 E | 27° NW | +6.8 |
| 4–5 | 187 | S 9 E | N 33 E | 29° NW | −24.6 |
| 5–6 | 298 | S 72 W | N 25 E | 30° NW | +23.5 |
| 6–7 | 320 | N 9 W | N 30 E | 29° NW | +21.4 |

Recasting into terms of equation A.1, we get

| Station | d | $\alpha$ | $\beta$ | h | t |
|---|---|---|---|---|---|
| 1–2 | +205 | 25° | 78° | +18.2 | +101 |
| 2–3 | +171 | 26° | 56° | +11.3 | +72 |
| 3–4 | −97 | 27° | 1° | +6.8 | +5 |
| 4–5 | −187 | 29° | 42° | −24.6 | −81 |
| 5–6 | +298 | 30° | 47° | +23.5 | +129 |
| 6–7 | +320 | 29° | 39° | +21.4 | +116 |

Thickness = 423 − 81 = 342 feet.

In order to solve problem 15 graphically, we must estimate an average dip and strike. A value of 28° to the N 52 W seems reasonable. Using this assumed average dip, the problem is solved graphically in figure A.22b. The vertical distance *ab* of 56.6 feet is the difference in elevation between the ends of the traverse. The measured thickness of 340 feet is close to the calculated thickness of 342 feet.

Unfortunately, it is not always possible to recognize data that cannot be averaged successfully. For example, it would seem that the thickness in the following problem could be determined accurately enough by a graphic solution, because the dips and strikes do not vary greatly.

## Problem 16

Along a winding road, on the side of a hill, the following data were taken across the entire outcrop of a formation. Determine the thickness.

| Station | Distance (in feet) | Traverse Bearing | Strike | Dip | Elevation Difference |
|---|---|---|---|---|---|
| 1–2 | 79 | S 34 E | N 67 E | 57° SE | +5.4 |
| 2–3 | 84 | S 34 E | N 67 E | 57° SE | +5.4 |
| 3–4 | 71 | S 45 E | N 67 E | 57° SE | +5.4 |
| 4–5 | 66 | N 70 E | N 67 E | 57° SE | −3.0 |
| 5–6 | 119 | N 5 W | N 67 E | 51° SE | +5.4 |
| 6–7 | 93 | North | N 67 E | 51° SE | +5.4 |
| 7–8 | 79 | N 12 E | N 67 E | 51° SE | +5.4 |
| 8–9 | 84 | N 25 E | N 67 E | 51° SE | +5.4 |
| 9–10 | 53 | N 51 E | N 67 E | 51° SE | −2.5 |
| 10–11 | 45 | S 70 E | N 60 E | 60° SE | +4.0 |
| 11–12 | 73 | S 30 E | N 60 E | 60° SE | −5.4 |
| 12–13 | 82 | S 3 E | N 70 E | 55° SE | −5.4 |
| 13–14 | 103 | S 3 E | N 70 E | 55° SE | +5.4 |

Recasting this into terms of equation A.1, we get

| Station | d | $\alpha$ | $\beta$ | h | t |
|---|---|---|---|---|---|
| 1–2 | +79 | 57° | 79° | +5.4 | +68 |
| 2–3 | +84 | 57° | 79° | +5.4 | +73 |
| 3–4 | +71 | 57° | 68° | +5.4 | +58 |
| 4–5 | +66 | 57° | 3° | −3.0 | +1 |
| 5–6 | −119 | 51° | 72° | +5.4 | −85 |
| 6–7 | −93 | 51° | 67° | +5.4 | −63 |
| 7–8 | −79 | 51° | 55° | +5.4 | −47 |
| 8–9 | −84 | 51° | 42° | +5.4 | −40 |
| 9–10 | −53 | 51° | 16° | −2.5 | −13 |
| 10–11 | +45 | 60° | 50° | +4.0 | +32 |
| 11–12 | +73 | 60° | 90° | −5.4 | +61 |
| 12–13 | +82 | 55° | 73° | −5.4 | +61 |
| 13–14 | +103 | 55° | 73° | +5.4 | +83 |

Thickness = 437 − 248 = 189 feet.

When estimating an average dip and strike for the graphic solution of problem 16, a dip of 55° to the S 23 E and strike of N 67 E appear to be reasonable averages. Yet, in a graphic solution made with this assumed dip and strike, the thickness will be 155 feet. This is about 20% less than the calculated thickness. The error is caused by inability to estimate a valid average dip and strike. Therefore, equation A.1 should be used wherever there is a significant variation in dip or strike.

## Practice Problems

### Problem 17

Using the nomogram in figure A.21, complete the data for each of the following observations:

| | Width of Outcrop (in feet) | Thickness (in feet) | Dip (in degrees) |
|---|---|---|---|
| a | — | 90 | 14 |
| b | — | 60 | 2 |
| c | 210 | — | 50 |
| d | 680 | — | 2 |
| e | 2,500 | 1,600 | — |
| f | 470 | 35 | — |

### Problem 18

In a stream bed, the base of a shale is exposed at an elevation of 1,845 feet and the top of the shale at 1,632 feet. The horizontal distance (map distance) between these contacts is 2,100 feet in a direction N 33 E. The average dip of the formation is 42° to the N 5 E. Determine the thickness.

### Problem 19

A red shale is dipping 35° to the S 60 E. The following traverse was made between the lower and the upper contacts. Determine the thickness.

| Station | Horizontal Distance | Traverse Bearing | Strike | Dip | Elevation Difference |
|---|---|---|---|---|---|
| 1–2 | 362 | N 60 E | N 30 E | 35° SE | +13.6 |
| 2–3 | 445 | N 80 E | N 30 E | 35° SE | +21.0 |
| 3–4 | 330 | S 10 W | N 30 E | 35° SE | −17.4 |
| 4–5 | 676 | S 75 E | N 30 E | 35° SE | +30.8 |

### Problem 20

A traverse between the upper and lower contacts of a formation yielded the following data. Determine the thickness of the formation.

| Station | Distance (in feet) | Traverse Bearing | Strike | Dip | Elevation Difference |
|---|---|---|---|---|---|
| 1–2 | 281 | N 30 W | N 18 E | 22° E | +16.1 |
| 2–3 | 222 | N 43 W | N 14 E | 28° E | +24.4 |
| 3–4 | 196 | N 61 W | N 8 E | 34° E | +38.5 |
| 4–5 | 182 | N 61 W | N 5 E | 42° E | −12.7 |
| 5–6 | 104 | N 36 W | N 2 E | 46° E | −21.3 |
| 6–7 | 392 | N 7 E | North | 47° E | −18.2 |

### Problem 21

A traverse between the top and the bottom of a bed yielded the following data. Determine the thickness of the bed.

| Station | Distance (in feet) | Traverse Bearing | Strike | Dip | Elevation Difference |
|---|---|---|---|---|---|
| 1–2 | 276 | N 20 W | N 15 E | 56° SE | +18.5 |
| 2–3 | 314 | N 41 W | N 20 E | 60° SE | −23.7 |
| 3–4 | 166 | N 60 W | N 12 E | 62° SE | −36.4 |
| 4–5 | 178 | N 60 W | N 5 E | 55° SE | −12.6 |
| 5–6 | 193 | N 22 W | N 8 E | 52° SE | +27.2 |
| 6–7 | 387 | N 15 E | N 2 E | 56° SE | +18.3 |

## Three-Point Problem

Any three points on a plane, not in a straight line, will determine the attitude of that plane. Therefore, from three known points on a geological surface, such as a bedding plane, a fault, or any plane contact, it is possible to determine the dip and strike, provided that the surface is a straight plane. If the surface is curved or warped, then an average strike and dip may be determined. Here is an example:

### Problem 22

Drill hole *a* is 1,600 feet N 46 E of drill hole *b;* and drill hole *c* is 1,300 feet N 26 W of *b*. The three drill holes are located on a plane, and the same horizon or "marker bed" is encountered at the following depths: *a*, 1,254 feet; *b*, 1,894 feet; and *c*, 1,494 feet. Determine the strike and dip.

**Solution 1.** The locations of the drill holes are plotted as shown in figure A.23. It is convenient, although not necessary, to reduce the shallowest drill hole to zero depth, and then reduce the other two drill holes the same amount. This makes *a* zero, *c* 240 feet deep, and *b* 640 feet deep. Obviously, at some place along line *ab*, the horizon will be found at a depth of 240 feet. A line connecting this point with *c* will give the strike of the dipping horizon, because a strike is a level line.

Draw *ak* at any angle to *ab*. The only purpose of this line is to make it easy to divide *ab* into any number of proportional parts. On *ak*, using any scale, mark the position of 640 and 240. Connect *b* with 640 and project, with parallel lines, all other division points that are desired onto *ab*, as shown in figure A.23. In this problem the important point is 240, and its projection onto *ab* will determine the strike of N 70 W. The dip of 24° to the S 20 W is found in a section perpendicular to the strike.

**Figure A.23**
Three-point problem. Solution 1.

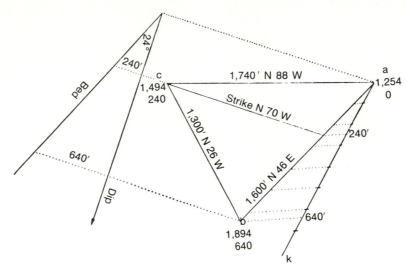

**Figure A.24**
Three-point problem. Solution 2.

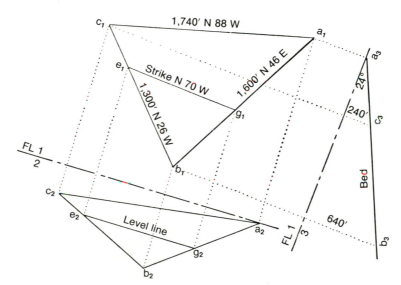

Note that it is not necessary to use an exaggerated vertical scale in the section along the dip, since the dip angle is large enough to be measured with a protractor. If necessary, an exaggerated vertical scale may be used, as illustrated in figure A.10. Note also that any scale may be used to divide line *ak;* often it is convenient to use a scale for the proportional division that is different from that used in the rest of the problem.

**Solution 2.** Draw the plan view of drill holes *a, b,* and *c,* as shown in view 1 (figure A.24). Show the depths to the horizon by drawing any section, such as view 2. Here $a_2$ is zero depth, $c_2$ is 240 feet deep, and $b_2$ is 640 feet deep. In this section, any level line, such as $e_2g_2$, when projected onto the plan, will give the strike of N 70 W. This is true because a strike line is a level line. A section perpendicular to the strike, such as view 3, will give a dip of 24° to the S 20 W. Note that the bed appears as an edge in this view.

**Solution 3.** Since sections along lines connecting any two of the drill holes will give apparent dips of the horizon or marker bed, the three-point problem may be treated as two apparent dips. From two apparent dips the true dip may be determined as already shown (pages 336 and 337).

## Practice Problems

### Problem 23

A vein is found by three vertical drill holes. *a* is 1,200 feet N 10 E of *c,* has an elevation of 1,045 feet, and found the top of the vein at a depth of 1,835 feet. *b* is 1,012 feet N 25 W of *c,* has an elevation of 907 feet, and found the top of the vein at a depth of 948 feet. *c* started drilling at an elevation of 857 feet and found the top of the vein at a depth of 418 feet. Determine the strike and dip of the vein by all three methods.

### Problem 24

In an area of gently dipping beds the bottom of a lignite bed has an elevation of 1,256 feet at point *A.* The bottom of the same lignite has an elevation of 1,185 feet at point *B,* 10 miles S 85 E of *A.* At point *C,* 3 miles S 72 E of *B,* the bottom of this lignite could not be seen, although lignite was found; however, an elevation of 1,090 feet was taken on a spring line directly below *C.* Determine the average regional strike and the average dip in feet per mile.

### Problem 25

A well-developed thin breccia zone marks a major fault. Three points were located in this zone of brecciation, as near to the top in each case as possible. Point *a* has an elevation of 1,485 feet; point *b,* 400 feet north of *a,* has an elevation of 1,745 feet; and point *c,* 630 feet S 60 E of *b,* has an elevation of 1,860 feet. Determine the dip and strike of the fault.

## Outcrop Completion

Patterns made by outcropping beds, dikes, veins, and faults are governed by their attitude and by the topography. If the topography has little relief, or if the dip is steep, then the outcrops trend in nearly straight lines or bands that parallel the strike direction. However, if the topography is rugged, or if the dip is gentle, then the outcrop pattern is fairly intricate. Fortunately, the determination of this pattern in covered ground is not difficult, provided that the direction and amount of dip remain nearly constant and a topographic map is available.

**Figure A.25**

Projection of an upper and a lower contact. The top of a limestone outcrops at *C* and dips 10° due west. The limestone is 300 ft. thick. The contours are shown by dashed lines, and the contour interval is 100 ft. The upper contact is the dotted line *CDEF.* The lower contact is the dotted line to the east. All the area between the dotted contact lines is underlain by limestone.

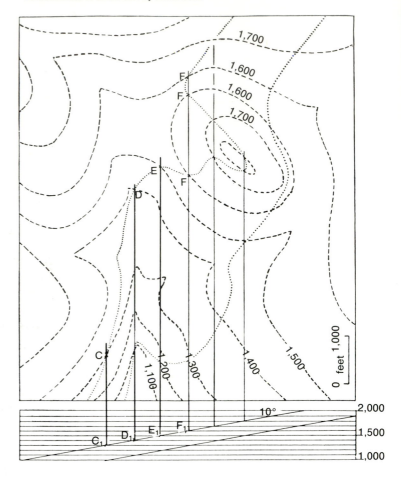

Geological maps, showing the attitude of the contacts between formations, are usually not made by "walking out" these contacts in the field. A few points accurately located on the contacts, together with the dip, generally give sufficient information to permit the projection of all in-between points of outcrop. The method of projection is merely a graphic solution in which the plan view and one or more section (elevation) views are used. Since the position of the folding line is obvious in the following projection examples, it has been omitted.

In figure A.25 the top of a limestone 300 feet thick outcrops at *C,* with an elevation of 1,300 feet, and dips 10° due west. The contours are shown as dashed lines

**Figure A.26**

Projection aided by half-interval contours. The top of a limestone 250 ft. thick outcrops at A and dips 20° due east. The contours are shown by dashed lines, and the contour interval is 250 ft. The outcrop pattern is determined by projection and is shown by solid, heavy lines. The half-interval contours are shown by dotted lines, wherever additional control is needed. Intersection points on these are found by projection from half-interval contours in the cross section.

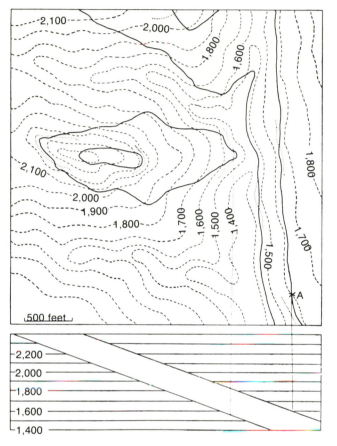

and the contour interval is 100 feet. In an east-west section, the top of the bed can be drawn through $C_1$ (elevation 1,300 feet) with a dip of 10°, provided that the vertical scale is not exaggerated. The bottom of the bed may then be drawn by scaling the thickness of 300 feet.

To determine the pattern of the upper contact, projection lines are drawn connecting points of equal elevation in both views. These projection lines are really structure contour lines, since they connect points of equal elevation, and hence are level lines. The upper contact may now be drawn through the intersection points as shown by the dotted line *CDEF*. *The dotted contact line must cross no contour line except at an intersection between topographic and structure contour lines.* In the same way, the bottom of the bed may be projected, as shown by the dotted contact line to the

east. To avoid confusion, no projection lines (structure contours) have been drawn from the bottom of the bed. The area between the dotted contacts is underlain by limestone.

In some cases the contour interval does not give sufficient control for accurate projection. Then half-interval contours may be interpolated for guidance. This is illustrated in figure A.26, where the top of a limestone 250 feet thick outcrops at *A* and dips 20° due east. The contours are shown by dashed lines. The projection of the outcrop pattern is shown by solid heavy lines. The half-interval contours are shown by dotted lines wherever additional control is desired. Intersection points on these are determined by projection lines from half-interval contours in the section.

**Figure A.27**
Projection with an exaggerated vertical scale. The bottom of a
limestone outcrops at points A, B, and C, and the top at D. Since A
and B are at the same elevation, a line through these points gives
the strike of the limestone. The actual dip is about 2°; hence, the
vertical scale must be exaggerated. Contours are shown by dotted
lines, and the contour interval is 20 ft. Streams are shown by light,
solid lines; contacts are heavy, dashed lines. The apparent thickness
of 150 ft is measured with the exaggerated vertical scale. The actual
thickness is 300 ft.

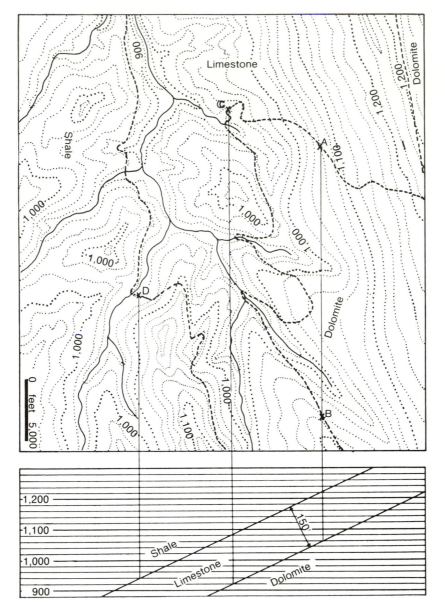

If the topographic map on which the outcrop pattern is to be drawn has a fairly small scale, such as 1 inch to the mile, and if the dips are gentle, then the vertical section may be exaggerated for accurate projection. This is illustrated in figure A.27, where the bottom of a limestone outcrops at *A, B,* and *C,* and the top at *D.* Since *A* and *B* are at the same elevation, 1,060 feet, a line through these points will give the strike of the limestone. A perpendicular surface distance from line *AB* to point *C,* which is also on the bottom of the limestone, will give the horizontal component of the dip; and the difference in elevation between point *C* and 1,060 feet will give the vertical component of the dip. The tangent of this ratio gives a dip of about 1°, which is too small to plot for projection purposes. Therefore, the vertical scale should be exaggerated. Alternatively, structure contour spacing may be computed by trigonometry.

A vertical section is set up at right angles to the strike line *AB,* and any convenient scale will do for the contour interval of 20 feet. In figure A.27 the vertical scale is 25 times the horizontal scale. Points *A* and *B* are located on the 1,060-foot contour and point *C* on the 920 contour. A line joining these points will represent the bottom of the limestone with a true dip of 1° on the exaggerated vertical scale. The top of the limestone is located along a projected strike line on the 940 contour, and the top of the bed is drawn parallel to the bottom contact. The outcrop pattern is determined by the usual projection. So that the map will be easy to read, the contour lines are dotted, the streams are shown by thin solid lines, and the contacts are heavy dashed lines. Areas underlain by dolomite, limestone, and shale are indicated.

To summarize, note the following:

1. Completing an outcrop pattern by projection is merely an application of the projection methods used in earlier exercises.
2. The cross section should be perpendicular to the strike to show the true dip; the projection lines are then strike lines (structure contours).
3. If the dip changes in direction or amount, reorient the cross section, or construct a new one.
4. As shown in previous graphical solutions, no difficulty arises if the vertical scale is exaggerated, provided that the dip is treated as a gradient (rate of slope) and not as an angle.

5. Contacts must not cross contours unless intersection points are present. When in doubt, use half-interval contours in *both* the plan *and* the cross section.
6. If the dip is not known, and if no two contact points are at the same elevation, it may be necessary to determine the strike by solving a three-point problem.
7. Emphasis has been on outcrop patterns of sedimentary formations; but the projection method is just as useful for veins, dikes, and faults.
8. Major faults may present a problem, because exposures of clean, defined surfaces are rare. However, if a fault is traced accurately by common field criteria (chapter 12), its dip may be found by the projection method.

## Practice Problems

### Problem 26
Figure A.28 is a topographic map with a contour interval of 20 feet. A series of formations dip due south. Outcrop *E* marks the top of a limestone. Outcrops *A* and *B* mark the bottom of this limestone; this is also the top contact of an underlying black shale. Outcrop *C* marks the bottom of the black shale; this is also the top contact of an underlying gray shale whose bottom crops out at *D.* Draw the geologic map, assuming the attitude of beds to remain constant. Also determine the thickness of each formation.

### Problem 27
Figure A.29 is a topographic map of a mining property with a contour interval of 100 feet. At point *A* a fault dips 50° to the S 43 W. At *D* a vein crops out and dips 35° due west. At *B* the bottom of a limestone 300 feet thick dips 20° due south. In the area southwest of the fault, the vein crops out at *D'* and dips 28° to the S 70 W. The bottom of the limestone outcrops at *B'* and dips 30° to the S 20 E. Draw the geologic map. (Save this exercise to later determine displacement along the fault.)

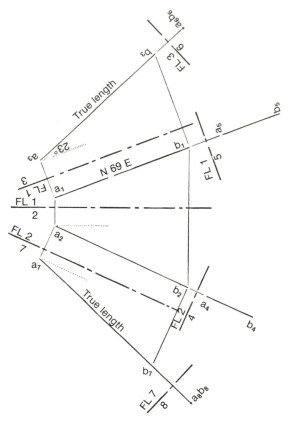

## Point, Line, and Plane Problems Using the Orthographic Projection Method

In the following examples, a line is assumed to be straight throughout its course, whether it represents a direction, a borehole, or a tunnel; a plane is assumed to be straight, whether it is a bed, a dike, a vein, or a fault.

### True Length of a Line

A line must lie at right angles to the lines of sight for any view, in order to appear in its true length in that view. That is, any level line will be seen in its true length in the plan view; any vertical line will appear in its true length in all elevation views.

Line *AB,* shown in figure A.30, slopes 23° to the N 69 E and is shown in plan view 1 and in elevation view 2. The true length, $a_3b_3$, is drawn in view 3, which lies at right angles to the lines of sight to $a_1b_1$. The same true length, $a_7b_7$, is shown in view 7, which lies at right angles to the lines of sight to $a_2b_2$.

### True Slope of a Line

The true slope of a line can be seen only in that elevation view which shows the line in its true length. View 3, figure A.30, shows the true slope of 23° in the direction N 69 E, because line *AB* appears in its true length and the view is an elevation. Note that elevation view 2 does not give the true slope, because the line does not appear in its true length. Note also that inclined view 7 does not give the true slope, even though line *AB* is seen in its true length.

### The Plane as an Edge

A plane appears as an edge or line in the view in which any line in that plane appears as a point. The true dip (direction and angle) of a plane can be seen only in an elevation view where the plane appears as an edge. To geologists, a more familiar statement would be that a strike is a level line, and that the true dip (direction and angle) is shown in an elevation view at right angles to the strike. Furthermore, a plane must appear as an edge before a view can be drawn showing the true shape and size of any object in the plane.

In figure A.31, the plane *ABC* is shown in the plan and front-elevation views. Any line such as $x_2y_2$ is drawn in view 2 and shown in its true length $x_1y_1$ in view 1. In view 3, this line appears as a point $x_3y_3$, and the plane as an edge $a_3b_3c_3$ with a true dip of 45°. View 4 shows the plane in its true shape and size.

Incidentally, since *XY* is a level line, $x_2y_2$, its projection into the plan view $x_1y_1$ is a strike line. A section at right angles to the strike, such as view 3, will give the true dip of the plane.

### Application

The above principles have numerous applications, among them the following:

1. The shortest distance, or the distance in any given direction, from a point to a line or to a plane
2. The shortest distance, or the distance in any given direction, between any two nonintersecting, nonparallel lines
3. True displacements along faults
4. True slopes of lines

In actual practice, the lines may be boreholes, tunnels, adits, etc., that are to be driven to veins, dikes, and faults, or to certain specified underground mining developments. The following examples illustrate the general method. Additional examples will be found in the Practice Problems.

## Distance in a Given Direction from a Point to a Line

In figure A.32a, line $AB$ and point $Z$ are given in the plan and front-elevation views. What is the true distance and slope of a line from point $Z$ to line $AB$, in a direction S 75 W from point $Z$.

In view 1, the dashed line bearing S 75 W from point $Z$ intersects line $a_1b_1$ at $y_1$. In view 2, the location of $y_2$ is determined by projection. View 3 shows the true distance $z_3y_3$ and the true slope of 10° to the S 75 W.

## Shortest Distance from a Point to a Line

In figure A.32b, line $AB$ and point $Z$ are given in plan and front-elevation views. What is the bearing, slope, and true length of the shortest distance from point $Z$ to line $AB$?

In view 3, $AB$ is shown in its true length as $a_3b_3$, and the perpendicular or shortest distance to point $Z$ is shown as $z_3y_3$. In view 4, the line $AB$ is shown as a point $a_4b_4$, and the true length of the shortest distance is shown as $z_4y_4$. Projecting the intersection point $y_3$ back to the plan view gives the true bearing of the shortest distance as $z_1y_1$. The true slope is shown in view 5 as 41° and also the true length of the shortest distance $z_5y_5$, which checks the true distance $z_4y_4$.

**Figure A.31**
The plane as an edge, and also its true shape and size.

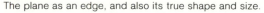
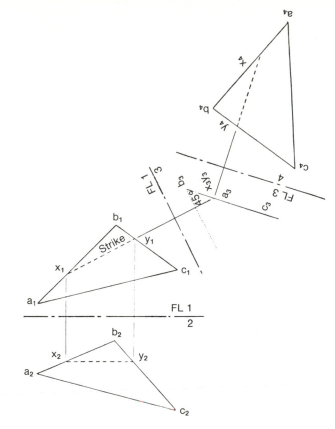

---

**Figure A.32**
Distance from a point to a line. (a) in a given direction (b) shortest distance.

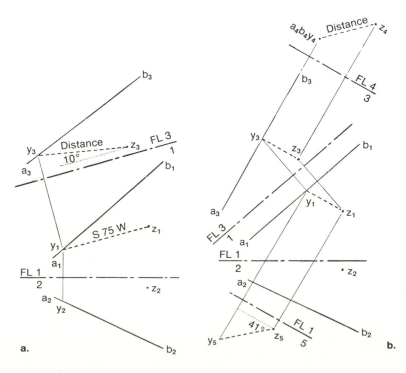

**Figure A.33**
Construction of shortest distances. (a) between any two non-intersecting, non-parallel lines; (b) from a point to a plane.

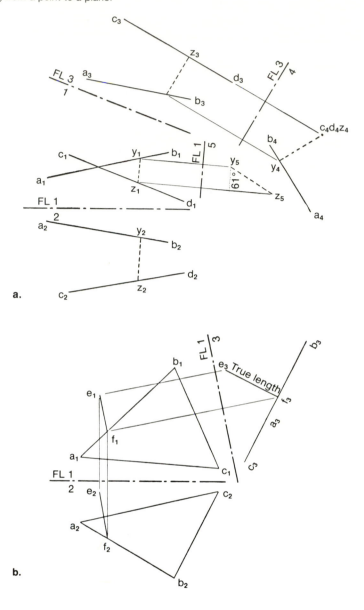

a.

b.

## Shortest Distance Between Any Two Nonintersecting, Nonparallel Lines

Figure A.33a shows two nonintersecting, nonparallel lines $AB$ and $CD$ in plan and front-elevation views. What is the true length, bearing, and true slope of the shortest distance between these two lines?

Either line must be shown in its true length and then projected as a point. In this example line $CD$ is chosen, and it is shown in its true length in view 3 as $c_3d_3$. In view 4, $CD$ is shown as a point, $c_4d_4$, and the dashed line $y_4z_4$, which is perpendicular to line $AB$, is the shortest true distance. The intersection point $y_4$ is projected back into view 3, and the dashed perpendicular from it to $c_3d_3$ gives $z_3$. The line $YZ$ is now fully determined. Its projection into the plan view gives the bearing $y_1z_1$. View 5 shows the true slope of $60°$ and the true distance, which checks the distance $y_4z_4$.

**Figure A.34**
True distance in any given direction from a point to a plane, the
piercing point of this line, and the true angle between this line and
the plane.

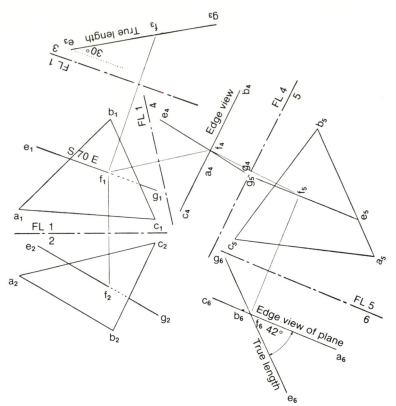

## Shortest Distance from a Point to a Plane

Figure A.33b shows point $E$ and plane $ABC$ in plan and front-elevation views. The plane is shown as an edge $a_3b_3c_3$ in view 3. The perpendicular distance from $e_3$ to the edge view of plane $e_3f_3$ is the true length of the shortest distance. Projection into the plan view gives $e_1f_1$, the bearing of the shortest distance.

## True Distance in Any Given Direction from a Point to a Plane, the Piercing Point of this Line, and the True Angle Between this Line and the Plane

Figure A.34 shows point $E$ and plane $ABC$ in plan and front-elevation views. A sloping shaft is to be dug on a bearing S 70 E from point $E$ to the fault plane $ABC$, with a slope of 30° to the southeast. What is the true length, the piercing point, and the true angle between the shaft and the fault plane?

View 3 is an elevation taken S 70 E, so as to show the 30° angle of slope. The position of point $e_3$ is determined by projection, and any convenient length, such as $e_3g_3$, is laid off on the slope. This is projected back into the plan view as $e_1g_1$.

View 4 shows the plane as an edge $a_4b_4c_4$ and also the line of the proposed shaft as $e_4g_4$. The intersection of this line with the edge view of the plane $f_4$ is the piercing point. This piercing point may now be projected to all views. In view 3, $e_3f_3$ is the true length of the shaft from point $E$ to the fault plane.

View 5 shows the fault plane in its true shape and size, and also the position of the shaft $e_5g_5$. Any elevation view from view 5, such as view 6, will give the true angle between the shaft and the fault plane. In this example it is 42°. Also the true length of the shaft from point $E$ to the fault plane is shown as $e_6f_6$, which should check the true length $e_3f_3$ in view 3.

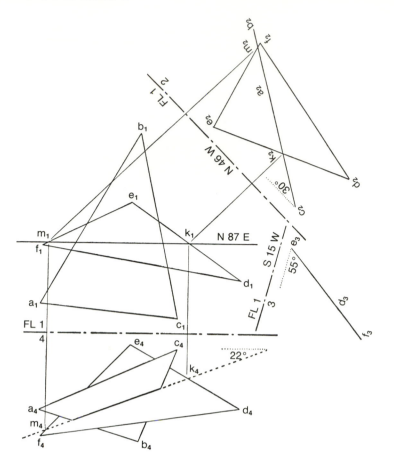

## Line of Intersection of Any Two Nonparallel Planes

Problem 28 explains the method for finding the bearing and plunge of the line of intersection of two planes.

### Problem 28

A bed dips 30° to the N 46 W, and a fault dips 55° to the S 15 W. What is the bearing and plunge of the intersection of these two planes?

Remember that these two planes are actually of indefinite extent. However, for convenience, any three points may be chosen on each plane, such as $a_2b_2c_2$ and $d_3e_3f_3$ in figure A.35, and projected so as to give a plan view of the two planes. Three methods of solving the problem will follow.

1. One method requires the drawing of a view so that one of the planes appears as an edge, cutting the other plane. In figure A.35, view 2 was chosen and bed *ABC* appears as an edge-cutting fault plane *DEF* at points $k_2$ and $m_2$. These two points lie on the intersection of the two planes and may be projected into the plan view to give the bearing of the line of intersection, N 87 E. Then, projection into view 4, which is an elevation taken S 87 W, will give the plunge of the line of intersection as 22°.

   Note that bed *ABC* could have been projected into view 3, where it would have been cut by the edge view of fault plane *DEF*, and the line of intersection could be determined in the same manner.

2. There is another way to solve problem 28. In figure A.36, the bed dipping 30° to the N 46 W is shown in view 2, and the fault plane dipping 55° to the S 15 W is shown in view 3. Any unit depth $x$ is laid off in these two views, and points $c_2$ and $c_3$ are determined. Projection of these points onto the plan view gives an intersection point $c_1$, which is common to both planes and hence lies on the line of intersection of the two planes. Connecting this point $c_1$ with the common point of origin gives the bearing of the intersection, S 88 W. Elevation view 6, taken S 88 W, will give the line of intersection in its true length and also the plunge of 22°.

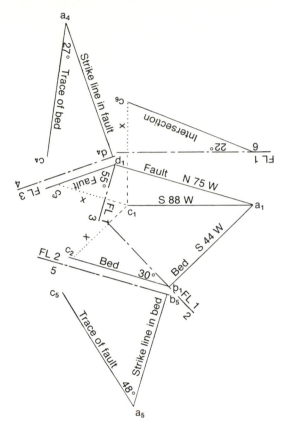

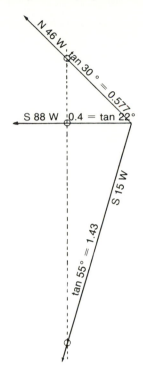

In addition, the trace of the bed $a_4c_4$ on the fault plane is shown in view 4. Here the fault plane will appear in its true shape and size, so that true distances and angles on the fault plane may be measured. The pitch of the trace of the bed is 27° on the fault plane. Likewise, the trace of the fault $a_5c_5$ on the bedding plane is shown in view 5. The pitch of the trace of the fault on the bedding is 48°.

3. The tangent vector method may also be used to solve problem 28. In figure A.37, the two dipping planes are laid off as tangent vectors. The bearing of the line of intersection is S 88 W, and the plunge is 22° in that direction.

It is useful to become familiar with all three methods of finding the intersection of two planes. Solution 1 is cumbersome, but it gives a good three-dimensional picture of the intersection. Solution 2 is simple and has the advantage of giving true distances along the plunge of the intersection. It may also be expanded to show the pitch of the trace of either plane on the other plane. Solution 3 is rapid, but it merely shows the bearing of the line of intersection and its plunge. In solution 2, the vertical scale may be exaggerated for cases of very low dips.

## Practice Problems

### Problem 29
An inclined shaft is driven a length of 400 feet from a point $A$, but 500 feet west and 260 feet north of $A$, a hole is to be drilled so as to intersect the lower end of the inclined shaft. Determine the bearing, slope, and true length of the drill hole.

### Problem 30
A vein dips 30° to the S 40 W. A drill hole is located 300 feet due south of the upper contact of the vein. The drill hole is inclined at an angle of 60° to the S 20 W. The total amount of vein material passed through is 85 feet. (a) What is the true thickness of the vein? With what slope should the hole have been drilled to reach the vein in the shortest distance? (b) What length of hole was drilled to reach the top of the vein? (c) In what direction and with what slope should the hole have been drilled to reach the vein in the shortest distance? (d) What is the shortest distance?

### Problem 31
An inclined shaft, sloping 50° and bearing S 58 E, comes to the surface at $A$. An abandoned tunnel, sloping 30° and bearing S 65 W, comes to the surface at $B$. Point $B$ is 290 feet east and 55 feet south of $A$ and is at the same elevation as $A$. Determine the bearing, true length, and slope of the shortest connecting entry. At what distances from the surface in the abandoned tunnel and in the inclined shaft should the new entry be started?

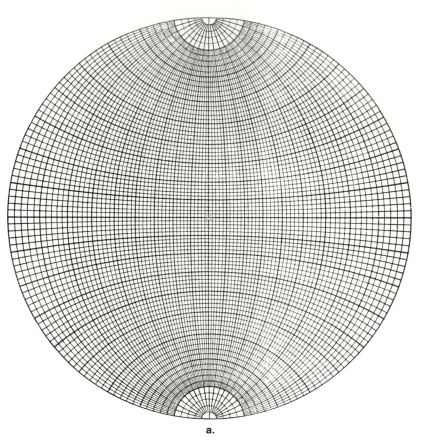

a.

## Problem 32

A vein dips 40° to the S 60 W. A prospect hole is located 200 feet west and 150 feet south of the outcrop of the vein. The hole is drilled on a slope of 55° to the S 70 E going through the entire vein but no farther. Assume the surface of the ground to be level. Find the piercing point, the length of the drill hole, and the angle between the drill hole and the vein. If the drill pierces 60 feet of vein material, how thick is the vein?

## Problem 33

Find the bearing and plunge of the intersection of a fault plane dipping 60° to the S 37 E and a bed dipping 45° to the S 40 W. Use two methods.

## Stereographic Projection

The stereographic projection has been used by mineralogists and crystallographers since the middle of the nineteenth century. Only much later, however, was it applied to problems of structural geology, in spite of the fact that a stereographic projection is often the best method for measuring angles and spatial relations between lines and planes. We shall consider only a few of the simpler stereographic solutions of structural geology problems. For a more extended treatment of this method, see Bucher (1944) and Phillips (1971).

The most useful stereographic base for the structural geologist is the Wulff net, illustrated in figure A.38a. Imagine that a transparent terrestrial globe is sliced through the poles along a meridian and that the

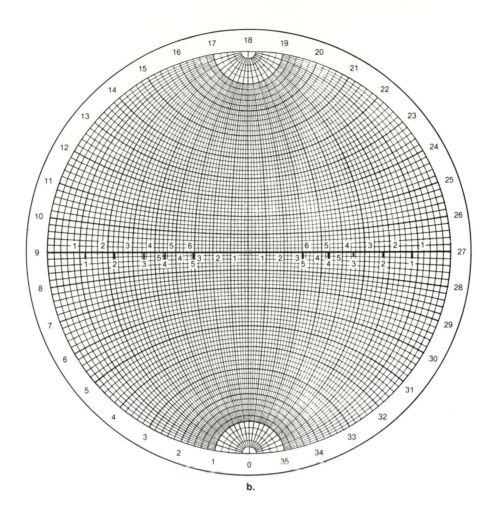

b.

plane surface of one resulting hemisphere is placed horizontally with the eye of the observer above its center. If the latitude and longitude circles are spaced at 2° intervals, this view reveals the Wulff (meridional) stereographic net of figure A.38a. The same net would be seen if the observation point was at the intersection of a meridian and the equator, on the surface of a transparent terrestrial globe.

## Procedure

Fasten the net to a tabletop or drawing board with adhesive tape. Place a sheet of tracing paper or other transparent overlay above the net, and drive a small pin into the center of the net, so that the tracing paper can be rotated. Or drive the sharpened point of a thumbtack through from the underside of the net, if it is on heavy cardboard. Mark the positions of the north and south poles, as well as east and west, *outside* the ends of the two diameters of the net.

In diagram *a,* figure A.39, the heavy lines are drawn on the tracing paper, and a 10° meridional net that lies underneath is shown by dots and lighter lines. The illustration represents a plane *AODB* dipping due east at 30°. Read off the strike at *A* or *D,* and the dip direction at *C. BC* is the amount of dip, the 30° being measured from the perimeter. *ABD* is the trace of the great circle made by this plane on the net.

In diagram *b,* figure A.39, a plane *AODB* is shown dipping 30° to the S 50 E. The direction of dip *C* is drawn to the S 50 E. Then the tracing paper is rotated until *C* coincides with the east diameter of the net, in which position the true dip of 30° can be measured. The great circle representing a 30° dip may now be traced from the net, exactly as in diagram *a.* Rotation of the tracing paper until the direction of dip *C* is properly oriented to the S 50 E is the final step.

**Figure A.39**
Stereographic projection. Diagram *a* represents a plane dipping 30°
due east. Diagram *b* represents a plane dipping 30° to the S 50 E.

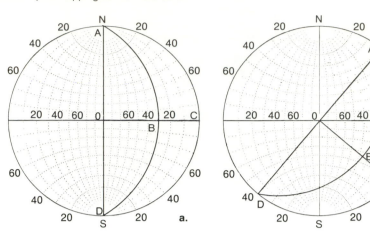

## Examples

### Apparent Dip from True Dip: Problem 5 (repeated).

A bed dips to the S 60 W at 30°. A cross section is to be drawn along a north-south line. What is the apparent dip of the bed in this cross section?

In diagram *a*, figure A.40, the true dip direction *OC* is laid off on the tracing paper to the S 60 W, and the apparent dip direction *OF* is laid off due south. In diagram *b*, figure A.40, the tracing paper has been rotated to the vertical angle position so that the true dip of 30°, measured by *CB*, may be traced as the great circle *ABD*. All lines drawn from *O* lying in the plane *AODB*, except the dip line *OBC* and the strike line *AOD*, will be apparent dip directions. In the direction *OF*, the amount of apparent dip is measured by *FE*. The amount of apparent dip is determined by turning *OEF* to the position where vertical angles are read (east or west) as shown in diagram *c*. If plotted on a 2° net the apparent dip can be read as 16°.

### True Dip from Two Apparent Dips: Problem 6 (repeated).

In a rock cut along a bend in a railroad track, two apparent dips are read, namely, 20° to the S 40 W and 35° to the S 20 W. Find the direction and amount of true dip.

In diagram *a*, figure A.41, the two apparent dip directions *OB* and *OD* are drawn on the tracing paper to the S 40 W and S 20 W, respectively. Rotate the apparent dip direction *OB* to the vertical angle position, and mark the apparent dip of 20° at *A*. Then rotate the apparent dip direction *OD* as shown in diagram *a*, and mark the apparent dip angle of 35° at *C*. Apparent dips are *lines;* therefore, their stereographic projections are *points* (*A* and *C*).

In diagram *b*, figure A.41, the tracing paper has been rotated until points *A* and *C* fall on the same great circle. The plane *GEH*, which is the dipping bed, may now be drawn. The amount of true dip is 40° measured by *FE*. The strike is measured at *G* or *H*.

The direction of true dip is shown in diagram *c*, figure A.41, where the apparent dip directions have been rotated to their proper orientation. On a 2° net, the direction of dip is read as S 24 E.

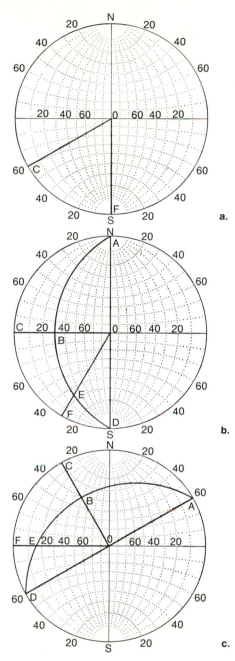

a.

b.

c.

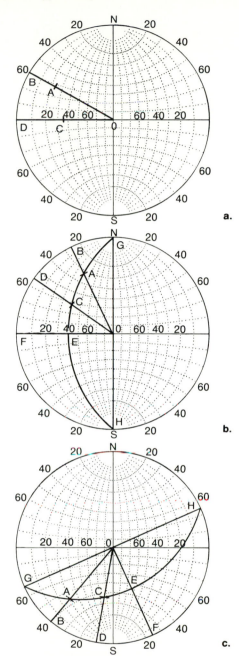

a.

b.

c.

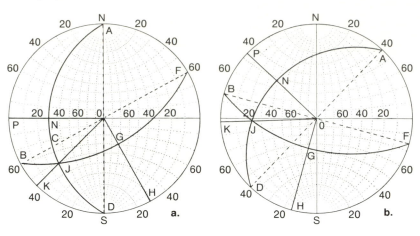

## Line of Intersection of Any Two Oblique Planes:
**Problem 28 (repeated).** A bed dips 30° to the N 46 W, and a fault dips 55° to the S 15 W. Find the bearing and the plunge of the intersection of these two planes.

In diagram *a*, figure A.42, the fault plane is shown as *BGF*, and the bed as *DA*. These two planes intersect at *J*. The plunge of the intersection is measured by *KJ* and is read as 22° by rotating the tracing paper so that *KJ* lies in the west position where vertical angles may be measured. Point *J* represents both the trace of the bedding on the fault plane *BGF* and the trace of the fault on the bedding plane *DNA*.

The pitch of the trace of the fault on the bedding is measured by *JD* and represents the angle with any horizontal line in the bedding. Since the strike *D* of the bedding is horizontal, the angle of pitch may be read from *D* to *J* as 48°, the tracing being in the correct position, with the strike of the bed along the polar diameter.

The pitch of the trace of the bedding on the fault is measured by *BJ*. To read the angle of pitch, the tracing is rotated until the strike line of the fault *BF* lies on the polar diameter. In this position the pitch angle from *B* to *J* is 27°.

To determine the trend of the plunge, rotate the tracing into its proper orientation, as shown in diagram *b*, figure A.42, so that the fault dips S 15 W and the bed dips N 46 W. The direction of plunge is shown by *OJK* and the bearing is S 88 W.

Compare this solution with the orthographic solution of the same problem on pages 362–363 and figure A.36.

## Displacements Caused by Nonrotational Faulting
Field data for the solution of fault problems should be obtained far enough from the fault zone to eliminate drag and local distortions. Of course, it is assumed that the fault plane is not warped appreciably throughout the area in which displacements are to be determined. Otherwise, the solution will be only approximate.

### Problem 34
A fault that dips 35° to the N 60 W has displaced a vein and a bed. The vein dips 58° to the S 75 W, and the bed dips 43° due south. The outcrops of the fault and the displaced bed and vein are shown in the outcrop map (diagram *a*) of figure A.43. Determine the components of the net slip along the fault.

On the stereonet, diagram *b*, figure A.43, the directions of dip of the fault, *F*, of the vein, *E*, and of the bed, *D*, are laid off as shown. The great circle *ABC* showing the 35° dip of the fault is drawn by the method previously described. The great circles showing the dips of the bed and the vein have not been drawn, but their intersections with the fault plane are shown at *H* (bed) and at *G* (vein). *CH* is the pitch of the bedding trace, and *AG* is the pitch of the vein in the fault plane. Since the angle of pitch is measured from a strike line, the overlay tracing is rotated until the strike of the fault falls along the polar diameter of the net. In this position, the bedding trace pitches 40° measured from the SW, and the vein trace pitches 73° measured from the NE, in the fault plane.

Diagram *c* of figure A.43 is a view in the *plane of the fault*. Points $f_1 g_1 k_1 h_1$ are taken from the outcrop map (diagram *a*). Draw the traces of the bed and of

**Figure A.43**
Determination of net slip and components of net slip along a nonrotational fault. Diagram a, outcrop map; diagram b, stereographic projection of angular relations; diagram c, fault-plane view; HW, hanging wall traces; FW, footwall traces.

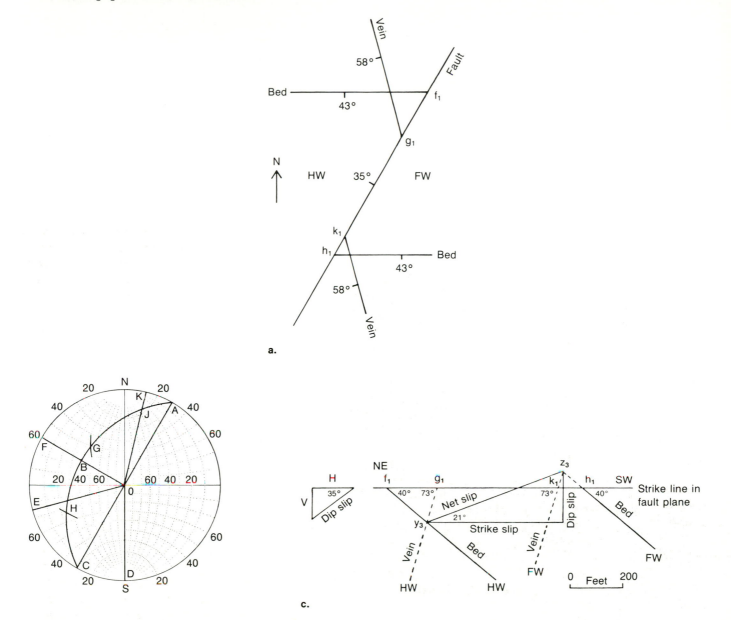

a.

c.

the vein using the pitches obtained from the stereonet, and intersection points $y_3$ and $z_3$ on the fault plane. Before displacement these two points coincided; hence, line $y_3z_3$ is the net slip. The diagram shows strike slip, dip slip, and the horizontal ($H$) and vertical ($V$) components of slip.

The pitch of the net slip on the fault plane is 21°. This pitch is transferred to the stereonet and marked as $J$ on the fault circle $ABC$. The line $OJK$ gives the direction of net slip as N 13 E, when properly oriented. The plunge of the net slip is $JK$ and, when read in the vertical-angle position, is found to be 12°.

## Problems Involving Rotation

It is sometimes necessary to restore faults, beds, joints, veins, and dikes that have been rotated, to their original attitudes. Problems involving rotation are easily solved by stereographic projection. The following two-tilt problem will serve as an example.

### Problem 35

Two formations are separated by an angular unconformity. The upper beds dip 30° to the N 40 E. The underlying beds dip 40° to the N 20 W. Determine the dip and strike of the underlying beds, when the upper beds were being deposited (horizontal).

In diagram *a,* figure A.44, the lower beds are shown by plane *GHF,* and the upper beds by plane *BDC.* These planes have a common point at *K.* The problem is to rotate the upper plane into a horizontal attitude and to determine the effect of this rotation upon the lower plane.

With the strike of the upper plane *B* moved to the north pole of the net, as shown in diagram *a,* this plane may be rotated 30° to horizontal. Then point *D* will move to the perimeter (point *E*), as will all other points on the great circle *BDC,* such as point *K.* Note that the paths of movement are along arcs of small circles. In the same way, all points on the lower bed *FHG* will also move 30° along paths of small circles, toward the perimeter, as shown by the dashed arcs.

In diagram *b,* figure A.44, the tracing has been moved so that point *L* will occupy the position corresponding to zero dip. A great circle is drawn through the ends of the dashed arcs; it represents the lower bed after a rotation of 30°. The angle of dip of the lower bed, when the upper bed is horizontal, is measured by *NP* and is 34°.

In diagram *c,* figure A.44, the tracing has been properly oriented and the direction of dip of the lower bed *ONP,* when the upper bed is horizontal, is N 61 W. The required answer is 34° to the N 61 W.

Problem 36 illustrates the usefulness of the stereographic projection in solving rotational displacements along faults.

### Problem 36

A fault dips 30° to the N 70 E. On the east side of the fault, the beds dip 20° to the N 30 W. Rotation of 40° has occurred about an axis perpendicular to the fault plane, in such a fashion that the beds on the west side have been displaced in a clockwise direction. Determine the dip and the strike of the beds to the west.

The solution is in three main steps: (1) rotate the fault plane vertically 30° to a horizontal attitude and determine the resulting attitude of the east bed;

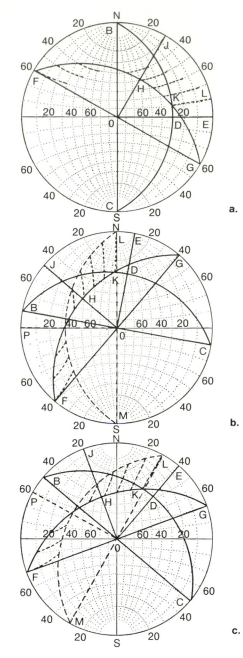

a.

b.

c.

(2) rotate the east bed horizontally 40° in a clockwise direction on the level fault plane; (3) return the fault plane to its original 30° dip. The attitude of the bed will then give the desired dip and strike of the beds to the west.

In diagram *a,* figure A.45, the fault plane is shown as *ABD,* and the beds to the east as *GEH.* The strike of the fault plane *AD* is placed on the north pole of the

## Figure A.45
Rotation around an axis perpendicular to a fault plane.

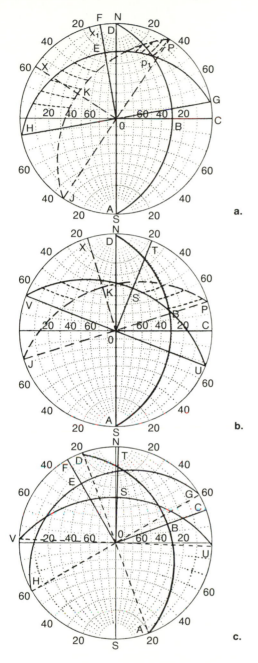

a.

b.

c.

## Figure A.46
Determination of rotation by the difference in pitch.

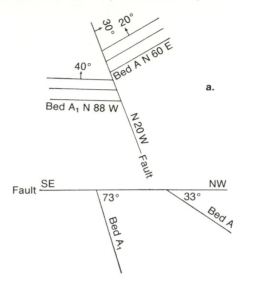

a.

b.

net. The fault may now be rotated 30° to horizontal, moving point *B* to the perimeter at point *C*. The east bed moves to *PKJ* because of this 30° rotation and dips in the direction *X*. Note that a 40° horizontal rotation clockwise will move point *X* to $X_1$.

In diagram *b*, figure A.45, this 40° horizontal rotation has been made, and *PKJ* is shown in its new position. The final step is to return the fault to its original

attitude by a vertical rotation of 30° from its present horizontal attitude. This rotation will move point *C* from the perimeter to point *B*. Also the plane *PKJ* will rotate to *USV*. This plane *USV* represents the beds to the west of the fault after they have been rotated 40° clockwise about an axis perpendicular to the fault plane.

In diagram *c*, figure A.45, *ABD* represents the fault, *GEH* the beds east of the fault, and *USV* the beds west of the fault, all properly oriented. The direction of dip of the west beds is N 2 E, as shown by *T*, and the amount of dip is 40°, as shown by *TS*.

More frequently the dips of the displaced beds, veins, or dikes on the two sides of the fault are known, and the problem is to find the direction and angle of rotation on the fault plane.

Diagram *a*, figure A.46, is a plan view showing the fault dipping 30° to the N 70 E, and the displaced beds dipping 20° to the N 30 W and 40° to the N 2 E. If the pitches of the beds in the fault plane are determined by the method shown on page 368, the amount and the direction of rotation are readily obtained.

Diagram *b*, figure A.46, is a section in the fault plane and shows the east beds pitching 33°, measured from the northwest, and the west beds pitching 73°, measured also from the northwest. It is probable that the west beds have been relatively rotated 40° clockwise, about an axis perpendicular to the fault plane. Note that sections in fault planes should be made as though looking directly down on the fault plane. Thus, in this problem, northwest should be to the right. Otherwise, the relative rotation (clockwise or counterclockwise) will not be correct.

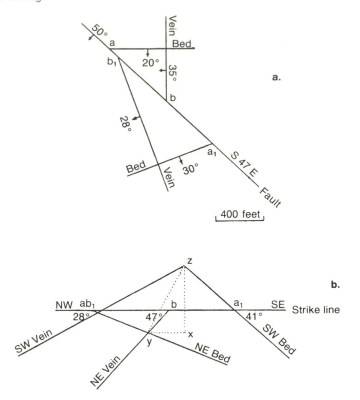

## Determination of Displacements
## Caused by Rotational Faulting

Figure A.29 and problem 27 show that the dips and strikes of the faulted vein and bed differ on the two sides of the fault. That is, straight lines on opposite sides of the fault, outside the zone of drag, are not parallel but have been rotated. In this case, the solution for the displacements is carried out in the manner already discussed on page 368. However, the amount and direction of rotation may also be required.

To avoid the effect of topography on the outcrops of the fault and the disrupted bed and vein, choose some uniform level in figure A.29 for the data. Diagram *a*, figure A.47, is a plan view of the outcrops, taken from figure A.29, as they would appear on a level plain at an elevation of 4,000 feet. Solve stereographically for the pitches of the vein and bed (bottom of limestone) in the plane of the fault. Then, on the northeast side, the bed pitches 22° from the southeast, and the vein pitches 47° from the northwest. On the southwest side, the bed pitches 41° from the southeast and the vein pitches 28° from the northwest.

Diagram *b*, figure A.47, is a view of the fault plane, on which the angles of pitch have been plotted. Point *y* is the intersection of the northeast bed and vein; and point *z* is the intersection of the faulted bed and vein to the southwest. Since *y* and *z* would be identical before faulting, the line *yz* is the true net slip. The dip slip is *xz*, and the strike slip is *xy*.

Note that 19° is the difference in the angles of pitch of the two parts of the disrupted bed and that the same 19° difference is true for the angles of pitch of the two parts of the disrupted vein. It is probable, therefore, that a rotation of 19° occurred during faulting.

The total net slip *yz* may be accounted for by a rotation of 19° about an axis perpendicular to the fault plane, located at the point of intersection of the two traces of the bed, provided that the northeast bed and vein are rotated clockwise. Or this 19° rotation may have been about any other axis perpendicular to the fault plane, and additional shifting along straight lines would bring about the total net slip. Unless the example represents part of a faulted plunging fold, it is fairly certain that the northeast side has been rotated 19° counterclockwise with respect to the southwest side.

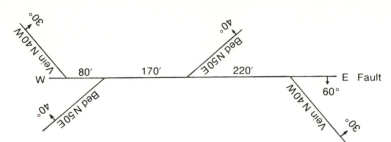

## Practice Problems

### Problem 37

A formation dips 60° to the S 15 E. What are the apparent dips in a section trending S 60 E and in a section trending due east?

### Problem 38

In a gorge, two apparent dips are read, namely 32° to the S 40 E and 48° to the N 70 E. Find the direction and amount of true dip.

### Problem 39

A series of beds are followed along their strike of N 45 E. In a north-south railroad cut, the apparent dip is 30° to the south. What is the true dip?

### Problem 40

Find the bearing and plunge of the intersection of a fault plane dipping 60° to the S 37 E and a bed dipping 45° to the S 40 W. What is the pitch of the trace of the bed on the fault plane? What is the pitch of the trace of the fault on the bed?

### Problem 41

The outcrops of a faulted vein and a thin key bed are shown in figure A.48. The vein dips 30° to the N 50 E, the bed dips 40° to the N 40 W, and the fault dips 60° due south. Determine the plunge and pitch of the net slip and the amounts of net slip, dip slip, and strike slip.

### Problem 42

The outcrops of a faulted vein and a thin key bed are shown in figure A.49. The vein dips 35° to the S 60 W on the north side of the fault and 42° to the S 48 W on the south side. The bed dips 40° due east on the north side and 30° to the N 70 E on the south side. The fault dips 30° due south. The differences in dip and strike on the two sides of the fault result from rotation on the fault plane. Determine the plunge and pitch of the net slip and the amounts of net slip, dip slip, and strike slip. Also determine the direction and the amount of rotation about an axis perpendicular to the fault plane.

**Problem 43**

Given a series of lower beds dipping 56° to the S 12 W and overlain unconformably by beds dipping 30° to the S 25 E, find the original dip and dip direction of the underlying beds.

**Problem 44**

A fault dips 40° due east. On the west side of the fault, the beds dip 30° to the N 45 E. Rotational movement of 30° about an axis perpendicular to the fault plane has displaced the beds to the east in a counterclockwise sense. Determine the dip and the strike of the beds to the east. Confirm your answer by plotting pitches of the east and the west beds in the fault plane and checking the rotation.

**Problem 45**

On a cleavage face, the lineation pitches 30° to the southwest. The cleavage dips 60° to the S 50 E. Determine the direction and the amount of plunge of the lineation.

**Problem 46**

On the northeast side of a fault, a vein dips 35° N 65 W, and the beds dip 50° S 30 E. On the southwest side of the fault, the beds dip 36° S 20 W, and the vein dips 54° N 30 W. Determine the amount and direction of rotation on the fault that dips 30° N 40 E.

## Structure Contour Maps

Geological structures are often shown by contour lines, drawn on the top or the base of any single bed. Usually a distinctive and easily recognized bed is chosen as the contour horizon. Contour lines on this key bed connect points of equal elevation, just as topographic contours connect points of equal elevation on the land surface.

Where the key bed appears at the surface, the elevations of as many points as desirable are determined along the outcrop. Usually, the regional structure soon carries the key bed either underground or above the surface. In that case, the elevations are continued on any overlying or underlying parallel bed, and the thickness interval is added or subtracted so as to bring the elevation to the original key horizon. For example, if the key bed becomes covered and the elevations are continued on a bed that is 60 feet higher, then these elevation figures should be reduced by 60 feet to be brought to the key bed that is being contoured. Oftentimes a dozen or more different beds must be used, in addition to the contour horizon, to outline the structure. Consequently, determination of thickness intervals, between the contour horizon and all supplementary beds used, is very important. Check frequently to be certain that any assigned thickness interval remains constant.

The contour interval selected for the structure map will depend on: (1) the amount of data available; (2) the accuracy of the elevations; (3) the steepness of the structure; and (4) the scale of the map. The contour interval should not be less than the overall errors involved. It would be misleading, for instance, to contour with a 10-foot interval, if the elevation points were scarce and questionable to an accuracy of less than 50 feet.

In drawing structure contours, interpolate proportionately between elevation points. Often this may be done accurately enough by eye; but, if there is any doubt, a proportional scale should be used. After the contours have been drawn in this mechanical way, the resulting structure usually needs modification. Sharp curves should be smoothed; where control is poor, the structure should be indicated by dashed, rather than by solid, contour lines. Also, known structural trends and styles may in themselves suggest changes. In any event, the final map should show the elevations of all control points, so that anyone may check the interpretation.

**Problem 47**

The outcrops of some 10 different beds are shown on figure A.50 by dotted lines and marked by capital letters. Draw a structure contour map with a 10-foot interval, on bed *J* as the key horizon. Beds *A, B, D, E, G,* and *I* are above the contour horizon at distances shown in the table below. Subtract the appropriate thickness interval from each elevation on these beds to bring the data to the contour horizon. Beds *K, L,* and *M* are below the contour horizon; raise elevations on these beds by the appropriate thickness interval shown in the table. The area is structurally a region of small domes on a regional dip to the west. Where control is poor, show contours by dashed lines. Make every fifth contour, such as 800 and 850, a heavy line. The entire area should be contoured.

| Bed | Distance Above or Below Contour Horizon |
|---|---|
| A | 132 feet above |
| B | 79 feet above |
| D | 67 feet above |
| E | 54 feet above |
| G | 43 feet above |
| I | 20 feet above |
| J | Contour horizon |
| K | 16 feet below |
| L | 36 feet below |
| M | 126 feet below |

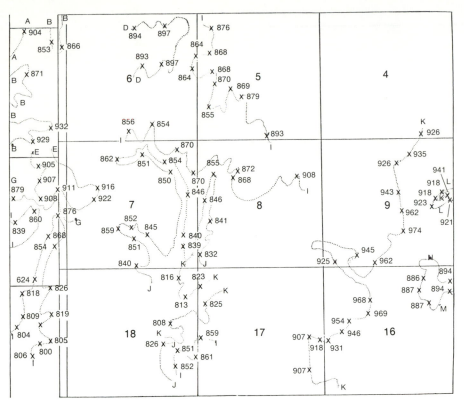

A structure map contoured on elevations of a buried key bed is often called a *subsurface map*. The elevations are usually obtained from well logs, mine records, and geophysical data. Unless otherwise specified, the elevations are referred to sea level as a datum.

For example, if the top of a key bed is found at a depth of 1,500 feet in a well, the surface elevation of which is 500 feet, the elevation figure for contouring will be —1,000 feet, or 1,000 feet below sea level. In this example, the minus sign is shown on the map not only for the elevation but also for the —1,000-foot contour line.

All elevations and contours above sea level are positive, although the plus sign is usually omitted. All elevations and contours below sea level are negative, and the minus sign should always be shown. On a map with positive contours, the structurally high areas always carry the larger contour numbers. With negative contours, the structurally high areas always carry the smaller contour numbers.

Contouring of subsurface maps follows the same principles as topographic surface contouring. Usually the data are not so complete as in surface contouring, and considerable judgment and experience may be required for the final structural interpretation.

Preliminary subsurface contouring begins with proportionate interpolation between each pair of elevation points. If the elevation points are a fair distance apart, interpolation should not be estimated but should be accurately scaled. Any later changes from this preliminary contouring, aside from smoothing sharp curves and corners, should be made only for sound reasons.

The following subsurface problems are not difficult with sufficient care in scaling proportionate interpolation.

**Figure A.51**
Base map for subsurface contouring of problem 48. (Copyright 1949 by Charles Merrick Nevin.)

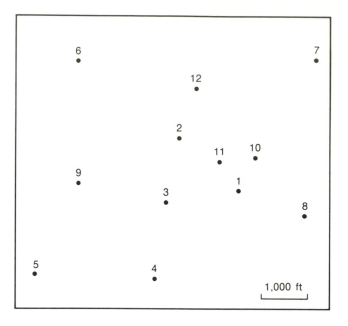

## Problem 48

The locations of 12 wells are shown in figure A.51. The area is to be contoured with a 50-foot interval. Well data for the subsurface structure are as tabulated.

| Well Number | Elevation of Top of Well | Depth to Key Horizon | Elevation of Key Horizon |
|---|---|---|---|
| 1 | 830 | 940 | −110 |
| 2 | 725 | 850 | −125 |
| 3 | 745 | 1000 | −255 |
| 4 | 615 | 1010 | −395 |
| 5 | 435 | 920 | −485 |
| 6 | 505 | 920 | −415 |
| 7 | 290 | 725 | −435 |
| 8 | 650 | 900 | −250 |
| 9 | 605 | 810 | −205 |
| 10 | 650 | 750 | −100 |
| 11 | 750 | 825 | −75 |
| 12 | 535 | 805 | −270 |

**Figure A.52**
Nomograms for converting dip angles to structure contours.

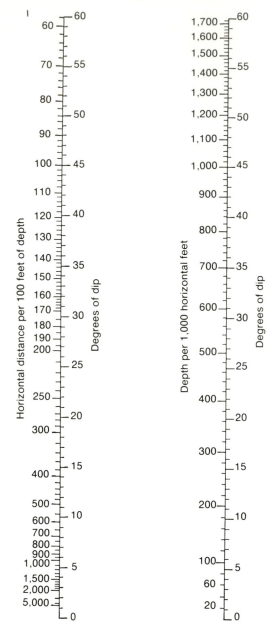

Sometimes it is necessary to construct a contour map of an area for which the data are shown by strikes and dips. Since dips can also be expressed as gradients, they may be readily converted to structure contours.

For instance, a dip of 20° will carry the bed to a depth of 364 feet in 1,000 feet of horizontal distance, perpendicular to the strike. Or, if a 100-foot contour interval is used, the contours would be 275 feet apart (horizontally) where the dip is 20°. These relations are shown by the two nomograms of figure A.52. Note that the nomogram scales can be enlarged or reduced. Thus, in nomogram *A,* if the contour interval is 500 feet, the horizontal distance between contours will be 1,375 feet; and, if the contour interval is 1,000 feet, the distance between contours will be 2,750 feet, where the dip is 20°. In nomogram *B,* the depth to the bed will be 182 feet for each 500 feet of horizontal distance, and 36 feet for each 100 feet of horizontal distance, where the dip is 20° and the measurement is perpendicular to the strike.

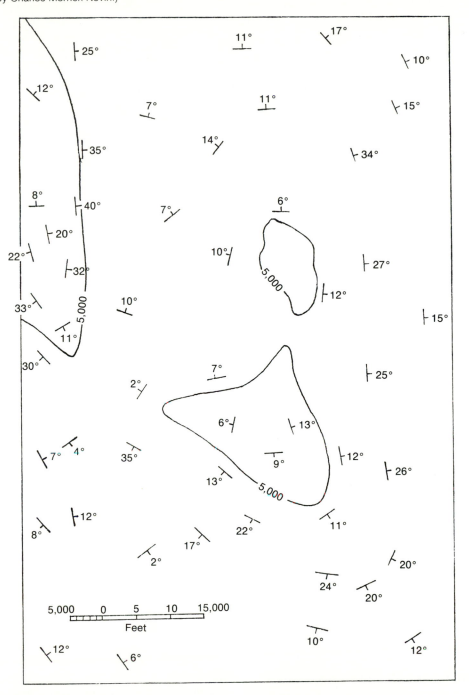

Nomogram *A,* figure A.52, is especially useful in determining the spacing of contours, once the contour interval has been chosen. Nomogram B is useful in checking the completed contour map for errors. Also, if the dip is small, this nomogram can be read more accurately because of the expanded scale.

*Problem 49*

The structure of an area in the Rocky Mountains is shown in figure A.53 by dip symbols. Draw a structure contour map, using a contour interval of 1,000 feet. The 5,000-foot contour is given as a starting datum. The horizontal scale can be transferred to a slip of paper to speed up the measurements. For example, where the dip is 12°, the 4,000-foot contour will lie 4,700 feet down-dip from the 5,000-foot contour and can be so scaled with the slip of paper.

**Figure A.54**
The effect of a regional dip illustrated by structure contours.
(a) Dome with a vertical closure of about 40 feet. (b) Regional dip to
the northwest, of 50 ft to the mile. (c) Regional dip superposed on
the dome and added to it. (d) The result is a nose plunging to the
northwest. Notice the migration of the high point of the structure.

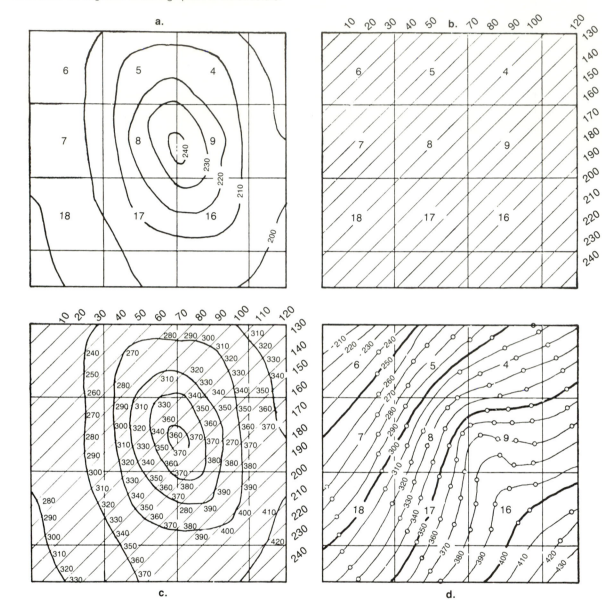

## Effect of Regional Dip and Thickening-Thinning of Formations on Flexures

We subconsciously tend to refer curvature in a vertical plane as seen and mapped at the surface, to a horizontal plane; we do this, whether its form is shown by dip symbols or by structure contours. In fact, measuring dips and measuring elevations for contouring both imply horizontal reference planes. However, where curvatures are gentle, or where vertical closure is small, this may lead to incorrect conclusions where a regional dip is present, and where subsurface thickening and thinning are appreciable. An example will explain.

Suppose that a dome had been formed and that, later, the entire area was tilted to the northwest. The net result, as seen and mapped at the surface, would be to open the dome to the southeast and steepen it to the northwest. If the vertical closure of the dome was small, and the amount of regional dip fairly large, the dome would be mapped as a nose plunging northwest (figure A.54). This is one reason why noses are so common in areas of gentle regional dip. In general, a regional tilt causes the high point of a structure to migrate, with depth, away from the direction of regional dip.

**Figure A.55**
The effect of southward thickening on a surface plunge to the north.
(a) Structure contours showing a plunge to the north, on which is
superposed a uniform thickening to the south. (b) The effect of
sedimentary thickening has been removed, and the structure shown
will be found below the thickened formation.

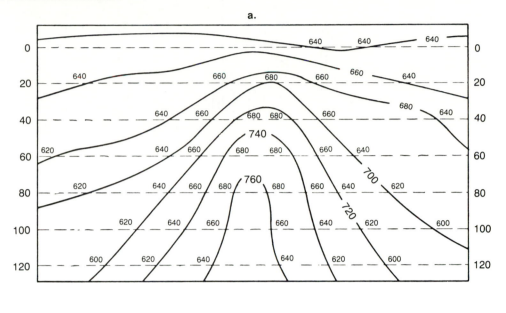

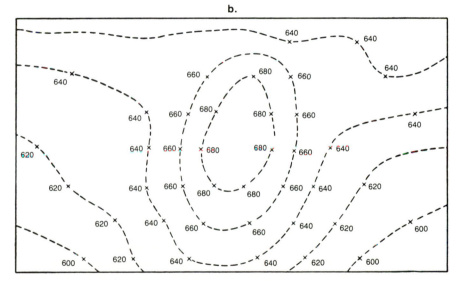

Thickening and thinning of formations (divergence and convergence) directly affect depth determinations. Usually the rate of convergence or divergence of a formation is rather uniform for a considerable distance, and thus, once known, its effect can be readily determined. Nevertheless, it is surprising how frequently this effect is either overlooked or incorrectly estimated.

In an area of mild deformation, surface flexures may plunge in several directions. Depending on the direction of stratigraphic thickening and thinning alone,

the subsurface structure may be very different from the surface flexure. Figure A.55 illustrates the effect of thickening when it occurs in a direction opposite to the plunge of the surface curvature. If this same method were used, what would happen if the thickening were in some other direction, or if it occurred at a different rate?

The method of evaluating the effects of regional dips and thickening and thinning is shown in figures A.54 and A.55. Use it to solve the following problem.

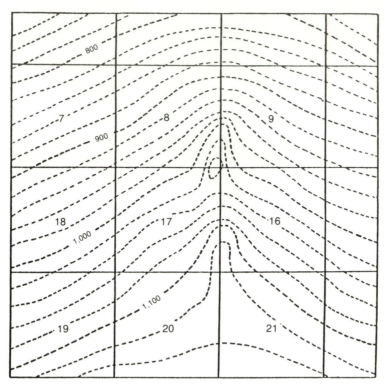

## Problem 50

In figure A.56 a surface nose, plunging to the north, is shown by dashed contours. The contour interval is 20 feet. There is a uniform thickening to the south of 100 feet per mile, at a depth of 1,500 feet below sea level. Recontour this map to show the effect of sedimentary thickening, at the general depth of 1,500 feet below sea level.

## The Busk Method of Reconstructing Folds

H. G. Busk (1929) devised a geometrical method for extrapolating folds at depth from known attitudes of folded layers at the surface. The assumptions are that all layers maintain uniform thickness and that the cross sections of fold hinges are perfect circles. Clearly, this method is applicable only where fold geometry approximates the assumptions.

Figure A.57 illustrates the method. It should be practiced on actual geological maps of folded regions. The steps for the general case (figure A.57a) are as follows:

1. Draw the cross section line on the map through the selected region.
2. Project the bedding outcrops along their strikes on to the cross section line, and measure the angles between the cross section line and the bedding strikes in order to determine apparent dips by using the alignment diagram on page 343.
3. Draw the topographical profile and locate on it the intersected and projected bedding traces, giving them the apparent dip values found in step 2 above.

**Figure A.57**
Construction of cross sections by the Busk method. See text for
procedure. (a) General case. (b) Adjustment of known key beds.
(Courtesy L. G. Duran.)

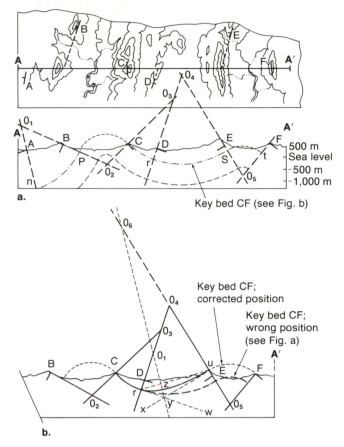

4. Draw perpendiculars to every (apparent) dip on
   the cross section, and extend them to intersect
   the perpendiculars to adjacent outcrops; this will
   give points $O_1$, $O_2$, $O_3$, $O_4$, $O_5$ (figure A.57a).
5. Using the above points as centers, and starting at
   any desired outcrop or key bed, such as $O_3$,
   construct the arcs $Cp$, $pn$, $Cr$, $rs$, and $st$.
6. The position of any desired horizon on the section
   may be found by drawing arcs with the same
   centers and giving the horizon the proper relative
   position according to thicknesses or
   stratigraphical intervals.

### Adjustments

If outcrops C and F, for example, (figure A.57a) are
found to belong to the same bed, it is obvious that the
section constructed according to the above method does
not fit the field information.

Figure A.57b illustrates the method to be used to
correct such a discrepancy. The method for the adjustment of known key beds can be summarized as follows:

1. Determine point $r$ with $O_3C$ as a radius and $O_3$ as
   center, and point $u$ with $O_5$ as center and $O_5F$ as
   a radius.
2. Draw perpendiculars $rw$ and $ux$ respectively to
   $O_4r$ and $O_4u$ at points $r$ and $u$.
3. Connect points $r$ and $u$ by the straight line $ru$;
   erect a perpendicular to it through $y$ (which is
   the intersection of perpendiculars $rw$ and $ux$),
   and prolong this perpendicular until it intersects
   the extension of $O_4O_5$ at point $O_6$.
4. With $O_7$ and $O_6$ as centers, construct arcs $rz$ and
   $zu$, connecting $Cr$ and $Fu$. The new line $CrzuF$
   represents the true position of the bed cropping
   out at $C$ and $F$.

# Proof of the Mohr Circle Construction

In a body under a triaxial state of stress, the principal stresses are $\sigma_1$, $\sigma_2$, and $\sigma_3$. We assume that, for an elementary treatment, we may neglect $\sigma_2$ and only consider the $\sigma_1$–$\sigma_3$ plane. Then, lines in that plane represent traces of planes perpendicular to it and parallel to $\sigma_2$.

We now consider an arbitrary plane within the body, parallel to $\sigma_2$ and at an angle $\theta$ with $\sigma_3$. It will make a trace $AB$ in figure B.1, and AB makes an angle $\theta$ with the $\sigma_3$-axis. Assume that the plane represented by AB has unit area (e.g., 1 sq cm), and length $AB$ has unit length (e.g., 1 cm). We can resolve $AB$ along $AC$ (parallel to $\sigma_1$) and along $BC$ (parallel to $\sigma_3$). Then, by simple trigonometry, we see that the area represented by $AC = 1 \times \cos \theta$, and the area represented by $BC = 1 \times \sin \theta$.

Next we consider the *forces* acting on each of the surface elements represented by *AB, BC,* and *AC.* Since force equals stress times the area over which it acts, we obtain

force on side $BC = \sigma_3 \times \sin \theta$;
force on side $AC = \sigma_1 \times \cos \theta$.

The force on side $AB$ consists of a normal force, $\sigma_\eta \times 1$ and a tangential (shear) force, $\tau_\eta \times 1$.

For equilibrium, the forces acting along $AB$ must balance, and so must the forces acting perpendicular to $AB$ (i.e., parallel to $CH$). Hence, resolving along $CH$:

Force acting $\perp AB$ = force $\perp AC$ resolved
along $CH$
+ force $\perp BC$ resolved
along $CH$
or $1 \times \sigma_n = \sigma_1 \cos \theta \times \cos \theta$
$+ \sigma_3 \sin \theta \times \sin \theta$
Simplifying: $\sigma_n = \sigma_1 \cos^2\theta + \sigma_3 \sin^2\theta$
Now $\cos^2\theta = \frac{1}{2}(1 + \cos 2\theta)$
and $\sin^2\theta = \frac{1}{2}(1 - \cos 2\theta)$

## Figure B.1

Trace *AB* of an arbitrarily oriented plane on a normal cross-section plane, and stress components by conventional symbols (see text).

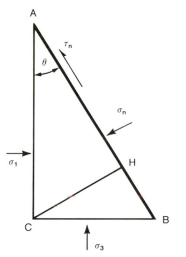

## Figure B.2

Mohr's circle for stress state in figure B.1.

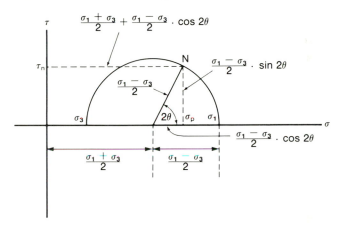

Hence, substituting and simplifying:
$$\sigma_n = \tfrac{1}{2}(\sigma_1 + \sigma_3) + \tfrac{1}{2}(\sigma_1 - \sigma_3) \cos 2\theta \quad \text{(B.1)}$$

and resolving along *AB*:

Force acting along *AB* = force $\perp$ *BC* resolved along *AB* + force $\perp$ *AC* resolved along *AB*

$$\text{or } 1 \times \tau_n = \sigma_1 \cos\theta \times \sin\theta - \sigma_3 \sin\theta \times \cos\theta$$

Simplifying: $\tau_n = (\sigma_1 - \sigma_3) \sin\theta \times \cos\theta$

Now $\sin\theta \times \cos\theta = \tfrac{1}{2}\sin 2\theta$

$$\therefore \qquad \tau_n = \tfrac{1}{2}(\sigma_1 - \sigma_3) \sin 2\theta \quad \text{(B.2)}$$

Rearranging equations B.1 and B.2, and squaring:

$$[\sigma_n - \tfrac{1}{2}(\sigma_1 + \sigma_3)]^2 = [\tfrac{1}{2}(\sigma_1 - \sigma_3)]^2 \cos^2 2\theta$$

$$\text{and } \tau_n^2 = [\tfrac{1}{2}(\sigma_1 - \sigma_3)]^2 \sin^2 2\theta \quad \text{(B.3)}$$

Adding equations B.3, we get

$$[\sigma_n - \tfrac{1}{2}(\sigma_1 + \sigma_3)]^2 + \tau_n^2 = [\tfrac{1}{2}(\sigma_1 - \sigma_3)]^2 \\ (\cos^2 2\theta + \sin^2 2\theta)$$

that is,

$$[\sigma_n - \tfrac{1}{2}(\sigma_1 + \sigma_3)]^2 + \tau_n^2 = [\tfrac{1}{2}(\sigma_1 - \sigma_3)]^2 \quad \text{(B.4)}$$

This is the equation of a circle with radius $\tfrac{1}{2}(\sigma_1 - \sigma_3)$ centered on the $\sigma = $ axis at $\tfrac{1}{2}(\sigma_1 + \sigma_3)$ as shown in figure B.2. Note that $\tfrac{1}{2}(\sigma_1 - \sigma_3) = \tau_{max}$, the maximum shear stress, and $\tfrac{1}{2}(\sigma_1 + \sigma_3) = \bar{\sigma}$, the mean normal stress. Equation B.4 is the equation of Mohr's circle for biaxial stress, a very useful graphic method of solving equations B.1 and B.2.

# Appendix C
# Analysis of Separation

We may represent the relationship between a fault surface and a faulted key horizon in different projected views. Analyzing these views is an important task of the subsurface geologist. In addition to the separation, the parameters include the angular relationship between three surfaces: the fault, the faulted horizon, and (since conventional measurements are referred to the horizontal), a horizontal surface. Nehm (1939) showed that there are four, and only four, basic patterns in this angular relationship. Figure C.1 gives these relationships in stereographic projection. Using the symbols in the caption of figure C.1, and designating the dips of the fault and the key horizon $\delta_F$ and $\delta_K$, respectively, we may characterize the four basic patterns as follows: In patterns 1–3, *KIF* is greater than 90°, and fault and key horizon dip in the same direction. For pattern 1, $\delta_K$ is greater than $\delta_F$; and for pattern 3, $\delta_K$ is less than $\delta_F$. For pattern 2, $\delta_K$ and $\delta_F$ are of comparable magnitude. The limiting condition for this pattern is that both *IK* and *IF* be acute angles. In pattern 4, *KIF* is less than 90°, and fault and key horizon dip in opposite directions. There are also three special limiting cases, those for which *IK, IF,* or *KF* are 90°.

These four basic intersection patterns yield eight separation patterns when each is combined with normal separation (figure C.2) and reverse separation (figure C.3). The patterns are shown in the four most commonly employed views: map view, view perpendicular to the fault surface, vertical cross section perpendicular to the fault surface, and vertical cross section perpendicular to the key horizon. Figures C.2 and C.3 also reveal the fundamental separation patterns for which beds are repeated or omitted; this depends not only on separation sense, but also on the angular relationship of the fault and the faulted surface. A fault that displaces a folded surface may, in fact, repeat or omit it in a given section, depending on the local dip of that surface (e.g., figure 12.11).

**Figure C.1**

The four basic intersection patterns between a fault and a faulted key horizon, referred to the horizontal in stereographic projection. $K$ = strike of key horizon; $F$ = strike of fault; and $I$ = trend of intersection between fault and faulted key horizon. $KIF$, $IK$, $IF$ are horizontal angles measured along the equator of the projection.

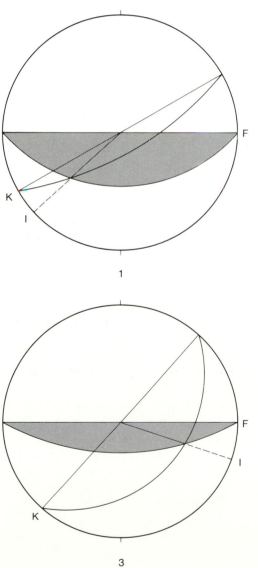

1

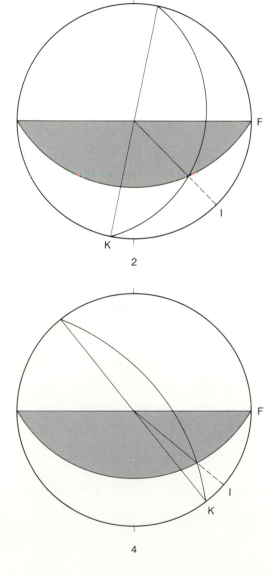

2

3

4

**Figure C.2**

Faulted surface in normal dip separation. The four basic separation patterns, each as seen in plan, in the fault surface, in vertical cross section perpendicular to the fault surface, and in vertical cross section perpendicular to the displaced surface. (Adapted from Nehm, 1939.)

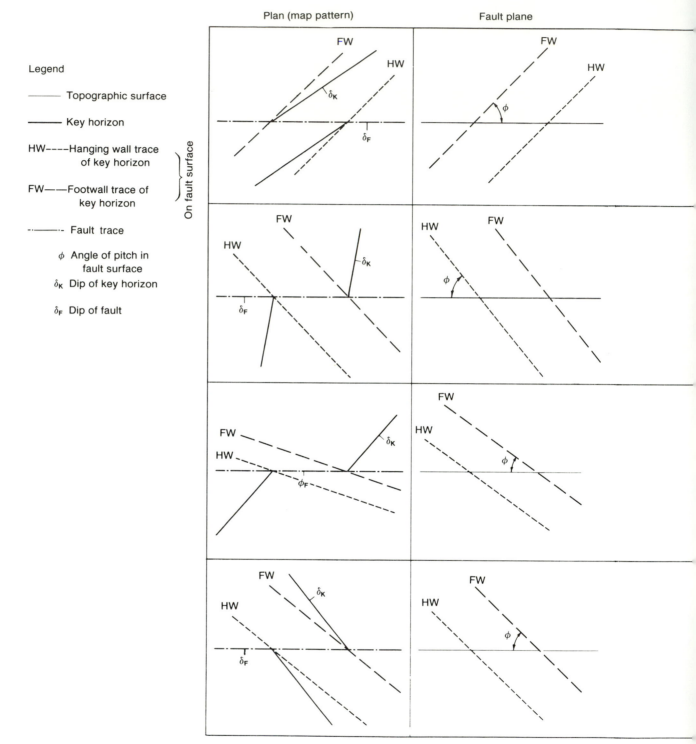

Plan (map pattern)

Fault plane

Legend

———— Topographic surface

———— Key horizon

HW————Hanging wall trace of key horizon

FW————Footwall trace of key horizon

—·——·— Fault trace

$\phi$ Angle of pitch in fault surface

$\delta_K$ Dip of key horizon

$\delta_F$ Dip of fault

On fault surface

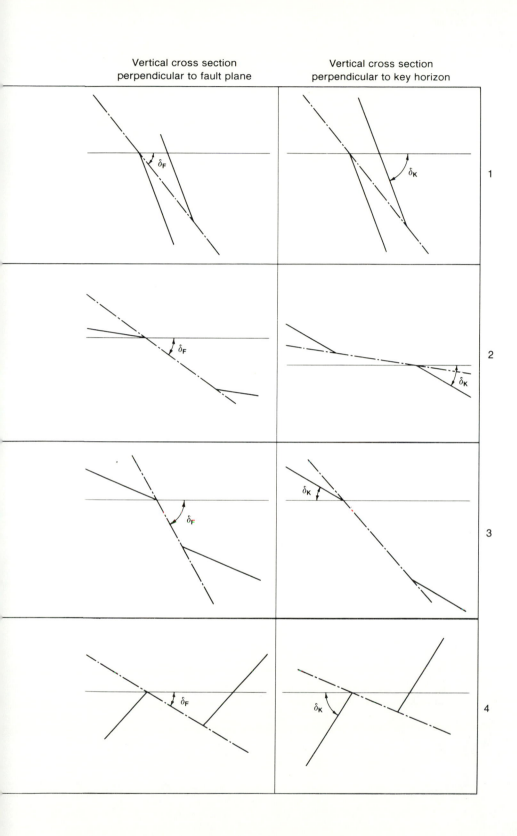

Vertical cross section
perpendicular to fault plane

Vertical cross section
perpendicular to key horizon

**Figure C.3**

Faulted surface in reverse dip separation. The four basic separation
patterns, each as seen in plan, in the fault surface, in vertical cross
section perpendicular to the fault surface, and in vertical cross
section perpendicular to the displaced surface. (Adapted from
Nehm, 1939.)

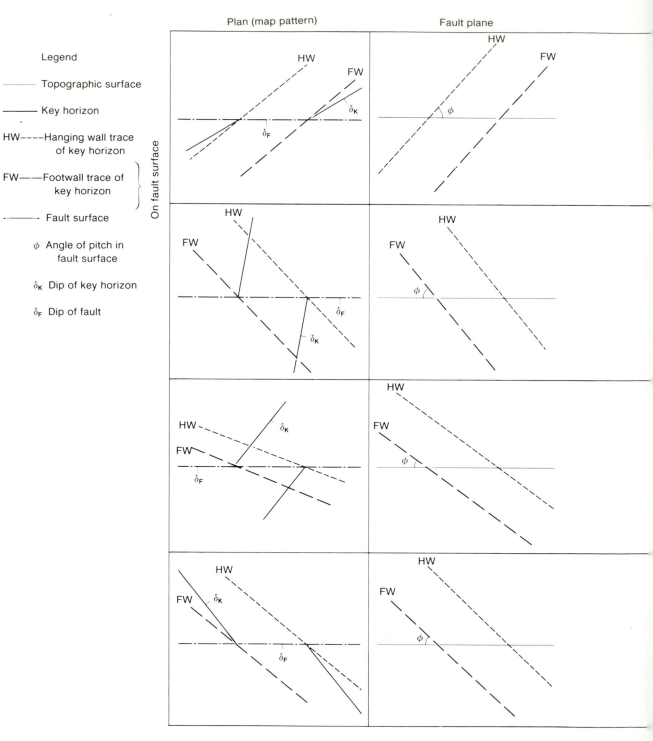

Legend

———— Topographic surface

———— Key horizon

HW ----Hanging wall trace
of key horizon

FW ———Footwall trace of
key horizon

—·——·— Fault surface

$\phi$  Angle of pitch in
fault surface

$\delta_K$  Dip of key horizon

$\delta_F$  Dip of fault

On fault surface

Plan (map pattern)

Fault plane

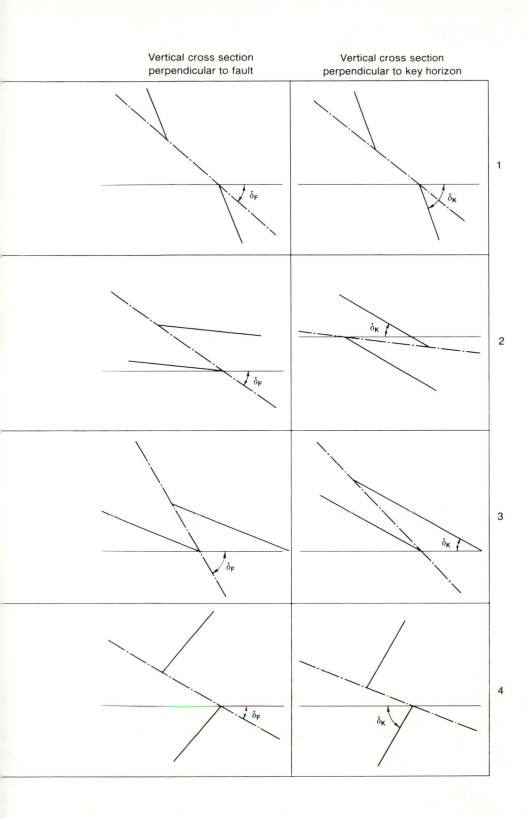

# Geologic Map Symbols[1]

## Bedding

| | |
|---|---|
| 25 ⟩ | Strike and dip of beds. |
| ⟩⟋ | Approximate strike and dip. |
| 25 ⟩ | Strike and dip of beds where upper bed can be distinguished, used only in areas of complex overturned folding. |
| 40 ⟫ | Generalized strike and dip of crumpled, plicated, crenulated, or undulating beds. |
| ⊕ | Horizontal beds. |
| ⟋ 90 | Strike of vertical beds. |
| ⤬ 60 | Strike and dip of overturned beds. |
| ⟵ 25 | Strike and dip of beds and plunge of slickensides. |
| 50 ⟶ | Apparent dip. |

## Foliation and Cleavage

The map explanation should always specify the kind of cleavage mapped.

| | |
|---|---|
| ⟋ 50 | Strike and dip of foliation. |
| ⤙ | Strike of vertical foliation. |
| ✦ | Horizontal foliation. |
| 75 ⟍ | Strike and dip of cleavage. |
| ⟍ | Strike of vertical cleavage. |
| ╫ | Horizontal cleavage. |
| ⫽ ⫽ ⫽ ⊢ | Alternative symbols for other planar elements. |

[1]Based on AGI Data Sheet 2 (Dietrich et al. 1982).

## Lineations

| | |
|---|---|
| 40 ↙ | Bearing and plunge of lineation. |
| ↗↙ | Horizontal lineation. |
| ↗ 25 / 10 | Strike and dip of beds and plunge of lineation. |
| ↗↙ 60 | Strike and dip of beds showing horizontal lineation. |
| 40 ⤢ 90 | Vertical beds, showing plunge of lineation. |
| ↗↙ 90 | Vertical beds, showing horizontal lineation. |
| ◆ | Vertical lineation. |
| ↗↓ | Double lineation. |
| 25 ↘ 60 | Strike and dip of foliation and plunge of lineation. |
| 60 ↗↘ | Strike and dip of foliation showing horizontal lineation. |
| ↘ 30 | Vertical foliation, showing plunge of lineation. |
| ↗↘ | Vertical foliation, showing horizontal lineation. |
| ∿ ↓40 | Generalized strike of folded beds, or foliation, showing plunge of fold axes. |

## Joints

| | |
|---|---|
| 60 ╱ | Strike and dip of joint. |
| ─■─ | Horizontal joint. |
| ◆ 90 | Strike of vertical joint. |
| 40 ■╱75 | Strikes and dips of multiple systems. |

| | Contacts |
|---|---|
| ⌒ | Definite contact. |
| - - - - - ⌒ | Inferred contact. |
| 50 ↘ | Contact, showing dip. |
| ⌒ ⌒ | Approximate contact. |
| ...... | Concealed contact. |
| 90 ⊢ | Vertical contact. |

| | Folds |
|---|---|
| 40 ◄ | Upright anticline, showing crest and sense of plunge. |
| 25 ↙ ↗ ⋔ | Overturned anticline, showing trace of axial surface, dip of limbs, and plunge. |
| 40 ◄ ⊂ | Upright minor anticline, showing sense of plunge. |
| ⋮ ≡ ⋮ | Approximate crests and troughs. |
| ⋮ ⋯ ⋮ | Concealed fold crests and troughs. |
| ↓ 25 | Upright syncline, showing trough and sense of plunge. |
| ⋔ | Overturned syncline, showing trace of axial surface and dip of limbs. |
| ⊂ ◄ 15 | Upright minor syncline, showing sense of plunge. |
| ⋮ ⋯ ⋮ | Inferred crests and troughs. |
| ⋮ ⌐ ⋮ ⌐ | Doubtful fold crests and troughs, dotted where concealed. |
| ◄───► | Horizontal fold axes. |
| ◄↕► | Dome. |

## Faults

|  |  |
|---|---|
| ⌢↓60 | Fault, showing dip. |
| – – – – | Approximate fault. |
| ·······.···· | Concealed fault. |
| ⌢⊥⊥⊥⊥⊥ | Normal fault, hachures on down side. |
| U / D | Dip-slip fault, showing relative displacement—U (up) and D (down). |
| ⌒→25 | Normal fault, showing bearing and plunge of relative displacement of downthrown block. |
| ⌒→40 | Fault, showing bearing and plunge of grooves, striations, or slickensides. |
| ═══ | Lineaments. |
| 90 ⌒⊥ | Vertical fault. |
| ·– – – – | Inferred fault. |
| – – –?·.···· | Doubtful fault, dotted where concealed. |
| ⇄ | Strike-slip fault, showing relative displacement. |
| – –⊣ | Thrust or slide fault; T, upper unit. |
| ▲▲▲ | Thrust or reverse fault, barbs on side of upper unit. |
| D ⟋25 | Reverse fault, showing bearing and plunge of relative displacement of downthrown block (D). |
| ⟋⟋↘40 | Fault zone, showing dip. |
| ∴∵ | Fault breccia. |

# Appendix E

# Selected Units and Conversion Factors

## Length

1 inch = 2.540 centimeters
1 foot = 30.480 centimeters
1 statute mile = 1.609 kilometers
1 meter = 39.370 inches = 3.281 feet
1 kilometer = 0.621 statute mile
1 international nautical mile = 1 great circle minute of arc
= 1.852 kilometers
1 fathom = 6 feet = 1.829 meters

## Area

1 square inch = 6.452 square centimeters
1 square mile = 2.590 square kilometers
1 acre = 4,047 square meters
1 hectare = $10^4$ square meters = 2.47 acres
1 square kilometer = 0.386 square mile

## Volume (wet and dry)

1 cubic inch = 16.387 cubic centimeters
1 fluid ounce = 28.413 cubic centimeters
1 U.S. gallon = 128 fluid ounces = 3.785 liters
1 imperial gallon = 160 fluid ounces = 4.546 liters
1 gallon = 4 quarts = 8 pints (in both U.S. and Imperial systems)
1 liter = 1,000.027 cubic centimeters = 1.057 U.S. quarts
= 0.264 U.S. gallon

## Weight

1 ounce avoirdupois = 28.3495 grams
1 pound avoirdupois = 0.454 kilogram
1 kilogram = 2.205 pounds avoirdupois
1 ton (long)  = 2,240 pounds avoirdupois = 1.016 tons (metric)
1 ton (short)  = 2,000 pounds avoirdupois = 0.907 ton (metric)
1 ton (metric) = 1,000 kilograms

**Acceleration**

1 gal = 1 centimeter per second per second
Free fall, standard = 980.665 centimeters per second per second

**Pressure, Stress**

1 pound per square inch = 0.070 kilogram per square centimeter

1 kilogram per square centimeter = 14.223 pounds per square inch

1 bar = $10^6$ dynes per square centimeter = 0.1 megapascals

$\qquad$ = 1.020 kilograms per square centimeter

$\qquad$ = 0.987 normal atmosphere

1 megapascal = 10 bars = .01 kilobars

$\qquad$ = 10.197 kilograms per square centimeter

$\qquad$ = 145 pounds per square inch

**Density**

1 gram per cubic centimeter = 0.036 pound per cubic inch

$\qquad$ = 62.430 pounds per cubic foot

**Viscosity**

1 poise = 1 dyne per square centimeter

**Magnetic Field Intensity**

1 gauss = $10^5$ gammas

# Glossary

## A

**allochthone**
A rock unit or thrust sheet overlying a tectonic surface that has been displaced from its original site of emplacement.

**allochthonous**
See *Allochthone.*

**anatexis**
The breakdown of crustal rocks to form granitic material.

**angular unconformity**
An unconformity (q.v.), truncating a lower sequence of layered rocks whose attitude is at an angle to the unconformity.

**anisotropic**
Not isotropic; a medium having some physical property that varies with direction.

**anticline**
A fold, normally, but not necessarily, convex upward with a core of stratigraphically older beds.

**antiform**
A fold which closes upward.

**antithetic fault**
A set of normal faults that tend to oppose or attenuate crustal flexure. The fault blocks are rotated in a sense opposite to that of the external shear that produced them.

**apparent dip**
The angle between the horizontal and the trace of a geological surface (or plane) measured in a vertical plane not perpendicular to the strike.

**appressed fold**
A fold with a very small interlimb angle.

**Archimedes' principle**
The buoyant force acting on an immersed body is equal to the weight of the fluid that it displaces.

**asthenosphere**
The more mobile, yielding layer of weakness in the Earth underlying the lithosphere.

**attitude**

The orientation in space of a structural surface, plane, or line measured by dip and strike or plunge and trend.

**aulacogen**

A rift at the edge of, and at a large angle to, a craton near a concave reentrant in the craton margin. It is bounded by convergent normal faults and is filled with sediments contemporaneous with those on adjacent continental shelves.

**autochthone**

The rock body at its original site of emplacement.

**autochthonous**

See *Autochthone*.

**axial culmination**

Convex upward curvature of a fold axis.

**axial depression**

Concave upward curvature of a fold axis.

**axial (hinge) surface**

A surface that contains all the hinge lines of the stacked surfaces in a fold.

**axial plane**

(1) An axial surface that is planar. (2) A penetrative planar element that is parallel to the axial hinge surface.

**axial plane cleavage**

Transverse cleavage that is parallel to the axial plane of the related fold.

**axial trace**

The intersection of the axial surface of a fold with the topographic surface of the Earth or with any other specified surface. (Dennis 1967, 7)

## B

**backlimb**

The longer, more gently dipping limb in asymmetrical folds other than recumbent folds.

**bedding (bedding-plane) cleavage**

Cleavage that is parallel or almost parallel to bedding.

**bending**

The transverse deflection of a layer due to forces at right angles to the layer (contrast *buckling*).

**Benioff zone**

A seismic zone within the upper portion of a subducting lithospheric slab, commonly extending to a depth of several hundred kilometers, and dipping toward a continent or arc.

**beta diagram**

A stereographic projection in which related s planes are plotted as great circles in order to determine by their intersection the attitude of the statistically common axis of intersection or rotation of the s planes. Also spelled: β diagram.

**blind thrust**

A thrust that does not break the ground surface but causes deformation by folding in overlying layers.

**body force**

Any force acting on every point within a body.

**boudinage**

Relatively less ductile rock layers enclosed in more ductile rock and separated laterally, by layer-parallel stretching, into elongate segments each of which has a barrel-shaped cross section. The "barrels" lie end to end and commonly are separated by vein material. Each section is a *boudin*.

**box fold**

A fold formed under shallow overburden with a more or less rectangular cross section including two hinge zones.

**brittle**

The property of a material that, while undergoing elastic strain, fails by fracture below the yield stress.

**buckling**

The transverse deflection of a layer by layer-parallel compression (contrast *bending*).

**bulk modulus**

The modulus of elasticity which relates change in volume of a body to the stress that causes it. Represented by the symbol: K.

# C

**caldera**
A circular physiographic depression formed by collapse, which surrounds many central volcanoes.

**cataclasite**
A cohesive fault rock that is structureless and has few or no fragments visible to the naked eye.

**cataclastic flow**
Flow of rock by intergranular movement, usually accompanied by fracture.

**chevron fold**
Fold of mesoscopic scale with angular hinges and straight limbs.

**cleats**
Joints in coal.

**cleavage (in rocks)**
A set of closely spaced secondary fabric planes that impart on the rock a tendency to split along the planes.

**cleavage fan**
A family of diverging cleavage planes.

**cleavage refraction**
The change in attitude of cleavage as it crosses a contact between beds of different ductilities.

**columnar jointing**
Roughly tautozonal joints, forming a pattern of polygons in the plane normal to the joints, resulting in a system of columns with polygonal bases. (Dennis 1967, 94)

**compaction**
Reduction in pore space within a rock due to overburden or tectonic stress.

**competent**
An adjective referring to the relative ductility of a layer in a sequence of layers. The more competent layer is the less ductile.

**cone sheet**
Thin, inward-dipping sheets of dike rock that form an inverted cone pattern about an igneous center and are normally a few meters or less in thickness.

**conjugate folds**
Two sets of related folds whose axial surfaces are inclined toward one another.

**conjugate (fracture) system**
A system of fractures at different attitudes that appear to have formed in one event, usually consisting of two sets.

**continental drift**
The original theory that explained the present positions of the continents and ocean basins by the breakup and dispersion of a single great continent called Pangea.

**continuous cleavage**
Cleavage due to continuous, penetrative dimensional parallelism of platy minerals throughout a rock.

**contour lines**
Lines that connect points of equal elevation on a curved surface, above or below a reference datum (e.g., sea level) projected orthographically on a reference plane, usually a map.

**convergent plate boundary**
A boundary between two lithospheric plates that are moving toward each other; manifested at the surface as oceanic trenches or as continent-continent collision sutures.

**core**
The central portion of the Earth lying below the Gutenberg discontinuity. It consists mainly of iron in a solid inner core with a 1,200 km radius, which is surrounded by a liquid outer core 2,300 km thick. Total core radius equals 3,500 km.

**craton**
A relatively stable segment of a continent undergoing no more than epeirogenic movements and deformation for a prolonged period of time. It consists of *platforms* and *shields*.

**creep**
The slow, continuous deformation of solid rock below the yield point under a continuous stress.
   Another type of creep, *mass-wasting*, is the gradual downslope movement of soil, minerals, rock particles, and other superficial accumulations. See also *Fault creep*.

**crenulation cleavage**
Spaced cleavage along which an earlier schistosity has been deformed by microfolding (crenulation).

**cross-bedding**
Stratification in which layers are deposited in regular succession at an angle to the original surface of deposition. (Dennis 1967, 11)

**cross joint**
A joint that is approximately perpendicular to strong linear structure in a rock.

**cross section**
A diagram showing geological units and structures as cut by a given, usually vertical, plane.

**crush breccia**
A cohesive fault breccia consisting of intensely fragmented, unoriented fragments, and a matrix that makes up less than 10% of the rock.

**crust**
The outermost layer of the Earth that overlies the mantle above the Mohorovičić discontinuity. It consists of *continental crust* of mainly granodioritic composition and *oceanic crust* of mainly basaltic composition.

# D

**décollement**
Faulting by detachment along a stratigraphic surface.

**deformation**
The process that results in a change in the shape or size of a body by particle displacement within the body.

**denudation fault**
A low angle normal fault that moves essentially like a giant landslide mainly along bedding planes and brings younger rocks to lie over older rocks.

**detachment fault**
A fault marking a surface along which a layer or sheet has been detached and displaced.

**diagenetic structure**
Secondary structural feature formed in rocks after deposition, near the surface of the Earth, due to diagenetic processes.

**diapir**
Fold or plug-like structure whose more mobile core pierces overlying, less mobile rocks. (Dennis 1967, 33)

**diapiric fold**
See *Diapir*.

**dike swarm**
Extensive set of parallel dikes that are evidence of an extensional crustal regime at the time of emplacement.

**dilation**
Deformation of a body by change in volume but not shape. The particles within the body tend to separate or crowd together along straight lines. Synonym: *Dilatation*.

**dilatational strain**
See *Dilation*.

**dip**
The angle between a geologic surface (or plane) and the horizontal, measured downward from the horizontal, in a vertical plane along the line of greatest slope and perpendicular to the strike.

**dip isogon**
A line joining points of equal dip in a fold cross section.

**dip joint**
A steep joint whose strike is approximately in the direction of the dip of layering.

**dip slip**
The component of slip along the fault dip.

**dip-slip fault**
A fault whose net slip is approximately parallel to the dip of the fault.

**disconformity**
An unconformity between two sets of parallel strata that marks an interval of erosion or nondeposition.

**discordant bedding**
Bedding in which laminations within each layer are inclined to the main (true) bedding surfaces.

**disharmonic folding**
Contrasting fold wavelengths in a sequence of folded layers; may lead to décollement.

**disjunctive cleavage**
Cleavage consisting of thin, distinct planar discontinuities in the preexisting rock fabric. The intervening domain appears totally unaffected, so the cleavage planes tend to look like fractures.

**displacement**
A general term used for the relative change in position of formerly adjacent features along a fault.

**distortion**
Deformation of a body by change in shape but not in volume.

**divergent plate boundary**
A boundary between two lithospheric plates that are moving apart creating new oceanic-type lithosphere along a rift. Manifested at the surface as rifts and crests of active oceanic ridges.

**dome**
An upwarp, or other convex upward structure, circular or elliptical in map pattern whose outer boundaries dip away from a center in all directions.

**drag**
The deflection of beds along a fault surface.

**drape fold**
A fold "draped" around a relatively raised fault block (descriptive term).

**ductile deformation**
Permanent, irreversible, time-dependent deformation of solids without loss of cohesion.

**ductility**
The ability of a solid to flow without fracture.

**duplex**
An imbricate structure enclosed by a roof thrust and a sole thrust.

# E

**elastic deformation**
Deformation which is instantaneous and reversible and in which stress is proportional to strength.

**en echelon**
An overlapping or staggered arrangement, in a zone, of geologic structures that are oriented obliquely to the trend of the zone as a whole. The individual structures, e.g. fractures, or folds, are short, compared to the length of the zone. (Dennis 1967, 39)

**epeirogeny**
The broader, slower, vertical movements of the Earth's crust.

**erosion thrust**
A thrust along an erosion surface. (Dennis 1967, 156)

**eustatic**
Worldwide, uniform changes in sea level.

**extension fractures**
Fractures resulting from outward separation of two formerly adjoining surfaces.

# F

**fabric**
The internal ordering, shape, and orientation of repetitive elements in a rock.

**fabric axes**
Orthogonal axes of reference for the orientation of fabric elements.

**fabric data**
Measurements specifying the attitude of fabric elements.

**fabric diagram**
A stereographic projection of fabric data.

**fabric domain**
A discrete part of a rock body that has a uniform fabric or subfabric.

**fabric elements**
Repetitive components or internal structures of a rock that make up its fabric and that form some definable pattern.

**face**
(1) A natural or artificial exposure-plane. (2) Beds *face* in the direction of the stratigraphic top of the succession. (3) Folds *face* in the direction of the stratigraphically younger rocks along their axial hinge surface. (4) Faults *face* in the direction of the structurally lower unit. (Dennis 1967, 43–45)

**fault**
A fracture or a zone of fractures in rock along which displacement of the sides relative to one another, parallel to the fracture, has taken place.

**fault breccia**
A non-cohesive fault rock consisting of largely angular rock fragments that make up over 30% of the rock and are visible to the naked eye.

**fault creep**
Slow continuous movement along a fault.

**fault gouge**
A non-cohesive fault rock consisting largely of a clay-like paste.

**fault line**
See *Fault trace*.

**fault-line scarp**
A scarp produced by differential erosion of two different rock types brought side by side as a result of faulting.

**fault scarp**
A line of relatively straight, steep slopes or steps in topography indicating a vertical component of displacement by faulting.

**fault surface**
The surface along which fracture and displacement by faulting have taken place.

**fault trace**
The trace of a fault on the surface of the Earth or on any specified surface (e.g. map or cross section).

**fault zone**
A zone of sheared, crushed, or foliated rock along which numerous small displacements have taken place, adding up to an appreciable total offset of the undeformed walls.

**fenster**
See *Tectonic window*.

**fissility**
A property of rocks that allows them to be easily split along approximately parallel surfaces.

**fissure**
A fracture with a component of displacement normal to the fracture surface. (Dennis 1967, 63) The resulting opening is commonly filled with vein material.

**flexural-flow fold**
A fold generated by flexure of layers with some slip along layers and some flow of material within layers causing thickening in the hinge areas and thinning in the limbs.

**flexural-slip fold**
A fold generated by flexure of layers and slip along layers; it normally produces a parallel fold conserving parallel layers.

**flow**
Any deformation that is time-dependent and not instantly recoverable and occurs without permanent loss of cohesion. (See also *Creep*).

**fold**
A flexure or curvature in a geologic surface or in a set of stacked geologic surfaces.

**fold axis**
The nearest approximation to the line which, moved parallel with itself in space along a curved path, generates a fold, or any cylindrical portion of it. (Dennis 1967, 7)

**fold crest**
The line which joins the highest points of successive vertical cross sections of a fold, as defined by any particular folded surface. (Dennis 1967, 30)

**fold trough**
The axial line in a synform that joins the lowest points of the fold as defined by a particular bed. (Dennis 1967, 158)

**foliation**
Closely spaced planar anisotropy in rocks.

**footwall**
The lower side (wall) of an inclined fault.

**forced fold**
A fold produced in layered strata by the relative upward movement of an underlying fault block (genetic term).

**forelimb**
The shorter more steeply dipping limb in asymmetrical folds other than recumbent folds.

**form**
The external shape of a rock body created by its boundaries.

**fracture**
A surface along which loss of cohesion has taken place in a solid.

**fundamental joint system**
Orthogonal joint system found in most undisturbed rocks.

# G

**girdle**
Concentration of points (*poles*) representing orientations of fabric elements and coinciding approximately with a great circle in stereographic projection.

**gjá**
Gaping surface fissures in rocks of the Icelandic rift system. Plural: *gjar.*

**graben**
A linear topographic depression due to subsidence along a system of nearly parallel normal faults.

**graded bedding**
A texture of sedimentary layers in which grain size grades upward from coarser to finer.

**growth fault**
A listric normal fault which displaces an active surface of sedimentation causing a sedimentary unit on the hanging wall side of the fault to be thicker than its equivalent on the footwall side.

# H

**hanging wall**
The upper side (wall) of an inclined fault.

**halokinesis**
Deformation of highly ductile rocks, such as in diapirism, resulting from solid flow along pressure gradients.

**Hartmann's rule**
Compressive normal stress on a body always bisects the acute angle between the two sets of conjugate fractures that theoretically form at failure.

**heave**
The horizontal component of the dip separation. (Dennis 1967, 87)

**heterolithic unconformity**
See *Nonconformity.*

**hinge**
The locus of maximum curvature in a folded surface which is usually a line but may be a zone.

**homocline**
A succession of parallel dipping beds with uniform attitudes over a large area.

**Hooke's law**
The strain in simple extension is linearly proportional to the applied stress. (Bates et al. 1980, 296)

**horse**
Each individual, fault-bounded slice in an imbricate zone; also an isolated block in a fault zone.

**horst**
A relatively raised block due to relative uplift along a normal fault or faults.

**hot spot**
A volcanic center thought to be derived from a persistent rising plume of hot mantle material.

**hydrostatic pressure**
Isotropic stress in a stationary fluid.

# I

**ignimbrite**
Ash flow deposit.

**imbricate (schuppen) structure**
A structure resulting from parallel thrust faults overlapping in a shingle pattern.

**incompetent**
An adjective indicating the relatively lower ductility of a layer in a sequence of layers with contrasting ductilities.

**infrastructure**
The highly mobile, deep tectonic level in the crust where granitic material is generated.

**intrafolial folds**
Scattered small-scale, isoclinal folds whose limbs are parallel with the dominant foliation.

**intrusion**
The process and the result of emplacement by flowage of one rock in another.

**inverted limb**
See *Reversed limb.*

**island arc**
A chain of volcanic islands bordered on one side by a deep sea trench. (Dennis et al. 1979, 44)

**isoclinal fold**
A fold whose limbs are parallel.

**isostasy**
The tendency of all components in the Earth's crust to reach and maintain gravitational equilibrium.

**isotropic**
The state of having a given physical property equal in all directions.

# J

**joint**
Rock fracture or fissure along which there has been little or no displacement. (Dennis 1967, 93)

# K

**kink band**
A tabular zone, normally mesoscopic, in which foliation is deflected along kinks in the foliation.

**kink fold**
A chevron-type fold formed by kinking thin, even layers in schistose rocks causing sharp deflections along tabular zones.

**klippe**
An isolated allochthonous unit overlying a tectonic surface surrounded in map pattern by tectonically lower rocks.

# L

**leaching**
The selective removal, separation, or dissolving out of soluble materials from rock by the natural action of percolating water.

**left-slip fault**
A strike-slip fault in which the block across the fault from an observer appears offset to the left.

**Liesegang banding**
Secondary color banding in a rock, which transects bedding and which is caused by diffusion in groundwater of colored impurities, usually iron oxides.

**limb**
The weakly curved or uncurved portions of a fold between adjacent fold hinges.

**lineament**
Regional alignment of geological or physiographic features inferred to reflect a crustal structural discontinuity.

**linear**
Straight or slightly curved line on the surface of the Earth as seen in aerial or satellite imagery.

**lineation**
All penetrative linear structures in a rock without regard to origin.

**listric fault**
A normal or reverse fault that flattens with depth and usually has a curved strike and a spoon-shaped fault surface.

**lithosphere**
The relatively strong, noncontinuous, outermost silicate layer of the Earth consisting of the crust and the upper mantle. It overlies the weaker *asthenosphere* (q.v.).

**longitudinal joints**
Steep joints parallel to lineation or fold axes.

# M

**mantle**
The layer in the interior of the Earth between the Mohorovičić discontinuity, at the base of the crust, and the Gutenberg discontinuity, above the core. It is 2,900 km thick and consists mainly of ultramafic rock.

**Maxwell relaxation time**
A material constant given by the ratio of the viscosity of a material to its shear modulus.

**mesosphere**
The zone of relative strength beneath the asthenosphere making up the lower mantle of the Earth.

**microlithon**
The rock material between spaced cleavage planes.

**migmatite**
A mixed rock with both igneous and metamorphic components.

**minor fold**
A small-scale subsidiary fold usually associated with or related to a major fold.

**Mohorovičić discontinuity**
The seismic velocity discontinuity that marks the contact between the crust and the subjacent mantle where $P$ wave seismic velocities abruptly increase when leaving the crust and entering the upper mantle.

**Mohr circle**
See *Mohr diagram*.

**Mohr diagram**
A diagram that relates graphically shear stress and normal stress on a given plane. The resulting curve is a circle, the *Mohr circle*.

**Mohr envelope**
A tangent curve to a family of Mohr circles representing the state of stress that causes failure in shear.

**monocline**
Local steepening of an otherwise uniformly gently dipping or horizontal sequence of beds.

**mullion structure**
Columnar, commonly cuspoid structures bounded by long cylindrical surfaces that may be polished, covered by mica films, or striated longitudinally. They usually mark a contact between layers of contrasting ductilities.

**mylonite**
A foliated, cohesive fault rock with the following characteristics: (1) grain size reduction; (2) occurrence in a relatively narrow planar zone; and (3) foliation more intense than in adjacent rock.

# N

**nappe**
A thrust sheet or (less commonly) a large recumbent fold.

**neotectonics**
The study of recent crustal movements of the Earth as far back in time as the geological record shows a coherent and continuous tectonic history into the present.

**net slip**
See *Slip*.

**Newtonian liquid**
An ideally viscous substance in which the rate of shear strain is proportional to shear stress. See also *Viscous deformation*.

**nonconformity**
An unconformity marked by younger sedimentary rocks overlying eroded older, non-stratified rocks.

**nonsystematic joints**
Joints that have the following geometric properties: (1) they meet but do not cross other joints; (2) they are generally strongly curved; (3) they commonly terminate at bedding surfaces; (4) surface structures on faces are not oriented. (Dennis 1967, 96)

**normal cross section**
See *Tectonic profile*.

**normal fault**
A fault that dips toward the faultblock that has been relatively lowered.

**normal joint**
A joint formed perpendicular to foliation or layering.

**normal stress**
The stress component perpendicular to a given plane. It can be tensile or compressive.

**nose**
A sharp bend in an outcrop pattern where a plunging fold intersects the topographic surface.

# O

**obduction**
The overriding of oceanic lithosphere onto continental lithosphere.

**onlap unconformity**
An unconformity along which the younger, overlying beds appear truncated by the unconformity. Synonym: *Buttress unconformity*.

**orogeny**
Intense, episodic, irreversible deformation of rock bodies along restricted, elongate zones of continental crust within a limited time interval.

**orthogonal joint system**
Two or three more or less vertical sets of joints that intersect nearly at right angles to one another. See also *Fundamental joint system*.

**orthographic projection**
Projection along straight parallel lines on a plane of projection perpendicular to the projection lines.

**outcrop**
Area over which a rock body intersects the topographic surface of the Earth. It refers to both exposed bedrock and bedrock covered by surficial deposits.

**overthrust**
See *Thrust*.

**overturned fold**
A fold, other than a steeply plunging fold that has one overturned limb.

**overturned limb**
See *Reversed limb*.

# P

**paleomagnetism**
The study of remanent magnetization in rocks and sediments.

**palingenesis**
The rebirth of granitic magma in situ following *anatexis* (q.v.) of crustal rocks.

**palinspastic cross section**
A pre-shortening cross section of a deformed crustal segment.

**parallel cleavage**
Cleavage which has developed along a preexisting foliation.

**parallel fold**
A fold in which the thickness of the folded layers remains constant and in which successive folded surfaces remain mutually parallel throughout.

**parallel unconformity**
See *Disconformity*.

**passive folding**
Folding in relatively ductile and mechanically isotropic rock units, usually forming similar-type folds, and in which bedding planes have no mechanical significance.

**penecontemporaneous deformation**
Deformation of sedimentary rocks during or immediately following their deposition but before complete consolidation.

**permeability**
The measure of relative ease of fluid flow through a porous material: the inverse of resistance to flow. It is a dynamic quantity, unlike *porosity* (q.v.), which is a geometric property.

**perpendicular separation**
The shortest distance between two displaced parts of a faulted surface measured at right angles to the faulted surface.

**phyllonite**
A mica-rich mylonite that has the mesoscopic appearance of a phyllite.

**pi diagram**
A stereographic projection in which related *s* planes are plotted as their normals (poles) to determine their statistically common axis of rotation, which is the axis of the resulting girdle. Also spelled: *π diagram.*

**pillow lava**
Lava extruded under water forming distorted globular "pillows."

**pinch and swell**
A structure formed by the stretching of a competent layer in a less competent matrix, with ductility contrast less than that required for boudinage, resulting in pinches and swells in the competent layer.

**pinnate (feather) joints**
Extension joints developed along a fault which are arranged *en echelon* and inclined to the fault surface. Their acute angles with the fault point arrowhead-fashion in the direction of relative displacement.

**pitch**
The angle between any line lying within a surface or plane and a horizontal line lying within the same surface or plane.

**plane strain**
Strain in which all displacements are parallel to one plane; the strain along the intermediate strain axis is zero.

**plasticity**
The property of crystals to deform permanently, without rupture, along lattice planes.

**plates (lithospheric)**
The horizontally rigid shell segments of lithosphere which are in constant motion relative to one another.

**plate boundary**
Zone of seismic and tectonic activity between adjoining lithospheric plates.

**plate tectonics**
A model of global tectonics in which the Earth's lithosphere is divided into a number of rigid plates that interact with one another at their boundaries causing seismic and tectonic activity along these boundaries.

**plumose structure**
A delicate tracery of featherlike markings on a joint surface. The plumes diverge from a central axis and, in the outermost parts of fracture, pass into a system of minor planes. (J. Roberts 1961, 481–482)

**plunge**
The vertical angle between a line in space and the horizontal measured along its trend in the down-plunge direction.

**pluton**
A non-volcanic body of igneous rock. (Dennis 1967, 122)

**pore pressure**
The isotropic stress or pressure transmitted by a fluid that fills the pore spaces of a rock.

**porosity**
The percentage of bulk volume of a rock occupied by pore space (void space).

**porphyroclasts**
The larger fragments of broken mineral grains or aggregates of grains in a finer matrix of a fault rock.

**preferred orientation**
Non-random orientation of fabric elements.

**pressure gradient**
The rate of variation of pressure in a given direction in space at a given time. (Bates et al. 1980, 497)

**pressure solution**
Dissolution, in sedimentary or metamorphic rocks, at grain to grain contacts where grain to grain pressure exceeds the pore pressure, selectively dissolving materials in zones of high pressure and redepositing them in zones of lower pressure.

**primary fabric**
Rock fabric developed at the time of original deposition or emplacement.

**primary flat joints**
Flat lying joints in a pluton that are usually, but not necessarily, parallel to flat lying planar fabrics.

**primary structures**
Structures that originate contemporaneously with the formation and/or emplacement of a rock. (Bates et al. 1980, 498)

**principal axes of stress**
Three mutually perpendicular lines drawn normal to each of the three principal planes of stress along which shear stresses are zero.

**protomylonite**
A cohesive fault rock in the early stages of mylonitization in which megascopically visible, commonly lenticular fragments make up about 50% of the rock.

**pseudotachylite**
An intensely sheared cohesive fault rock that has melted and looks like tachylite.

**ptygmatic fold**
Closely appressed folds in veins or dikes with axial planes parallel to the dominant foliation.

**pull-apart basin**
A graben or basin structure formed by an extension component across a strike-slip zone.

**pure shear**
Irrotational strain that consists of uniform extension of a body in one direction and uniform contraction at right angles to this direction. No volume change or change along the intermediate strain axis is associated with pure shear.

# R

**recumbent fold**
An overturned fold with a more or less horizontal axial surface.

**reticulate cleavage**
Cleavage along surfaces that branch and rejoin in a network fashion.

**reversed limb**
The forelimb of a fold that has been rotated or tilted beyond the vertical so that the sequence of strata appears reversed.

**reverse fault**
A fault that dips toward the fault block which has been relatively raised.

**rheid**
A substance, below its melting point, which deforms by viscous flow during the time of applied stress at an order magnitude at least three times that of elastic deformation under similar conditions. (Carey 1954, 71)

**rheidity**
The Maxwell relaxation time of a substance multiplied by 1,000.

**rheology**
The study of the relationships between stress, strain, time, and rate of strain of a material.

**Riecke's principle**
Dissolution of a mineral tends to occur most readily at points where external pressure is greatest while crystallization of a mineral occurs most readily at points where external pressure is least. (Bates et al. 1980, 537–538)

**rift**
A long, narrow, topographic depression of regional or global extent bounded by roughly parallel striking normal faults; a graben of regional extent associated with seismic and volcanic activity due to extension in the lithosphere under this zone.

**ring complex**
A complex system of igneous intrusions, mainly dikes of circular and radial outcrop patterns.

**ring dikes**
Concentric dikes dipping at a small angle from the vertical, either toward or away from a common center, and normally hundreds of meters thick.

**right-slip fault**
A strike-slip fault in which the block across the fault from an observer appears offset to the right.

**rodding**
A family of monomineralic, thin, rod-shaped parallel bodies in deformed metamorphic rocks.

**roof thrust**
A flat-lying thrust surface overlying an imbricate zone or duplex.

**rotational slide**
A landslide that creates its own sliding surface which is listric in form.

**rough cleavage**
Continuous cleavage in rocks with a high proportion of granular minerals and some oriented platy minerals. Thus, cleavage surfaces tend to be rough.

**rupture**
Deformation characterized by loss of cohesion in a material.

# S

**salt dome**
A dome or plug-shaped salt diapir, circular or elliptical in plan, formed as a result of density contrast between a mother salt bed and the overlying country rock.

**schistosity**
(1) A planar fabric element in metamorphic rocks resulting from the parallel orientation of platy or elongated mineral grains. (2) A coarsely foliated continuous cleavage.

**schlieren**
Segregations of minerals in igneous bodies.

**secondary fabric**
Rock fabric developed during deformation of a rock.

**separation (fault)**
The distance between the displaced parts of a faulted surface measured in any specified direction.

**serial folding**
Folding at high crustal levels in which folds form in sequence, one fold piling up against the next.

**set of joints**
A family of parallel joints. (Dennis 1967, 97)

**shear fractures**
Fractures resulting from displacement (faulting) along the fracture surface.

**shear modulus**
The modulus of elasticity in shear. Also called the modulus of rigidity. Represented by the symbol: G.

**shear strain**
Strain in which particles in a body move past one another.

**shear stress**
The stress component within and parallel to a given plane. It tends to move particles past one another.

**shear zones**
Tabular zones of failure that localize shear strain.

**sheeting**
The division of massive, plutonic rocks into rectangular slabs or sheets by nearly horizontal joints which are generally parallel with the exposed surface of the rock body. (Dennis 1967, 134)

**shift**
The true relative displacement measured along a fault after eliminating the effect of drag.

**similar fold**
A fold in which successive surfaces form congruent or similar curves. (Dennis 1967, 75)

**similitude**
The ratio of model dimensions to corresponding dimensions of a prototype.

**simple shear**
A strain in which planes parallel with one another and with the intermediate principal axis of strain slide parallel to an arbitrary, fixed plane. Displacements are proportional to their perpendicular distance from the fixed plane. No volume change or change along the intermediate strain axis is associated with simple shear.

**slab slide**
A landslide that slides along preexisting surfaces of weakness, such as lubricated joints or bedding surfaces.

**slaty cleavage**
Cleavage characteristic of slates. It is usually a homogeneous, fine-grained cleavage with very smooth and continuous or very closely spaced parallel cleavage planes.

**sled-runner thrust**
A thrust that develops by detachment along a flat lying surface terminating in a listric, concave upward toe.

**slickensides**
The smoothed, polished, and in some cases, fibrous, striations along a fault surface.

**slip (fault)**
The distance between two originally adjacent points situated on opposite sides of a fault measured in the fault surface.

**smooth cleavage**
Continuous cleavage in rocks that are made up almost entirely of fine-grained, usually micaceous, minerals, and thus have smooth cleavage surfaces.

**spaced cleavage**
Cleavage surfaces that are spaced at finite intervals at the scale under consideration.

**s plane**
See *s surface*.

**sole fault**
A low-dip, basal master fault that is a detachment surface tangent to a family of listric faults, either normal or reverse.

**sole thrust**
The lowest, flat-lying thrust surface underlying an imbricate zone.

**s surface**
Any kind of penetrative planar fabric element.

**strain**
The change in shape or volume of a body due to stress. The relative displacement of constituent particles as a result of stress.

**strain ellipsoid**
A three-dimensional, graphic representation of strain. The X semi axis normally represents maximum extension, and the Z semi axis normally represents maximum shortening.

**stratification**
The deposition of materials in layers. This normally refers to bedding in sedimentary rocks, but it can include depositional layering in igneous rock. It refers to both the process and its result.

**stratigraphic separation**
Synonym of perpendicular separation where the separated surface is a bedding surface.

**strength**
The limiting stress before failure of a specific kind under specified conditions. (Dennis 1967, 143)

**stress**
The intensity of force per unit area on a plane in a body.

**stress corrosion**
Propagation of cracks by circulation of solutions along networks of incipient cracks.

**stress field**
The aggregate of stresses in a body, homogeneous or varying from point to point, in a given domain.

**stress trajectories**
Lines of equal value for a stated stress component.

**strike**
The direction of the intersection between a geological surface or plane and a horizontal plane.

**strike joint**
A joint whose strike is nearly coincident with the strike of strata or layering and generally inclined at a large angle to the layers. (Dennis 1967, 97)

**strike slip**
The component of slip along the fault strike.

**strike-slip fault**
A fault with predominately horizontal displacement whose net slip is almost parallel to the strike of the fault.

**structural terrace**
A shelf-like flattening of the dip in inclined strata. (Dennis 1967, 145)

**structure contour**
A contour line on a geological surface; it represents the strike at any point.

**stylolites**
Irregular, suture-like seams in soluble rocks filled with insoluble residues resulting from solution along surfaces and grossly perpendicular to contemporaneous compression.

**subduction**
The process of one lithospheric plate descending under another into the asthenosphere.

**superstructure**
The tectonic level in which deformation is predominantly brittle.

**supratenuous fold**
An anticline in which constituent layers thin over the crest. (Dennis 1967, 75)

**surface force**
Any force acting on the external surface of a body.

**symmetrical fold**
A fold whose axial surface is a plane of symmetry (i.e., a fold whose limbs make equal angles with the axial surface). A vertical axial surface is sometimes postulated.

**symmetry**
Correspondence in size, shape, and position of parts that are on opposite sides of a dividing line or center. (The Merriam-Webster Dictionary, © 1974, by G&C Merriam Co.)

**syncline**
A fold, normally but not necessarily, concave upward with a core of stratigraphically younger beds.

**synform**
A fold that closes downward.

**synthetic fault**
A set of normal faults that tend to enhance crustal flexure. The rotation of the fault blocks is in the same sense as that of the external shear that produced them.

**systematic joints**
Joints that occur in one or more regular sets and with the following geometric properties: (1) they have straight or broadly curved traces or surfaces; (2) they are about perpendicular to the upper and lower surfaces of the rock unit in which they are present; (3) surface structures on the faces are oriented; (4) they cut across other joints. (Dennis 1967, 98)

**system of joints**
Two or more sets of joints that intersect at a more or less constant angle.

## T

**tachylite**
Basaltic glass.

**tear fault**
A strike-slip fault formed within a detachment sheet or nappe due to the sheet moving at unequal rates over a flat shear or detachment surface.

**tectonic level**
A level in the Earth's crust having a distinct tectonic style for a given tectonic episode.

**tectonic outlier**
See *Klippe.*

**tectonic profile**
A cross section perpendicular to fold axes in which the attitudes and thicknesses of beds are undistorted.

**tectonics**
The science of the regional structural and deformational features of the Earth's crust and the movements and forces which have produced them.

**tectonic unit**

A rock unit defined by tectonic contacts.

**tectonic window**

An isolated outcrop of a tectonic unit surrounded in map pattern by tectonically higher rocks.

**tephra**

Solid products of volcanism.

**throw**

The vertical component of the dip separation. (Dennis 1967, 154)

**thrust**

A fault with low average dips extending over a large area along which there is a large overlap of rocks that have moved from their original site of emplacement. In an *overthrust* the upper unit is assumed to have been the predominantly active unit; in an *underthrust* the lower unit is assumed to have been the predominantly active unit.

**thrust fault**

A synonym of reverse fault, often implies low dips. See also *Thrust*.

**thrust sheet**

The rock unit overlying a thrust surface.

**tilting**

A continuous crustal displacement resulting in differential changes in elevation.

**traction**

The stress component acting in any direction across a particular plane in a body.

**transform fault**

A strike-slip fault separating two lithosphere plates.

**transform plate boundary**

A boundary between two lithosphere plates along which relative displacement is strike slip.

**transition level**

The site of regional metamorphism between the *infrastructure* and the *superstructure*.

**transpression**

Convergence across strike-slip faults.

**transtension**

Extension across strike-slip faults.

**transverse cleavage**

Cleavage that transects older foliation, usually bedding. See also *Axial plane cleavage*.

**trench (oceanic)**

A long, narrow, depression in the deep seabed extending at least 2,000 meters below the adjacent seafloor, oriented normally parallel and adjacent to a continental margin or island arc. Associated with subduction zones.

**trend**

The bearing (direction) of a linear structure.

**triangular facets**

An aligned series of slopes truncating ridge spurs having a broad base and an apex pointing upward. Normally, this feature is associated with a fault plane located at the base of the facets, but it can also be due to stream or glacial erosion.

**triple junction**

A point or small area where three lithosphere plates meet.

**truncation**

The act and result of cutting or covering the top or end of a geologic structure or landform.

**turbidite**

A sedimentary rock deposited by turbidity currents; it commonly exhibits graded bedding, moderate sorting, and well developed primary structures.

**turbidity current**

A fast moving density current in water consisting of dense suspensions of sediments moving down subaqueous slopes under the influence of gravity and deposited horizontally where its speed is checked. These currents can originate from mud-slumps, submarine slides, and storm flooded streams.

## U

**ultramylonite**
A cohesive fault rock that has been sheared beyond the stage of mylonite *sensu stricto* and in which most fragments have been reduced to streaks; some recrystallization may have taken place but it appears aphanitic.

**unconformity**
A discontinuity in the geologic record represented by a surface of erosion or nondeposition that separates younger strata from older rocks.

**underthrust**
See *thrust*.

**uniformitarianism**
The fundamental principle in geology that assumes geological processes acting upon the Earth's crust at present represent processes that have acted upon the Earth's crust throughout geologic time, although not necessarily at the same rate or intensity. "The present is the key to the past." (Geikie 1882, 3)

## V

**valley bulges**
Where a valley is cut in a more competent bed overlying a less competent bed, this may cause the less competent, more ductile bed to flow up into the opening forming folds in the valley floor.

**varves**
Cyclic, seasonal, layers of sediment deposited in a body of still water annually. A glacial varve usually contains a lower, coarse-grained, light-colored, summer layer overlain by and grading up to a thinner, very fine-grained, dark-colored winter layer.

**vergence**
The direction towards which planes of folds are tilted (are *facing*); also the direction of displacement of a thrust or thrust system.

**viscosity**
Internal resistance to flow of a substance. The coefficient of viscosity is the ratio of shear stress to the rate of shear strain.

**viscous deformation**
Deformation in which the rate of shear strain is proportional to shear stress.

**volcanic pipe**
A volcanic vent filled with consolidated volcanic rock.

**volcanic vent**
A pipe-shaped opening through which products of volcanism can reach the surface.

## Y

**yield point**
The stress at which ductile, irreversible deformation first occurs following elastic deformation.

**yield strength**
See *Yield point*.

**yield stress**
See *Yield point*.

**Young's modulus**
The modulus of elasticity in tension or compression, involving change in length (Bates et al. 1980, 709). Represented by the symbol: E.

# References

Adams, F. D., and Nicholson, J. T. 1901. An experimental investigation into the flow of marble. *R. Soc. London, Philos. Trans.* A195: 363–401.

Airy, G. B. 1855. On the contribution of the effect of the attraction of mountain-masses. *R. Soc. London, Philos. Trans.* 145: 101–104.

Ampferer, O. 1906. Über das Bewegungsbild von Faltengebirgen. *K.u.K. Geol. Reichsanst. Vienna Jahrb.* 56: 539–622.

Amstutz, G. C. 1960. Polygonal and ring tectonic patterns in the Precambrian and Paleozoic of Missouri, U.S.A. *Ecologae Geol. Helv.* 1959 52: 904–913.

Anderson, D. L., Ben-Menahem, A., and Archambeau, C. B. 1965. Attenuation of seismic energy in the upper mantle. *J. Geophys. Res.* 70: 1441–1448.

Anderson, E. M. 1905. The dynamics of faulting. *Edinburgh Geol. Soc. Trans.* 8: 387–402.

——. 1942. *The Dynamics of Faulting*. Edinburgh: Oliver and Boyd. 191 pp.

Anderson, T. B. 1964. Kink-bands and related geological structures. *Nature* 202: 272–274.

Ardell, A. J., Christie, J. M., and Tullis, J. A. 1973. Dislocation substructures in deformed quartz rocks. *Cryst. Lattice Defects* 4: 275–285.

Argand, E. 1924. La tectonique de l'Asie. *Int. Geol. Congr., 13th, Compt. Rend.* 5: 171–372.

——. 1977. *Tectonics of Asia*. Trans. and ed. A. V. Carozzi. New York: Hafner. 218 pp.

Armstrong, R. L., and Dick, H. J. B. 1974. A model for the development of thin overthrust sheets of crystalline rock. *Geology* 2: 35–40.

Atwater, T. 1970. Implications of plate tectonics for the Cenozoic tectonics of western North America. *Geol. Soc. Am. Bull.* 81: 3513–3536.

Bailey, E. B. 1935. *Tectonic Essays, Mainly Alpine*. Oxford: Clarendon Press. 200 pp.

Bailey, E. B., and McCallien, W. J. 1937. Perthshire tectonics; Schiehallion to Glen Lyon. *R. Soc. Edinburgh, Trans.* 5: 79–117.

Bakewell, R. 1813. *An Introduction to Geology*. London: Harding. 362 pp.

Balk, R. 1937. Structural behavior of igneous rocks. *Geol. Soc. Am. Mem.* 5. 177 pp.

———. 1953. Salt structure of Jefferson Island salt dome, Iberia and Vermilion Parishes, Louisiana. *AAPG Bull.* 37: 2455–2474.

Bankwitz, P. 1965. Über Klüfte, Beobachtungen im thüringischen Schiefergebirge. *Geologie* 14: 242–253.

———. 1966. Über Klüfte, II. *Geologie* 15: 896–941.

———. 1978. Über Klüfte, III & IV. *Z. Geol. Wiss.* Berlin 6: 285–311.

Bannert, D. 1969. Luftbildkartierung des Lineationsnetzes vom Ries und seiner Umgebung. *Geol. Bavarica* 61: 379–384.

Barrell, J. 1914. The strength of the earth's crust, Parts 4 and 5. *J. Geol.* 22: 289–314, 441–468, and 655–683.

Becker, G. F. 1893. Finite homogeneous strain, flow and rupture of rocks. *Geol. Soc. Am. Bull.* 4: 13–90.

Behrmann, R. B. 1949. Geologie und Lagerstätten des Oelfeldes Reitbrook bei Hamburg: *Erdöl und Tektonik*. Hannover: Amt. f. Bodenf., pp. 190–221.

Beloussov, V. V. 1961. The origin of folding in the earth's crust. *J. Geophys. Res.* 66: 2241–2254.

Bernauer, F. 1939. Island und die Frage der Kontinentalverschiebungen. *Geol. Rundsch.* 30: 357–358.

———. 1943. Junge Tektonik auf Island und ihre Ursachen. In Niemcyk, Oskar, *Spalten auf Island, 14–64*. Stuttgart: Wittwer.

Berthelsen, A. 1960. Structural studies in the Pre-Cambrian of western Greenland. II, Geology of Tovqussap Nûna. *Meddel. Grønland*, v. 123–1, 223 pp.

———. 1979. Recumbent folds and boudinage structures formed by subglacial shear: An example of gravity tectonics. *Geol. Mijnbouw* 58: 253–260.

Beutner, E. C., and Diegel, F. A. 1985. Determination of fold kinematics from syntectonic fibers in pressure shadows, Martinsburg slate, New Jersey. *Am. J. Sci.* 285: 16–50.

Biot, M. A. 1961. Theory of folding of stratified viscoelastic media and its implications in tectonics and orogenesis. *Geol. Soc. Am. Bull.* 72: 1595–1620 and 1621–1631.

Bischoff, G. 1966. Statische Gesetzmässigkeiten des basischen Deckenvulkanismus und deren Hinweise auf Vorgänge im oberen Erdmantel; mit Beispielen von Analogien aus Südamerika und Afrika. *Z. Dtsch. Geol. Ges.* 116 (1964): 813–831.

Blyth, F. G. H. 1952. *Geology for Engineers*. 3d ed. London: Arnold. 336 pp.

Bock, H. 1976a. *Geometrische Eigenschaften von Kluftflächen und ihr Einfluss auf die Festigkeit geologischer Körper*. Ph.D. thesis, University of Bochum. 201 pp.

———. 1976b. Einige Beobachtungen und Überlegungen zur Kluftentstehung in Sedimentgesteinen. *Geol. Rundsch.* 65: 83–101.

———. 1980. Das fundamentale Kluftsystem. *Z. Dtsch. Geol. Ges.* 131: 627–650.

Bolsenkötter, H. 1955. Feintektonische Untersuchungen an Schlechten und Klüften in Steinkohlenflözen des Ruhrgebietes. *Geol. Rundsch.* 44: 443–472.

Boyer, S. E., and Elliott, D. 1982. Thrust systems. *AAPG Bull.* 66: 1196–1230.

Brace, W. F. 1960. An extension of the Griffith theory of fracture to rocks. *J. Geophys. Res.* 65: 3477–3480.

Branco, W., and Fraas, E. 1905. Das kryptovulkanische Becken von Steinheim. *Kön. Preuss. Akad. Wiss. Abh.* 1905: 1–64.

Bruce, C. H. 1973. Pressured shale and related sediment deformation. Mechanism for development of regional contemporaneous faults. *AAPG Bull.* 57: 878–886.

Bruhn, R. L., Yusas, M. R., and Huertas, F. 1982. Mechanics of low-angle normal faulting: An example from Roosevelt Hot Springs geothermal area, Utah. *Tectonophysics* 86: 343–361.

Bucher, W. H. 1920, 1921. The mechanical interpretation of joints. Pt. 1: *J. Geol.* 28: 707–730; Pt. 2: *J. Geol.* 29: 1–28.

———. 1944. The stereographic projection, a handy tool for the practical geologist. *J. Geol.* 52: 191–212.

———. 1956. Role of gravity in orogenesis. *Geol. Soc. Am. Bull.* 67: 1295–1318.

———. 1963. Cryptoexplosion structures caused from without or from within the earth? *Am. J. Sci.* 261: 597–649.

Buddington, A. F. 1959. Granite emplacement with special reference to North America. *Geol. Soc. Am. Bull.* 70: 671–747.

Burchfiel, B. C., Wernicke, B., Willemin, J. H., Axen, G. J., and Cameron, C. S. 1982. A new type of décollement thrusting. *Nature* 300: 513–515.

Burke, K. 1979. Two problems of intra-continental tectonics: Re-elevation of old mountain belts and subsidence of intracontinental basins. In *Proc. Int. Res. Conf. Intra-Continental Earthquakes,* ed. J. Petrouski and C. R. Allen, 157–163. Ohrid, Yugoslovia: Skopje Inst. Earthquake Eng. Earthquake Seismol.

Busk, H. G. 1929. *Earth Flexures.* Cambridge University Press. Cambridge, England. 106 pp.

Butler, R. W. H. 1982. The terminology of structures in thrust belts. *J. Struct. Geol.* 4: 239–245.

Carey, S. W. 1954. The rheid concept in geotectonics. *Geol. Soc. Australia J.* 1: 67–117.

———. 1958. A tectonic approach to continental drift. In *Continental Drift, a Symposium,* conv. S. W. Carey, 177–355. Hobart: University of Tasmania.

———. 1962. Folding. *J. Alberta Soc. Petrol. Geol.* 10: 95–144.

Carter, N. L. 1976. Steady state flow of rocks. *Rev. Geophys. Space Phys.* 14(3): 301–360.

Carter, N. L., Christie, J. M., and Griggs, D. T. 1964. Experimental deformation and recrystallization of quartz. *J. Geol.* 72: 687–733.

Carter, N. L., and Raleigh, C. B. 1969. Principal stress directions from plastic flow in crystals. *Geol. Soc. Am. Bull.* 80: 1231–1264.

Chappell, B. W., and White, A. J. R. 1974. Two contrasting granite types. *Pac. Geol.* 8: 173–174.

Chapple, W. M. 1970. The finite-amplitude instability in the folding of layered rocks. *Can. J. Earth Sci.* 7: 457–465.

———. 1978. Mechanics of thin-skinned fold-and-thrust belts. *Geol. Soc. Am. Bull.* 89: 1189–1198.

Christensen, M. N. 1965. Late Cenozoic deformation in the central Coast Ranges of California. *Geol. Soc. Am. Bull.* 76: 1105–1124.

Cloos, E. 1936. Der Sierra-Nevada-Pluton in Californien. *Neues Jahrb. Mineral.* 76 (Beilage Bd.): 356–450.

———. 1946. Lineation. *Geol. Soc. Am. Mem.* 18: 122 pp.

———. 1947a. Oolite deformation in the South Mountain fold, Maryland. *Geol. Soc. Am. Bull.* 58: 843–918.

———. 1947b. Boudinage. *Am. Geophys. Union Trans.* 28: 626–632.

———. 1955. Experimental analysis of fracture patterns. *Geol. Soc. Am. Bull.* 66: 241–256.

Cloos, H. 1922. Über Ausbau und Anwendung der granittektonischen Methode. *Preuss. Geol. Landesanst. Abh.* 89: 1–18.

———. 1928. Experimente zur inneren Tektonik. *Zentralbl. Mineral. Geol. Paleontol.* 1928-B: 609–621.

———. 1929. Künstliche Gebirge. *Nat. Mus.* 59: 225–272; 60: 258–269.

———. 1936. *Einführung in die Geologie.* Berlin: Borntraeger. 503 pp.

———. 1941. Bau und Tätigkeit von Tuffschloten. *Geol. Rundsch.* 32: 711–800.

———. 1948. Gang und Gehwerk einer Falte. *Z. Dtsch. Geol. Ges.* 100: 290–303.

———. 1954. *Gespräch mit der Erde.* 2d ed. Munich: Piper & Co. 389 pp.

Cloos, H., and Cloos, E. 1927. Die Quellkuppe des Drachenfels am Rhein; ihre Tektonik und Bildungsweise. *Z. Vulkanol.* 11: 33–40.

Clüver, D. 1700. *Geologia.* Hamburg: Liebezeit. 223 pp.

Cook, F. A., Albaugh, D. S., Brown, L. D., Kaufman, S., Oliver, J. E., and Hatcher, R. D., Jr. 1979. Thin-skinned tectonics in the crystalline southern Appalachians; COCORP seismic reflection profiling of the Blue Ridge and Piedmont. *Geology* 7: 563–567.

Coulomb, C. S. 1773. Sur une application des règles maximis et minimis à quelques problèmes de statique relatifs à l'architecture. *Acad. Sci. Paris Mém. Math. Phys.* 7: 343–382.

Cox, A., Doell, R., and Dalrymple, G. 1964. Reversals of the earth's magnetic field. *Science* 144: 1537–1543.

Crittenden, D. 1963. New data on the isostatic deformation of Lake Bonneville. *U.S. Geol. Surv. Prof. Pap.* 454–E. 31 pp.

Crowell, J. C. 1974. Origin of late Cenozoic basins in southern California. *Tectonics and Sedimentation, SEPM Spec. Publ.* 22: 189–204.

Curnelle, R., and Marco, R. 1983. Reflection profiles across the Aquitaine basin. In *Seismic Expression of Structural Styles: A Picture and Work Atlas,* ed. A. W. Bally, vol. 2, 3.2–11. AAPG Studies in Geology Series, no. 15. Tulsa, Okla.: American Association of Petroleum Geologists.

Currie, J. B., Patnode, H. W., and Trump, R. P. 1962. Development of folds in sedimentary strata. *Geol. Soc. Am. Bull.* 73: 655–674.

Dahlstrom, C. D. A. 1969. Balanced cross sections. *Can. J. Earth Sci.* 6: 743–757.

Daly, R. A. 1940. *Strength and Structure of the Earth.* Englewood Cliffs, N.J.: Prentice-Hall. 434 pp.

Daubrée, G. A. 1879. *Etudes Synthétiques de Géologie Expérimentale.* Paris: Dunod. 828 pp.

De Jong, K. A., and Scholten, R. 1973. *Gravity and Tectonics.* New York: John Wiley & Sons. 502 pp.

De Terra, H. 1931. Structural features in gliding strata. *Am. J. Sci.* Ser. 5. 21: 204–213.

Dennis, J. G. 1956. The geology of the Lyndonville area, Vermont. *Vermont Geol. Surv. Bull.* 8. 98 pp.

———. 1964. The geology of the Enosburg area, Vermont. *Vermont Geol. Surv. Bull.* 23. 56 pp.

———. 1967. International tectonic dictionary, English terminology. *Am. Assoc. Pet. Geol. Mem.* 7: 196 pp.

———. 1969. Zur genetischen Unterscheidung von gemeinen Klüften und Verschiebungen. *Geol. Rundsch.* 59: 222–228.

Dennis, J. G., and Kelley, V. C. 1980. Antithetic and homothetic faults. *Geol. Rundsch.* 69: 186–193.

Dennis, J. G., Murawski, H., and Weber, K. 1979. *International Tectonic Lexicon.* Stuttgart: Schweizerbart. 153 pp.

Dewey, J. F. 1965. Nature and origin of kink-bands. *Tectonophysics* 1: 459–494.

Dewey, J. F., and Bird, J. M. 1970. Mountain belts and the new global tectonics. *J. Geophys. Res.* 75: 2615–2647.

Dewey, J. F., and Burke, K. 1974. Hot spots and continental breakup: Implications for collisional orogeny. *Geology* 2: 57–60.

Dickinson, W. R., and Seely, D. R. 1979. Stratigraphy and structure of forearc regions. *AAPG Bull.* 63: 2–31.

Dieterich, J. H. 1969. Origin of cleavage in folded rocks. *Am. J. Sci.* 267: 155–165.

———. 1970. Computer experiments on mechanics of finite amplitude folds. *Can. J. Earth Sci.* 7: 467–476.

Dieterich, J. H., and Carter, N. L. 1969. Stress history of folding. *Am. J. Sci.* 267: 129–154.

Dietrich, R. V., Dutro, J. T., Jr., and Foose, R. M., compilers. 1982. *AGI Data Sheets.* Falls Church, Virginia: American Geological Institute.

Dietz, R. 1959. Shatter cones in cryptoexplosion structures (meteorite impact?). *J. Geol.* 67: 496–505.

———. 1961. Continent and ocean basin evolution by spreading of the sea floor. *Nature* 190: 854–857.

Donath, F. A., and Parker, R. B. 1964. Folds and folding. *Geol. Soc. Am. Bull.* 75: 45–62.

Du Toit, A. L. 1937. *Our Wandering Continents*. Edinburgh: Oliver and Boyd. 366 pp.

Edelglass, S. M. 1966. *Engineering Materials Science: Structure and Mechanical Behavior of Solids*. New York: Ronald. 499 pp.

Egyed, L. 1957. A new dynamic conception of the internal constitution of the earth. *Geol. Rundsch.* 46: 101–120.

Elliott, D. 1976. The motion of thrust sheets. *J. Geophys. Res.* 81: 949–963.

————. 1983. The construction of balanced cross sections. *J. Struct. Geol.* 5: 101.

Engelen, G. B. 1963. Gravity tectonics in the northwestern Dolomites (N. Italy). *Geol. Ultraiectina* 13: 92 pp.

Eric, J. H., and Dennis, J. G. 1958. Geology of the Concord-Waterford area, Vermont. *Vermont Geol. Surv. Bull.* 11: 66 pp.

Erxleben, A. W., and Carnahan, G. 1983. Slick Ranch area, Starr County, Texas Gulf coast. In *Seismic Expression of Structural Styles,* ed. A. W. Bally, vol. 2, 3.1–22. AAPG Studies in Geology Series, no. 15. Tulsa, Okla.: American Association of Petroleum Geologists.

Escher, A., Escher, J. C., and Watterson, J. 1975. The reorientation of the Kangamiut Dike swarm, West Greenland. *Can. J. Earth Sci.* 12: 158–173.

Escher, Arnold. 1841. Gebirgsprofil von St.-Triphon. *Leonhard und Bronn, N. Jahrb.* 1841: 342–346.

Eskola, P. E. 1949. The problem of mantled gneiss domes. *Geol. Soc. London Q. J.* 104: 461–476.

Ewing, M., and Ewing, J. 1962. Rate of salt-dome growth. *AAPG Bull.* 46: 708–709.

Fairbairn, H. W. 1936. Elongation in deformed rocks. *J. Geol.* 44: 670–680.

Falcon, N. L. 1969. Problems of the relationship between surface structure and deep displacements illustrated by the Zagros Range. In *Time and Place in Orogeny,* ed. P. E. Kent, G. E. Satterthwaite, and A. M. Spencer, 9–22. London: Geol. Soc. 311 pp.

Fermor, E. M. R., and Price, R. A. 1976. Imbricate structures in the Lewis thrust sheet around the Cate Creek and Haig Brook windows, southeastern British Columbia. *Geol. Surv. Can. Pap.* 76–1B: 7–10.

Feugueur, L. L. 1953. Comportement du charbon dans une nappe helvétique des Alpes françaises, le gisement d'Araches en Haute-Savoie. *Congr. Géol. Int. Compt. Rend. 19th Sess.* Pt. 3: 163–172.

Fleuty, M. J. 1964. The description of folds. *Geol. Assoc. London, Proc.* 75: 461–492.

Flinn, D. 1962. On folding during three-dimensional progressive deformation. *Geol. Soc. London Q. J.* 118: 385–433.

Friedman, M., and Conger, F. B. 1964. Dynamic interpretation of calcite twin lamellae in a naturally deformed fossil. *J. Geol.* 72: 361–368.

Friedman, M., and Sowers, G. M. 1970. Petrofabrics: A critical review. *Can. J. Earth Sci.* 7: 477–495.

Funk, H., Labhart, T., Milnes, A. G., Pfiffner, A. O., Schaltegger, U., Schindler, C., Schmid, S. M., and Trümpy, R. 1983. Bericht über die Jubiläumsexkursion ''Mechanismus der Gebirgsbildung'' der Schweizerischen Geologischen Gesellschaft in das ost- und zentralschweizerische Helvetikum und in das nördliche Aarmassiv vom 12. bis 17. September 1982. *Eclogae Geol. Helv.* 76: 91–123.

Gangel, L., and Murawski, H. 1977. Rasterelektronenmikroskopische (Stereoscan) Untersuchungen zur genetischen Interpretation von Schlechten und Klüften im Steinkohlengebirge. *N. Jahrb. Geol. Paleontol. Monatsh.* 12: 705–719.

Gass, I. G., and Gibson, I. L. 1969. Structural evolution of the rift zones in the Middle East. *Nature* 221: 926–930.

Geikie, A. 1882. *Textbook of geology*. London: Macmillan and Co. 971 pp.

Gilbert, G. K. 1890. Lake Bonneville. *U.S. Geol. Surv. Monogr.* 1. 438 pp.

Gill, J. E. 1935. Normal and reverse faults. *J. Geol.* 43: 1071–1079.

————. 1941. Fault nomenclature. *R. Soc. Can. Trans.* 35: 71–85.

————. 1971. Continued confusions in the classification of faults. *Geol. Soc. Am. Bull.* 82: 1389–1392.

Gray, D. R. 1979. Microstructure of crenulation cleavages. An indicator of cleavage origin. *Am. J. Sci.* 279: 97–128.

Greene, G. W., and Hunt, C. B. 1960. Observations of current tilting of the earth's surface in the Death Valley, California area. *U.S. Geol. Survey Prof. Pap.* 400-B. p. 275–276.

Griffith, A. A. 1921. The phenomena of rupture and flow in solids. *R. Soc. London Philos. Trans.* A221: 163–198.

Griggs, D. T. 1939. Creep of rocks. *J. Geol.* 47: 225–251.

Griggs, D. T., and Handin, J. 1960. Observations of fracture and a hypothesis of earthquakes. In *Rock Deformation, a Symposium. Geol. Soc. Am. Mem.* 79: 347–364.

Groschopf, P., and Reiff, W. 1969. Das Steinheimer Becken. Ein Vergleich mit dem Ries. *Geol. Bavarica* 61: 400–412.

Gross, W. W., and Hillemeyer, F. L. 1982. Geometric analysis of upper-plate fault patterns in the Whipple-Buckskin detachment terrane, California and Arizona. In *Mesozoic-Cenozoic Tectonic Evolution of the Colorado River region, California, Arizona, and Nevada,* ed. E. G. Frost and D. L. Martin, 257–265. San Diego: Cordilleran Publ. 608 pp.

Gutenberg, B. 1931. *Handbuch der Geophysik,* vol. 2. Berlin: Borntraeger. 539 pp.

———. 1955. Wave velocities in the earth's crust. In *Crust of the Earth—A Symposium,* ed. A. Poldevaart. *Geol. Soc. Am. Spec. Pap.* 62: 19–34.

Halbouty, M. T., and Hardin, G. C., Jr. 1959. A geological appraisal of present and future exploration techniques on salt domes of the Gulf region of the United States. *5th World Pet. Congr., Sect. L,* Pap. 5. 13 pp.

Hall, Sir James. 1815. On the vertical position and convolutions of certain strata and their relation with granite. *R. Soc. Edinburgh Trans.* 7: 79–108.

Haller, J. 1956. Die Strukturelemente Ostgroenlands zwischen 74° und 78°N. *Medd. Groenland* 154(3): 153.

Ham, W. E., and Wilson, J. L. 1967. Paleozoic epeirogeny and orogeny in the central United States. *Am. J. Sci.* 265: 332–407.

Hamberg, A. 1932. Struktur und Bewegungsvorgänge im Gletschereise nebst Beiträgen zur Morphologie der arktischen Gletscher. *Naturwiss. Unters. d. Sarekgebirges' Schwedisch-Lappland,* vol. 1. Abt. 3: 69–129.

Hammond, R. H., Buck, C. P., Rogers, W. B., Walsh, G. W., and Ackert, H. P. 1971. *Engineering Graphics.* 2d ed. New York: Ronald. 648 pp.

Handin, J. 1957. Experimental deformation of rocks and minerals. *Colo. Sch. Mines Q.* 52: 75–98.

———. 1966. Strength and ductility. In *Handbook of Physical Constants,* ed. S. P. Clark, Jr. *Geol. Soc. Am. Mem.* 97: 223–290.

Harding, T. P. 1974. Petroleum traps associated with wrench faults. *AAPG Bull.* 58: 1290–1304.

Harding, T. P., and Lowell, J. D. 1979. Structural styles, their plate-tectonic habitats, and hydrocarbon traps in petroleum provinces. *AAPG Bull.* 63: 1016–1058.

Harland, W. B. 1971. Tectonic transpression in Caledonian Spitsbergen. *Geol. Mag.* 108: 27–42.

Harris, L. D., and Milici, R. C. 1977. Characteristics of thin-skinned style of deformation in the southern Appalachians, and potential hydrocarbon traps. *U.S. Geol. Surv. Prof. Pap.* 1018. 40 pp.

Harrison, J. V., and Falcon, N. L. 1934. Collapse structures. *Geol. Mag.* 71: 529–539.

Hartmann, L. 1896. *Distribution des déformations dans les métaux soumis à des efforts.* Paris: Berger-Leurault.

Heard, H. C. 1960. Transition from brittle fracture to ductile flow in Solenhofen [*sic*] limestone as a function of temperature, confining pressure and interstitial fluid pressure. *Geol. Soc. Am. Mem.* 79: 193–226.

———. 1963. Effect of large changes in strain rate in the experimental deformation of Yule marble. *J. Geol.* 71: 162–195.

Heezen, B. C., and Tharp, M. 1965. Tectonic fabric of the Atlantic and Indian oceans and continental drift. *R. Soc. London, Philos. Trans.* 258: 90–106.

Heezen, B. C., Tharp, M., and Ewing, M. 1959. The floors of the oceans, 1, the North Atlantic. *Geol. Soc. Am. Spec. Pap.* 65. 122 pp.

Heim, A. 1878. *Untersuchungen über den Mechanismus der Gebirgsbildung.* Basle: Schwabe. Vol. 1, 346 pp.; vol 2, 246 pp.

Heirtzler, J. R., Dickson, G. O., Herron, E. M., Pitman, W. C. III, and Le Pichon, X. 1968. Marine magnetic anomalies, geomagnetic field reversals, and motions of ocean floor and continents. *J. Geophys. Res.* 73: 2119–2136.

Hendricks, H. E. 1947. Geologic map of the Steelville quadrangle, Mo. *In The geology of the Steelville Quadrangle.* Pl. 8 *Mo. Geol. Surv. 2nd Ser.* 36.

Hess, H. H. 1962. History of ocean basins. In *Petrographic Studies, A Volume in Honor of A. F. Buddington,* ed. M. L. James and B. F. Leonard, 599–620. Geol. Soc. Am. 660 pp.

Hill, M. L. 1947. Classification of faults. *AAPG Bull.* 31: 1669–1673.

———. 1959. Dual classification of faults. *AAPG Bull.* 43: 217–221.

Hill, M. L., and Dibblee, T. W., Jr. 1953. San Andreas, Garlock, and Big Pine faults, California. *Geol. Soc. Am. Bull.* 64: 443–458.

Hills, E. S. 1963. *Elements of Structural Geology.* New York: Wiley. 483 pp.

———. 1972. *Elements of Structural Geology.* 2d ed. New York: Wiley. 502 pp.

Hodgson, R. A. 1961. Regional study of jointing in Comb Ridge-Navajo mountain area, Arizona and Utah. *AAPG Bull.* 45: 1–38.

Hoeppener, K. E., and Schrader, P. 1969. Zur physikalischen Tektonik. Bruchbildung bei verschiedenen affinen Deformationen im Experiment. *Geol. Rundsch.* 59: 179–193.

Hoeppener, R. 1955. Das tektonische Inventar eines Aufschlusses in den Orthoceras-Schiefern bei Dillenburg. *Geol. Rundsch.* 44: 93–98.

———. 1964. Zur physikalischen Tektonik. *Felsmech. Ingenieurgeol.* 2: 22–44.

Holder, M. T. 1979. An emplacement mechanism for post-tectonic granites and its implications for their geochemical features. In *Origin of Granite Batholiths, Geochemical Evidence,* ed. M. P. Atherton and J. Tarney, 116–128. Orpington, England: Shiva Publishing Co.

Hollingworth, S. E., Taylor, J. H., and Kellaway, G. A. 1945. Large scale superficial structures in the Northampton ironstone field. *Geol. Soc. London Q. J.* 100 (1944): 1–34.

Holmes, A. 1931. Radioactivity and earth movements. *Trans. Geol. Soc. Glasgow.* 18: 559–606.

———. 1945. *Principles of Physical Geology.* New York: Ronald. 532 pp.

———. 1965. *Principles of Physical Geology.* 2d ed. New York: Ronald. 1288 pp.

———. 1978. *Principles of Physical Geology.* 3d ed. New York: Halsted Press. 730 pp.

Hossack, J. R. 1968. Pebble deformation and thrusting in the Bygdin area (Southern Norway). *Tectonophysics,* 5: 315–339.

———. 1979. The use of balanced cross sections in the calculation of orogenic contraction. A review. *J. Geol. Soc. London* 136: 705–711.

Hubbert, M. K. 1937. Theory of scale models as applied to the study of geologic structures. *Geol. Soc. Am. Bull.* 48: 1459–1519.

———. 1951. Mechanical basis for certain familiar geologic structures. *Geol. Soc. Am. Bull.* 62: 355–372.

Hubbert, M. K., and Rubey, W. W. 1959. Role of fluid pressure in mechanics of overthrust faulting. *Geol. Soc. Am. Bull.* 70: 115–166.

Hubbert, M. K., and Willis, G. 1957. Mechanics of hydraulic fracturing. *Shell Oil Co. Tech. Services Division, Publ. 415.* Also in: Hubbert, M. K.1972. *Structural Geology.* New York: Hafner. 329 pp.

Hutton, J. 1788. Theory of the earth. *R. Soc. Edinburgh Trans.* 1: 209–304.

Illies, H. 1965. Bauplan und Baugeschichte des Oberrheingrabens. *Oberrheinische Geol. Abh.* 14: 1–54.

———. 1970. Graben tectonics as related to crust-mantle interactions. In *Graben Problems: International Upper Mantle Project, Scientific Report 27,* ed. J. H. Illies and St. Mueller, 4–27. Stuttgart: Schweizerbart.

Inoue, E. 1960. Land deformation in Japan. *Geogr. Surv. Inst. (Japan) Bull.* 6: 73–134.

Irvine, T. N. 1965. Sedimentary structures in igneous intrusions with particular reference to the Duke Island ultramafic complex. *SEPM Spec. Publ.* 12: 220–232.

Isacks, B., Oliver, J., and Sykes, L. R. 1968. Seismology and the new global tectonics. *J. Geophys. Res.* 73: 5855–5899.

Jackson, M. P. A., and Talbot, C. J. 1986. External shapes and dynamics of salt structures. *Geol. Soc. Am. Bull.* 96: 305–323.

Jukes, J. B. 1859. Geology of the Staffordshire coalfield. *Geol. Surv. Great Britain Mem.* 215 pp.

Kehle, R. O. 1970. Analysis of gravity sliding and orogenic translation. *Geol. Soc. Am. Bull.* 81: 1641–1664.

Kennedy, W. Q. 1946. The Great Glen fault. *Geol. Soc. London Q. J.* 102: 41–76.

Kerr, J. W., and Christie, R. L. 1965. Tectonic history of the Boothia uplift and Cornwallis fold belt, Arctic Canada. *AAPG Bull.* 49: 905–926.

King, G. C. P. 1978. Geological faults: fracture, creep and strain. *R. Soc. London Philos. Trans.* A288: 197–212.

Koenigsberger, G., and Morath, O. 1913. Theoretische Grundlagen der experimentellen Tektonik. *Dtsch. Geol. Ges. Z.* 65: 65–86.

Korn, H., and Martin, H. 1959. Gravity tectonics in the Naukluft Mountains of South West Africa. *Geol. Soc. Am. Bull.* 70: 1047–1078.

Krumbein, W. C. 1942. Criteria for subsurface recognition of unconformities. *AAPG Bull.* 26: 36–62.

Kuenen, P. H. 1953. Significant features of graded bedding. *AAPG Bull.* 37: 1044–1066.

Lapworth, C. 1883. The secret of the Highlands. *Geol. Mag.* 10: 120–128; 193–199; and 337–344.

Laubscher, H. P. 1961. Die Fernschubhypothese der Jurafaltung. *Eclog. Geol. Helv.* 54: 221–282.

———. 1977. Fold development in the Jura. *Tectonophysics* 37: 337–362.

Lauerma, R. 1964. On the structure and petrography of the Ipernat Dome, Western Greenland. *Bull. Comm. Géol. Finlande* No. 215, 88 pp.

Le Pichon, X. 1968. Seafloor spreading and continental drift. *J. Geophys. Res.* 73: 3661–3697.

Le Pichon, X., Francheteau, J., and Bonnin, J. 1976. *Plate Tectonics.* Amsterdam: Elsevier Scientific Publishing Company. 311 pp.

Leith, C. K. 1905. Rock cleavage. *U.S. Geol. Surv. Bull.* 239. 216 pp.

Lensen, G. J. 1968. Analysis of progressive fault displacement during downcutting at the Branch River Terraces, South Island, New Zealand. *Geol. Soc. Am. Bull.* 79: 545–556.

Lovering, T. S. 1932. Field evidence to distinguish overthrusting from underthrusting. *J. Geol.* 40: 651–663.

Lowman, P. D., Jr. 1976. Geologic structure of California: Three studies with Landsat—1. *Calif. Geol.* 29: 75–81.

Lyell, C. 1865. *Elements of Geology.* 6th ed. London: John Murray, 2 vols. 669 pp.

———. 1872. *Principles of Geology.* Vol. 1, 11th ed. London: John Murray. 669 pp.

Macgregor, A. M. 1951. Some milestones in the Precambrian of Southern Rhodesia. *Geol. Soc. S. Afr. Proc.* 54: 27–71.

Mackin, J. H. 1950. The down-structure method of viewing geologic maps. *J. Geol.* 58: 55–72.

———. 1960. Structural significance of Tertiary volcanic rocks in southwestern Utah. *Am. J. Sci.* 258: 81–131.

Marr, J. E. 1900. Geology of the English Lake district. *Geol. Assoc. London Proc.* 16: 449–483.

Mason, R. G. 1958. A magnetic survey off the west coast of the United States between latitudes 32° and 36°N, longitudes 121° and 128°W. *R. Astron. Soc. Geophys. J.* 1: 320–329.

McKenzie, D. P. 1969. Speculations on the consequences and causes of plate motions. *R. Astron. Soc. Geophys. J.* 18: 1–32.

McKenzie, D. P., and Morgan, W. J. 1969. Evolution of triple junctions. *Nature* 224: 125–133.

Mead, W. J. 1925. The geologic role of dilatancy. *J. Geol.* 33: 685–698.

Means, W. D. 1976. *Stress and Strain—Basic Concepts of Continuum Mechanics for Geologists.* New York: Springer-Verlag, Inc. 339 pp.

Merla, G. 1951. Geologia dell Appennino settentrionale. *Soc. Geol. Ital. Boll.* 70: 95–382.

Meyerhoff, A. A. 1968. Arthur Holmes: Originator of spreading ocean floor hypothesis. *J. Geophys. Res.* 73: 6563–6565 (and discussion).

Mohr, O. 1882. Über die Darstellung des Spannungszustandes und des Deformationszustandes eines Körperelementes und über die Anwendung derselben in der Festigkeitslehre. *Civilingenieur* 28: 113–156.

———. 1900. Welche Umstände bedingen die Elastizitätsgrenze und den Bruch eines Materials? *Ver. Dtsch. Ing. Z.* 44: 1524.

Möller, D., and Ritter, B. 1980. Geodetic measurements and horizontal crustal movements in the rift zone of NE-Iceland. *J. Geophys.* 47: 110–119.

Montadert, L., de Charpal, O., Roberts, D., Guennoc, P., and Sibuet, J. C. 1979. Northeast Atlantic passive continental margins: Rifting and subsidence processes. *Am. Geophys. Union, Maurice Ewing Ser.* 3: 154–186.

Morgan, W. J. 1968. Rises, trenches, great faults, and crustal blocks. *J. Geophys. Res.* 73: 1959–1982.

Muehlberger, W. R. 1961. Conjugate joint sets of small dihedral angle. *J. Geol.* 69: 211–219.

Mueller, S., Peterschmitt, E., Fuchs, K., and Ansorge, J. 1969. Crustal structure beneath the Rhine graben from seismic refraction and reflection measurements. *Tectonophysics* 8: 529–542.

Müller, W. H., Schmid, S. M., and Briegel, U. 1981. Deformation experiments on anhydrite rocks of different grain sizes: Rheology and microfabric. *Tectonophysics* 78: 527–543.

Murawski, H. 1976. Raumproblem und Bewegungsablauf an listrischen Flächen, insbesondere bei Tiefenstörungen. *N. Jahrb. Geol. Paleont. Monatsh.* 1976: 209–220.

Murawski, H., and Merkel, D. 1983. Cleat examinations in the Ishikari coal region, Hokkaido/Japan. Some remarks about the connection between micro- and macrotectonic pattern. *N. Jahrb. Geol. Paleont. Monatsh.* 1983: 181–192.

———. 1833. Über den linearen Parallelisms mancher Felsarten. *N. Jahrb. Geol. Paleont. Mineralog.* 383–391.

Naumann, C. F. 1849. *Lehrbuch der Geognosie,* vol 1. Leipzig: Wilhelm Engelmann. 1000.

Nehm, W. 1939. *Markscheiderische Erwägungen zum Störungsproblem.* Mitteil. Markscheidewesen 50: 1–25.

Nettleton, L. L. 1934. Fluid dynamics of salt domes. *AAPG Bull.* 18: 1175–1204.

———. 1955. History of concepts of Gulf coast salt-dome formation. *AAPG Bull.* 39: 2373–2383.

Nevin, C. M. 1949. *Principles of Structural Geology.* 4th ed. New York: Wiley. 410 pp.

Nickelsen, R. P., and Hough, V. N. D. 1967. Jointing in the Appalachian plateau of Pennsylvania. *Geol. Soc. Am. Bull.* 78: 615; 609–630.

Nur, A. 1983. Accreted terranes. *Rev. Geophys. Space Phys.* 21(8): 1779–1785.

O'Connor, M. J., and Gretener, P. E. 1974. Quantitative modeling of the processes of differential compaction. *Bull. Can. Pet. Geol.* 22: 241–268.

Odé, H. 1956. A note concerning the mechanism of artificial and natural hydraulic fracture systems. *Colo. Sch. Mines Q.* 51: 19–29.

Oertel, G. 1955. Der Pluton von Loch Doon in Süd Schottland. *Geotekton. Forsch.* Heft 11: 83.

———. 1961. Stress, strain and fracture in clay models of geologic deformation. *Geotimes* 6: 26–31.

———. 1962. Extrapolation in geologic fabrics. *Geol. Soc. Am. Bull.* 73: 325–342.

Oldham, R. D. 1884. Notes on a graphic table of dips. *Geol. Mag.* 1: 412–415.

Oliver, J., and Isacks, B. 1967. Deep earthquake zones, anomalous structures in the upper mantle, and the lithosphere. *J. Geophys. Res.* 72: 4259–4275.

Oxburgh, E. R., and Turcotte, D. L. 1970. Thermal structure of island arcs. *Geol. Soc. Am. Bull.* 81: 1665–1688.

Page, B. M. 1963. Gravity tectonics near Passo della Cisa, Northern Apennines, Italy. *Geol. Soc. Am. Bull.* 74: 655–672.

Park, W. C., and Schot, E. H. 1968. Stylolites; their nature and origin. *J. Sediment. Petrol.* 38: 175–191.

Paterson, M. S. 1978. *Experimental Rock Deformation—The Brittle Field.* Berlin: Springer-Verlag.

Paterson, M. S., and Weiss, L. E. 1961. Symmetry concepts in the structural analysis of deformed rocks. *Geol. Soc. Am. Bull.* 72: 841–882.

———. 1962. Experimental folding in rocks. *Nature* 195: 1046–1048.

Pávai Vajna, F. von 1926. Über die jüngsten tektonischen Bewegungen der Erdrinde. *Foldt. Kozl.* 5: 282–297.

Pedersen, S. A. S. 1980. Regional geology and thrust fault tectonics in the southern part of the North Greenland fold belt, North Peary Land. *Groenl. Geol. Unders. Rapp.* 99: 79–87.

Peterson, M. S., Rigby, J. K., and Hintze, L. F. 1980. *Historical Geology of North America.* Dubuque, Iowa: Wm. C. Brown Co. 232 pp.

Pettijohn, F. J., and Potter, P. E. 1964. *Atlas and Glossary of Primary Sedimentary Structures.* Berlin: Springer-Verlag. 370 pp.

Pfiffner, O. A. 1980. Strain analysis in folds (Infrahelvetic complex, central Alps). *Tectonophysics* 61: 337–362.

Phakey, P., Dollinger, G., and Christie, J. M. 1972. Transmission electron microscopy of experimentally deformed olivine crystals. In *Flow and Fracture of Rocks,* ed. M. C. Heard, I. Y. Borg, and C. B. Raleigh. *Am. Geophys. Union, Geophys. Monogr.* 16: 117–138.

Phillips, F. C. 1971. *The Use of Stereographic Projection in Structural Geology.* 3d ed. London: Edward Arnold. 90 pp.

Pierce, W. G. 1966. Jura tectonics as a décollement. *Geol. Soc. Am. Bull.* 77: 1265–1276.

Pitcher, W. S. 1979. The nature, ascent and emplacement of granitic magmas. *Geol. Soc. London J.* 136: 627–662.

Plafker, G. 1965. Tectonic deformation associated with the 1964 Alaska earthquake. *Science* 148: 1675–1687.

Plessmann, W. 1965. Gesteinslösung, ein Hauptfaktor beim Schieferungsprozess. *Geol. Mitt.* 4: 69–82.

Plummer, C. C., and McGeary, D. 1985. *Physical Geology.* Dubuque, Iowa: Wm. C. Brown Co. 513 pp.

Powell, C. McA. 1979. A morphological classification of rock cleavage. *Tectonophysics* 58: 21–34.

Pratt, J. 1855. On the attraction of the Himalaya mountains, and of the elevated regions beyond them, upon the plumb line in India. *R. Soc. London, Philos. Trans.* 145: 53–100.

———. 1860. On the deflection of the plumb line in India, caused by the attraction of the Himalaya mountains and of the elevated regions beyond; and its modification by the compensation effect of a deficiency of matter below the mountain mass. *R. Soc. London, Philos. Trans.* 149: 745–778.

Price, N. J. 1959. Mechanics of jointing in rocks. *Geol. Mag.* 96: 149–167.

———. 1966. *Fault and Joint Development in Brittle and Semi-brittle Rock.* Oxford: Pergamon. 176 pp.

———. 1975. Rates of deformation. *Geol. Soc. London J.* 131: 553–575.

Price, N. J., and Johnson, M. R. W. 1982. A mechanical analysis of the Keystone-Muddy Mountain thrust sheet in southeast Nevada. *Tectonophysics* 84: 131–150.

Price, R. A. 1972. The distinction between displacement and distortion in flow, and the origin of diachronism in tectonic overprinting in orogenic belts. *Int. Geol. Congr. 24th Montreal.* Sect. 3: 545–551.

Price, R. A., and Mountjoy, E. W. 1970. Geologic structure of the Canadian Rocky Mountains between Bow and Athabasca Rivers—A progress report. *Geol. Assoc. Can. Spec. Pap.* 6: 7–25.

———. 1981. The Cordilleran foreland thrust and fold belt in the southern Canadian Rocky Mountains. *Geol. Soc. London Spec. Publ.* 9: 427–448.

Pridmore, C. L., and Craig, C. 1982. Upper-plate structure and sedimentation of the Baker Peaks area, Yuma County, Arizona. In *Mesozoic-Cenozoic Tectonic Evolution of the Colorado River Region, California, Arizona and Nevada,* ed. E. G. Frost and D. L. Martin, 357–375. San Diego: Cordilleran Publ. 608 pp.

Proffett, J. M., Jr. 1977. Cenozoic geology of the Yerington district, Nevada, and implications for the nature and origin of Basin and Range faulting. *Geol. Soc. Am. Bull.* 88: 247–266.

Pumpelly, R., Wolff, J. E., and Dale, T. N. 1894. Geology of the Green Mountains in Massachusetts. *U.S. Geol. Surv. Monogr.* 23. 206 pp.

Raleigh, C. B., Healy, J. H., and Bredehoeft, J. D. 1972. Faulting and crustal stress at Rangely, Colorado. In *Flow and Fracture of Rocks. Am. Geophys. Union, Geophys. Monogr.* 16 (Griggs volume): 275–284.

Ramberg, H. 1955. Natural and experimental boudinage and pinch-and-swell structures. *J. Geol.* 63: 512–526.

———. 1963a. Fluid dynamics of viscous buckling applicable to folding of layered rocks. *AAPG Bull.* 47: 484–505.

———. 1963b. Evolution of drag folds. *Geol. Mag.* 100: 97–106.

———. 1964. Note on model studies of folding of moraines in Piedmont glaciers. *J. Glaciol.* 5: 207–218.

———. 1977. Some remarks on the mechanism of nappe movement. *Geol. Foeren. Stockholm Foerh.* 99: 110–117.

———. 1982. *Gravity, Deformation and the Earth's Crust.* London: Academic Press. 452 pp.

Ramberg, H., and Ghosh, S. K. 1968. Deformation structures in the Hovin Group schists in the Hommelvik-Hell region (Norway). *Tectonophysics* 6: 311–330.

Ramsay, J. G. 1962. The geometry and mechanics of formation of similar type folds. *J. Geol.* 70: 309–327.

———. 1967. *Folding and Fracturing of Rocks.* New York: McGraw-Hill. 568 pp.

Ramsay, J. G., and Huber, M. I. 1983. *The Techniques of Modern Structural Geology. Vol. 1, Strain Analysis.* London: Academic Press. 306 pp.

Read, H. H. 1948. A commentary on place in plutonism. *Geol. Soc. London Q. J.* 104: 155–206.

———. 1949. A contemplation of time in plutonism. *Geol. Soc. London Q. J.* 105: 101–152.

Reid, H. F., Davis, W. M., Lawson, A. C., and Ransome, F. L. 1913. Report of the Committee on the Nomenclature of Faults. *Geol. Soc. Am. Bull.* 24: 163–186.

Reynolds, D. L. 1954. Fluidization as a geological process, and its bearing on the intrusive granites. *Am. J. Sci.* 252: 577–614.

Rickard, M. J. 1971. A classification diagram for fold orientations. *Geol. Mag.* 108: 23–26.

———. 1972. Fault classification—Discussion. *Geol. Soc. Am. Bull.* 83: 2545–2546.

Roberts, J. C. 1961. Feather-fracture, and the mechanics of rock-jointing. *Am. J. Sci.* 259: 481–492.

Robison, B. A. 1983. Low-angle normal faulting, Mary's River Valley, Nevada. In *Seismic Expression of Structural Styles,* ed. A. W. Bally, vol. 2, 2.2–12. AAPG Studies in Geology Series, no. 15. Tulsa, Okla.: American Association of Petroleum Geologists.

Ross, C. S., and Smith, R. H. 1961. Ash-flow tuffs: their origin, geologic relations, and identification. *U.S. Geol. Surv. Prof. Pap.* 366: 1–77.

Rubey, W. W., and Hubbert, M. K. 1959. Role of fluid pressure in mechanics of overthrust faulting. *Geol. Soc. Am. Bull.* 70: 167–206.

Runcorn, S. K. 1956. Palaeomagnetic comparisons between Europe and North America. *Geol. Assoc. Can. Proc.* 8: 77–85.

Rutter, E. H. 1976. The kinetics of rock deformation by pressure solution. *R. Soc. London, Philos. Trans.* A283: 203–219.

———. 1986. On the nomenclature of mode of failure transitions in rocks. *Tectonophysics* 122: 381–387.

Sales, R. 1914. Ore deposits of Butte, Montana. *Am. Inst. Min. Met. Eng. Trans.* 46: 11–30.

Sander, B. 1930. *Gefügekunde der Gesteine, mit besonderer Berücksichtigung der Tektonite.* Vienna: Springer. 352 pp.

———. 1948. *Einführung in die Gefügekunde der geologischen Körper.* Vol 1, *Allgemeine Gefugekunde und Arbeiten im Bereich Handstück bis Profil.* Vienna: Springer. 215 pp. [F. C. Phillips and G. Windsor, trans. 1970. *An Introduction to the Study of Fabrics of Geological Bodies,* pp. 1–215. Oxford and New York: Pergamon. 641 pp.]

Sanford, A. R. 1959. Analytical and experimental study of simple geologic structures. *Geol. Soc. Am. Bull.* 70: 19–52.

Schardt, H. 1893. Origine des préalpes romandes. *Eclog. Geol. Helvet.* 4: 129–142.

Schmidt, W. 1925. Gefügestatistik. *Tschermaks Mineral. Petrogr. Mitt.* 38: 392–423.

———. 1932. *Tektonik und Verformungslehre.* Berlin: Borntraeger. 208 pp.

Schneider, H. J. 1953. Der Bau des Arnspitzstockes und seine tektonische Stellung zwischen Wetterstein-und Karwendelgebirge. *Geol. Bavarica* 17: 17–55.

Schroeder, E. 1966. Beiträge zur Schiefergebirgstektonik in Ost-Thüringen. *Dtsch. Akad. Wiss., Abh., Kl. Chem. Geol. Biol.* (for 1965). 93 pp.

Scrope, G. P. 1825. *Considerations on Volcanoes.* London: W. Phillips. 270 pp.

Seager, W. R. 1970. Low-angle gravity glide structures in the Northern Virgin Mountains, Nevada and Arizona. *Geol. Soc. Am. Bull.* 81: 1517–1538.

Secor, D. J., Jr. 1965. Role of fluid pressure in jointing. *Am. J. Sci.* 263: 633–646.

Şengör, A. M. C. 1979. The North Anatolian transform fault: Its age, offset and tectonic significance. *Geol. Soc. London J.* 136: 269–282.

Sharp, R. P. 1958. Malaspina glacier, Alaska. *Geol. Soc. Am. Bull.* 69: 617–646.

Shiki, T., and Misawa, Y. 1982. Forearc geological structure of the Japanese Islands. In *Trench-Forearc Geology,* ed. J. K. Leggett. *Geol. Soc. London Spec. Publ.* 10: 63–73.

Shrock, R. R. 1948. *Sequence in Layered Rocks.* New York: McGraw-Hill. 552 pp.

Sibson, R. H. 1977. Fault rocks and fault mechanisms. *Geol. Soc. London J.* 133: 191–214.

Smith, A. G. 1981. Subduction and coeval thrust belts, with particular reference to North America. *Geol. Soc. London Spec. Publ.* 9: 111–124.

Smith, D. E., Kolenkiewicz, R., Dunn, P. J., and Torrence, M. H. 1979. The measurement of fault motion by satellite laser ranging. *Tectonophysics* 52: 59–67.

Smith, R. L., Bailey, R. A., and Ross, C. S. 1961. Structural evolution of the Valles caldera, New Mexico, and its emplacement of ring dikes. *U.S. Geol. Surv. Prof. Pap.* 424D: 145–149.

Smithson, S. B., Brewer, J. A., Kaufman, S., Oliver, J. E., and Hurich, C. 1978. Nature of the Wind River thrust, Wyoming, from COCORP deep reflection data and from gravity data. *Geology* 6: 648–652.

Smoluchowski, M. S. 1909. Versuch über Faltungserscheinungen schwimmender elastischer Platten. *Anz. Akad. Wiss. Krakow. Math. Phys. Kl.* 727–734.

Sorby, H. C. 1853. On the origin of slaty cleavage. *Edinburgh New Philos. J.* 55: 137–148.

Stearns, D. W. 1969. Fracture as a mechanism of flow in naturally deformed layered rocks. *Geol. Surv. Can. Pap.* 68–52: 79–89.

———. 1972. Structural interpretation of the fractures associated with the Bonita fault. In *Guidebook of East-Central New Mexico,* ed. V. C. Kelley and F. D. Trauger, 161–164. New Mexico Geological Society.

———. 1978. Faulting and folding in the Rocky Mountain forelands. In *Laramide Folding Associated with Basement Block Faulting in the Western United States,* ed. V. Matthews *Geol. Soc. Am. Mem.* 151: 1–37.

Steno, N. 1669. *De solido intra solidum naturaliter contento* (dissertationis prodromus). Florence: Typographia sub signo Stellae.

Stevens, G. R. 1975. The anatomy of a Marlborough fault line: The Wairau fault at Branch River. *Geol. Soc. N.Z. Guidebook.* 12 pp.

Stille, H. 1930. Über Einseitigkeiten der germanotypen Tektonik Nordspaniens und Deutschlands: *Nachrichten der Gesellschaft für Wissenschaften, Göttingen,* Math-Phys. Klass, 379–397.

Stockwell, C. H. 1950. The use of plunge in the construction of cross sections of folds. *Geol. Assoc. Can. Proc.* 3: 97–121.

Suess, E. 1875. *Die Entstehung der Alpen.* Vienna: Braumüller, 168 pp.

———. 1885. *Das Antlitz der Erde.* Vol. 1. Prague: Tempsky. 778 pp.

Sykes, L. R. 1967. Mechanisms of earthquakes and nature of faulting on the mid-oceanic ridges. *J. Geophys. Res.* 72: 2131–2153.

Sylvester, A. G. 1964. The Precambrian rocks of the Telemark area in south central Norway. III. *Nor. Geol. Tidsskr.* 44: 445–482.

Sylvester, A. G., and Christie, J. M. 1968. The origin of crossed-girdle orientations of optic axes in deformed quartzites. *J. Geol.* 76: 571–579.

Sylvester, A. G., and Smith, R. R. 1976. Tectonic transpression and basement-controlled deformation in San Andreas fault zone, Salton Trough, California. *AAPG Bull.* 60: 2081–2102.

Tapponnier, P., Peltzer, G., Le Dain, A. Y., and Armijo, R. 1982. Propagating extrusion tectonics in Asia: New insights from simple experiments with plasticine. *Geology* 10: 611–616.

Taylor, F. 1910. Bearing of the Tertiary mountain belt on the origin of the earth's plan. *Geol. Soc. Am. Bull.* 21: 179–226.

Te Punga, M. T. 1957. Live anticlines in western Wellington. *N.Z. J. Sci. Techn. Sect. B.* 38: 433–446.

Ter-Stepanian, G. 1962. Klassifizierung der Erdrutschrisse. *Geol. Bauwes.* 28: 53–54.

Tomkeieff, S. I. 1962. Unconformity—An historical study. *Geol. Assoc. London, Proc.* 73: 383–418.

Törnebohm, A. E. 1888. Om fjäll problemet. *Geol. Foeren. Stockholm Foerh.* 10: 328–336.

Trümpy, R. 1980. An outline of the geology of Switzerland. In *An Outline of the Geology of Switzerland (Guide-Book), 26th Int. Geol. Congr. Paris, 1980.* Basel: Wepf & Co. 334 pp.

Trusheim, F. 1957. Über Halokinese und ihre Bedeutung für die strukturelle Entwicklung Norddeutschlands. *Z. Dtsch. Geol. Ges.* 109: 111–158. Engl. transl. in *AAPG Bull.* 44 (1960): 1519–1541.

Tullis, J., Snoke. A. W., and Todd, V. R. 1982. Significance and petrogenesis of mylonitic rocks. *Geology* 10: 227–230.

Turner, F. J. 1953. Nature and dynamic interpretation of deformation in calcite of three marbles. *Am. J. Sci.* 251: 276–298.

Turner, F. J., and Weiss, L. E. 1963. *Structural Analysis of Metamorphic Tectonites.* New York: McGraw-Hill. 545 pp.

Van Hise, C. R. 1896. Principles of North American Precambrian geology. *U.S. Geol. Annu. Rep. 16th* (1894–1895) Pt. 1: 571–843.

Vine, F. J. 1966. Spreading of the ocean floor; new evidence. *Science* 154: 1405–1415.

Vine, F. J., and Matthews, D. H. 1963. Magnetic anomalies over oceanic ridges. *Nature* 199: 947–949.

Vollbrecht, K. 1964. Die Diabasvorkommen des Amazonasgebietes und das Problem des Intrusionmechanismus. *Geol. Rundsch.* 53: 686–706.

Von Kármán, T. 1911. Festigkeitsversuche unter allseitigem Druck. *Ver. Dtsch. Ing. Z.* 55: 1749–1757.

Walker, C. T., and Dennis, J. G. 1966. Explosive phase transitions in the mantle. *Nature* 209: 182–183.

Walker, G. P. L. 1965. Evidence of crustal drift from Icelandic geology. *R. Soc. London, Philos. Trans.* 258: 199–204.

Weber, K. 1976. Gefügeuntersuchungen an transversal geschieferten Gesteinen aus dem östlichen Rheinischen Schiefergebirge. *Geol. Jahrb.* D15: 3–98.

———. 1980. *Guide to Excursion, Rheinisches Schiefergebirge.* 1980 International Conference on the Effect of Deformation on Rocks. Göttingen. 65 pp.

———. 1981. Kinematic and metamorphic aspects of cleavage formation in very low-grade metamorphic slates. Pages 291–306 in: G. S. Lister, H. J. Behr, K. Weber, and H. J. Zwart, eds. The Effect of Deformation on Rocks. *Tectonophysics* 78: 291–306.

Wegener, A., 1915. *Die Entstehung der Kontinente und Ozeane.* Brunswick: Vieweg. 108 pp.

Wegmann, C. E. 1923. Zur Geologie der St. Bernhard Decke im Val d'Hérens (Wallis). *Soc. Neuchâtel. Sci. Natur. Bull.* 47.

———. 1929. Beispiele tektonischer Analysen des Grundgebirges in Finnland. *Comm. Géol. Finlande Bull.* 87: 98–127.

———. 1932. Note sur le boudinage. *Soc. Géol. France. Compt. Rend.* 5 Pt. 2: 477–489.

———. 1935. Zur Deutung der Migmatite. *Geol. Rundsch.* 26: 305–350.

———. 1965. Tectonic patterns at different levels. *Geol. Soc. S. Afr. Proc.* 66 (1963); Annex (A. L. du Toit Mem. Lect. no. 8), 78 pp.

Weiss, L. E. 1959a. Geometry of superposed folding. *Geol. Soc. Am. Bull.* 70: 91–106.

———. 1959b. Structural analysis of the basement system at Turoka, Kenya. *Overseas Geol. Miner. Resour. London,* 7: 3–35 and 123–153.

Wellman, H. W. 1955. New Zealand Quarternary tectonics. *Geol. Rundsch.* 43: 248–257.

———. 1971. Holocene tilting and uplift on the Glenburn coast, Wairarapa, New Zealand. *R. Soc. N.Z. Bull.* 9: 221–223.

Wilcox, R. E., Harding, T. P., and Seely, D. R. 1973. Basic wrench tectonics. *AAPG Bull.* 57: 74–96.

Williams, G. D., and Chapman, T. J. 1979. The geometrical classification of noncylindrical folds. *J. Struct. Geol.* 1: 181–185.

Willis, B. 1893. The mechanics of Appalachian structure. *U.S. Geol. Surv. 13th Annu. Rep. 1891–1892* Pt. 2: 212–281.

Wilson, G. 1961. Tectonic significance of small scale structures. *Soc. Geol. Belg. Ann.* 84: 423–548.

———. 1982. *Introduction to Small-Scale Geological Structures.* London: George Allen & Unwin. 128 pp.

Wilson, J. T. 1965. A new class of faults and their bearing on continental drift. *Nature* 207: 343–347.

Windley, B. F. 1969. Evolution of the early Precambrian complex of southern West Greenland. *Geol. Assoc. Can. Spec. Pap.* 5: 155–161.

Wood, D. S. 1973. Patterns and magnitudes of natural strain in rocks. *R. Soc. London Philos. Trans.* A274: 373–382.

Woodward, H. P. 1959. The Appalachian region. *5th World Pet. Congr. Pap.* 1–59, 19 pp.

Woodworth, J. B. 1896. On the fracture system of joints, with remarks on certain great fractures. *Boston Soc. Nat. Hist. Proc.* 27: 163–184.

Wright, F. E. 1911. *The Methods of Petrographic Microscope Research.* Publ. 168. Washington, D.C.: Carnegie Institute. 204 pp.

Wright, T. O., and Platt, L. B. 1982. Pressure dissolution and cleavage in the Martinsburg Shale. *Am. J. Sci.* 282: 122–135.

Wulff, G. 1902. Untersuchungen im Gebiete der optischen Eigenschaften isomorpher Krystalle. *Z. Kristallogr. Mineral.* 36: 1–28.

Ziegler, P. A. 1982. Faulting and graben formation in western and central Europe. *R. Soc. London Phil. Trans.* A305: 113–143.

Ziony, J. I. 1966. *The analysis of systematic jointing in part of Monument Upwarp, SE Utah.* Ph.D. thesis. Los Angeles: University of California. 112 pp.

Zwart, H. J. 1961. The chronological succession of folding and metamorphism in the central Pyrenees. *Geol. Rundsch.* 1960. 50: 203–218.

# Credits

## Chapter 1

**Figures 1.1 and 1.16:** From Isacks, B., J. Oliver, and L. R. Sykes, "Seismology and the New Global Tectonics" in *Journal of Geophysical Research, 73,* pp. 5855–5899, 1968. Copyright American Geophysical Union.

**Figure 1.2:** From Plummer, Charles and David McGeary, *Physical Geology,* 3d ed. © 1985 Wm. C. Brown Publishers, Dubuque, Iowa. All Rights Reserved. Reprinted by permission.

**Figure 1.4:** From Oliver, J. and B. Isacks, "Deep earthquake zones anomalous structures in the Upper Mantle, and the Lithosphere" in *Journal of Geophysical Research, 73,* pp. 4259–4275, 1967. Copyright 1967 American Geophysical Union.

**Figure 1.5:** From Katsumata, M. and L. R. Sykes, "Seismicity and tectonics of the Western Pacific: Izu-Mariana-Caroline and Ryukyu-Taiwan" in *Journal of Geophysical Research, 74,* pp. 5923–5948, 1969. Copyright 1969 American Geophysical Union.

**Figure 1.6:** From Meyerhoff, A. A., "Arthur Holmes: Originator of spreading ocean floor hypothesis" in *Journal of Geophysical Research, 73,* pp. 6563–6565 (and discussion), 1968. Copyright 1968 American Geophysical Union.

**Figures 1.7 and 1.14:** From Peterson, M. S., J. K. Rigby, and L. F. Hintze, *Historical Geology of North America, 2d ed.* © 1973, 1980 Wm. C. Brown Publishers, Dubuque, Iowa. All Rights Reserved. Reprinted by permission.

**Figures 1.8, 1.9, and 1.10:** From Vine, F. J., "Spreading of the ocean floor; New evidence" in *Science, 154,* pp. 1405–1415, 1966. Copyright © 1966 American Association for the Advancement of Science. Reprinted by permission.

**Figure 1.11:** From Dennis, John G., *International Tectonic Dictionary, English Terminology, Memoir 7.* © 1967 American Association of Petroleum Geologists. Reprinted by permission.

**Figure 1.13:** From Morgan, W. J., "Rises, Trenches, Great Faults, and Crustal Blocks" in *Journal of Geophysical Research, 73,* pp. 1959–1982, 1969. Copyright American Geophysical Union.

**Figure 1.15:** From Dickenson, W. R. and D. R. Seely, "Stratigraphy and structure of forearc regions" in *AAPG Bulletin, 63,* pp. 2–31, 1979. Copyright 1979 American Association of Petroleum Geologists. Reprinted by permission.

**Figure 1.17:** From Oxburgh, E. R. and D. L. Turcotte, "Thermal structure of island arcs" in *Geological Society of America Bulletin, 81,* pp. 1665–1688, 1970. Copyright © 1970 E. R. Oxburgh and D. L. Turcotte. Reprinted by permission.

**Figure 1.18:** From Gass, Ian G. and I. L. Gibson, "Structural evolution of the rift zones in the Middle East" in *Nature,* Vol. No. 221, pp. 926–930. Copyright © 1969 Macmillan Journals Limited.

**Figures 1.19 and 1.20:** From McKenzie, D. P. and W. J. Morgan, "Evolution of triple junctions" in *Nature,* Vol. No. 224, pp. 125–133. Copyright © 1969 by Macmillan Journals Limited.

**Figure 1.21:** From Dewey, J. F. and K. Burke, "Hot spots and continental breakup: Implications for collisional orogeny" in *Geology, 2,* pp. 57–60, 1974. Copyright © 1974 J. F. Dewey and K. Burke. Reprinted by permission.

**Figure 1.22:** From Ham, W. E. and J. L. Wilson, "Paleozoic epeirogeny and orogeny in the Central United States" in *American Journal of Science, 25,* pp. 332–407, 1967. Copyright © 1967 American Journal of Science. Reprinted by permission.

**Figure 1.23:** From Dennis, John G., H. Murawski, and K. Weber, *International Tectonic Lexicon,* p. 153. Copyright © 1979 Schweizerbart, Stuttgart, W. Germany. Reprinted by permission.

## Chapter 2

**Figures 2.1 and 2.2:** From Inoue, E., "Land deformation in Japan" in *Geographical Survey Institute (Japan), 6,* pp. 73–134, 1960. Copyright © 1960 Geographical Survey Institute (Japan). Reprinted by permission.

**Figure 2.4:** From Smith, D. E., R. Kolenkiewicz, P. J. Dunn, and M. H. Torrence, "The measurement of fault motion by satellite laser ranging" in *Tectonophysics, 52,* pp. 59–67, 1979. Copyright 1979 Elsevier Scientific Publishing, Amsterdam. Reprinted by permission.

**Figure 2.8:** From Wellman, H. W., "Holocene Tilting and uplift on the Glenburn coast, Wairarapa, New Zealand" in *Royal Society of New Zealand Bulletin, 9,* pp. 221–223, 1971. Copyright © 1971 Royal Society of New Zealand. Reprinted by permission.

**Figure 2.13:** From Stevens, G. R., "The Anatomy of a Marlborough Fault Line: The Wairau Fault at Branch River" in *Geological Society of New Zealand Guidebook,* p. 12, 1975. Copyright © 1975 Geological Society of New Zealand. Reprinted by permission.

**Figure 2.14:** From Bernauer, F., "Junge Tektonik auf Island und ihre Ursachen" in Niemcyk, Oskar, *Spaltin auf Island,* pp. 14.64, 1943. Copyright © 1943 Verlag Wittwer, Stuttgart, W. Germany. Reprinted by permission.

**Figures 2.16 and 2.17:** From Christensen, M. N., "Late Cenozoic Deformation in the Central Coast Ranges of California" in *Geological Society of America Bulletin, 76,* pp. 1105, 1124, 1965. Copyright © 1965 M. N. Christensen. Reprinted by permission.

## Chapter 3

**Figure 3.2:** From Zumberge, J. H. and C. A. Nelson, *Elements of Physical Geology.* Copyright © 1976 John Wiley & Sons, New York. Adapted by permission.

## Chapter 4

**Figure 4.2:** From Hills, E. S., *Elements of Structural Geology,* p. 502. © 1927 John Wiley & Sons, Inc. Reprinted by permission of Chapman & Hall, London.

**Figures 4.6 and 4.15:** From Kuenen, P. H., "Significant features of graded bedding" in *AAPG Bulletin, 37,* pp. 1044–1066, 1953. Copyright 1953 American Association of Petroleum Geologists. Reprinted by permission.

**Figure 4.11:** From Park, W. C. and E. H. Schot, "Stylolites: Their nature and origin" in *Journal of Sedimentary Petrology, 38,* pp. 175–191, 1968. Copyright © 1968 Society of Economic Paleontologists and Mineralogists. Reprinted by permission.

## Chapter 5

**Figure 5.5:** Courtesy Bradley Erskine.

**Figure 5.6:** From Hoeppener, R., *Felsmechanik & Ingenieurgeologie, Vol. 2.* Copyright © 1964 Springer-Verlag, New York. Reprinted by permission.

**Figure 5.7:** From Fairbairn, "Elongation in deformed rocks" in *Journal of Geology, 44,* pp. 670–680, 1936. Copyright © 1936 University of Chicago. Reprinted by permission.

**Figure 5.12:** From Handin, John W., "Strength and ductility" in *Geological Society of America Memoir 97,* pp. 223–290, 1966. Copyright © 1966 John W. Handin. Reprinted by permission.

**Figure 5.13:** From Handin, John W., "Behavior of Materials in the Earth's Crust" in *Colorado School of Mines Quarterly.* Vol. 52, No. 3, pp. 75–98, 1957. Copyright 1957 Colorado School of Mines Press. Reprinted by permission.

## Chapter 6

**Figure 6.2:** From Dennis, John G., *International Tectonic Dictionary, English Terminology, Memoir 7.* © 1967 American Association of Petroleum Geologists. Reprinted by permission.

**Figure 6.5:** From Handin, John W., "Strength and ductility" in *Geological Society of America Memoir 97,* pp. 223–290, 1966. Copyright © 1966 John W. Handin. Reprinted by permission.

**Figure 6.6:** From Walker, C. T. and John G. Dennis, "Explosive Phase Transitions in the Mantle" in *Nature,* Vol. No. 209, pp. 182–183, 1966. Copyright 1966 Macmillan Journals Limited.

**Figure 6.7:** From Price, N. J., "Rates of Deformation", in *Geological Society of London Quarterly Journal, 131,* pp. 553–575, 1975. Copyright © 1975 Geological Society of London. Reprinted by permission.

**Figures 6.8 and 6.9:** From Heard, H. C., "Transition from Brittle Fracture to Ductile Flow in Solenhogen Limestone as a Function of Temperature, Confining Pressure and Interstitial Fluid Pressure" in *Geological Society of America Memoir 79,* pp. 193–226, 1960.

**Figures 6.10 and 6.12:** From Heard, H. C., "Effect of Large Changes in Strain Rate in the Experimental Deformation of Yule Marble" in *Journal of Geology, 71,* pp. 162–195, 1963. Copyright © 1963 University of Chicago. Reprinted by permission.

**Figure 6.11:** Courtesy D. T. Griggs and J. D. Blacic.

**Figures 6.14 and 6.15:** From Edelglass, S. M., *Engineering Materials Science: Structure and Behavior of Solids.* Copyright © 1966 Ronald Press, New York. Reprinted by permission.

**Figure 6.17:** From Rutter, E. H., "The Kinetics of Rock Deformation by Pressure Solution" in *Royal Society of London Philosophical Transactions, A283,* pp. 203–219, 1976. Copyright © 1976 Royal Society of London. Reprinted by permission.

**Figures 6.19 and 6.20:** From Mueller, W. H., S. M. Schmid, and U. Briegel, "Deformation Experiments on Anhydrite Rocks of Different Grain Sizes: Rheology and Microfabric" in *Tectonophysics, 78,* pp. 527–543, 1981. Copyright © 1981 Elsevier Scientific Publishing, Amsterdam. Reprinted by permission.

**Figure 6.24:** From Jackson, M. P. A., "External Shapes and Dynamics of Salt Structures" in *Geological Society of America Bulletin, 96,* 1985. Copyright © 1985 M. P. A. Jackson. Reprinted by permission.

**Figure 6.25:** From Curnelle, R. and R. Marco, "Reflection Profiles Across the Aquitane Basin" in A. W. Bally, ed., *Seismic Expression of Structural Styles: A Picture and Work Atlas, 2: AAPG Studies in Geology, Series #15,* 2–3, 2–11, 1983. Copyright © 1983 American Association of Petroleum Geologists. Reprinted by permission.

**Figure 6.26:** From Balk, R., "Salt Structure of Jefferson Island Salt Dome, Iberia and Vermillion Parishes, Louisiana" in *AAPG Bulletin, 37,* pp. 2455–2474, 1953. Copyright © 1953 American Association of Petroleum Geologists. Reprinted by permission.

**Figure 6.28:** From Hollingworth, S. E., J. H. Taylor, and G. A. Kellaway, "Large Scale Superficial Structures in the Northampton Ironstone Field" in *Geological Society of London Quarterly Journal, 100,* pp. 1–34, 1945. Copyright © 1945 Geological Society of London. Reprinted by permission.

**Figure 6.29:** From Feugueur, "Comportment du charbon dans une nappe helvetique des Alpes francaises, le gisement d'Araches in Haute-Savoire". International Geological Congress, Comptes Rendus, 19th Session, Pt. 3, pp. 163–172, 1953.

## Chapter 7

**Figures 7.1 and 7.17:** From Wilson, Gilbert, *Introduction to Small-Scale Geological Structures.* Copyright 1982 by Allen & Unwin, London. Reprinted by permission.

**Figures 7.2, 7.3, and 7.19:** From Dennis, John G., *International Tectonic Dictionary, English Terminology, Memoir 7.* Copyright © 1967 American Association of Petroleum Geologists. Reprinted by permission.

**Figure 7.4:** From Richard, M. J., "A Classification Diagram for Fold Orientations" in *Geological Magazine, 108,* pp. 23–26, 1971. Copyright 1971 Cambridge University Press. Reprinted by permission.

**Figure 7.5:** Reprinted with permission from *Journal of Structural Geology, 1,* Williams, G. D. and T. J. Chapman, "The Geometrical Classification of Non-cylindrical Folds". Copyright 1979 Pergamon Journals.

**Figure 7.6:** From Ramsey, J. G., "The Geometry and Mechanics of Formation of Similar Type Folds" in *Journal of Geology, 70,* pp. 309–327, 1962. Copyright © 1962 University of Chicago. Reprinted by permission.

**Figure 7.7:** From Ramsey, J. G., *Folding and Fracturing of Rocks.* Copyright © 1979 McGraw-Hill Book Company, New York. Reprinted by permission.

**Figure 7.15:** From Wegmann, Eugene (C. E.), "Zur Geologie der St. Bernard Dieke in Val d'Herens (Wallis)" in *Soc. Neuchatel Sciences Natur. Bull., 47,* 1923.

**Figure 7.16:** Reproduced with permission of the Geological Association of Canada from Stockwell, C. H., "The Use of Plunge in the Construction of Cross Sections of Folds" in *Geological Association of Canada Proceedings, 3,* pp. 97–121, 1950. Copyright © 1950 Geological Association of Canada.

## Chapter 8

**Figure 8.1:** From Dennis, John G., *International Tectonic Dictionary, English Terminology, Memoir 7.* Copyright © 1967 American Association of Petroleum Geologists. Reprinted by permission.

**Figure 8.2:** From Ramberg, H. and Gosch, "Deformation Structures in the Hovin Group Schists in the Hommelvik-Hill Region (Norway)" in *Tectonophysics, 6,* pp. 311–330, 1968. Copyright © 1968 Elsevier Scientific Publishing, Amsterdam. Reprinted by permission.

**Figure 8.4:** From Laubscher, "Die Fernschubleypothese der Jurafalting" in *Ecologae Geologiae Helvetiae, 54,* pp. 221–282, 1961. Copyright © 1961 Ecologae Geologiae Helvetiae. Reprinted by permission.

**Figure 8.6:** From Carey, S. W., "The Rheid Concept in Geotechnics" in *Geological Society of Australia Journal, 1,* pp. 67–117, 1954. Copyright © 1954 Geological Society of Australia. Reprinted by permission.

**Figure 8.7:** From Ramberg, H., "Note on Model Studies of Folding of Moraines in Piedmont Glaciers" in *Journal of Glaciology, 5,* pp. 207–218, 1964. Copyright © 1964 International Glaciological Society. Reprinted by permission.

**Figure 8.8:** From Pfiffner, O. A., "Strain Analysis in Folds" in *Tectonophysics, 61,* pp. 337–362, 1980. Copyright © 1980 Elsevier Scientific Publications, Amsterdam. Reprinted by permission.

**Figures 8.10 and 8.14:** From Biot, M. A., "Theory of Folding of Stratified Viscoelastic Media and its Implications in Tectonics and Orogenesis" in *Geological Society of America Bulletin, 72,* pp. 1595–1620 and 1621–1631, 1961.

**Figure 8.11:** From Currie, J. B., H. W. Patnode, and R. P. Trump, "Development of Folds in Sedimentary Strata" in *Geological Society of America Bulletin, 73,* pp. 655–674, 1962. Copyright © 1962 J. B. Currie, H. W. Patnode, and R. P. Trump. Reprinted by permission.

**Figure 8.13:** From Dieterich, James H., "Origin of Cleavage in Folded Rocks" in *American Journal of Science, 267,* 155–165, 1969. Copyright © 1969 American Journal of Science. Reprinted by permission.

**Figure 8.15:** From Chapple, W. M., "The Finite-amplitude Instability in the Folding of Layered Rocks" in *Canadian Journal of Earth Sciences, 7,* pp. 457–465, 1970. Copyright © 1970 Canadian Journal of Earth Sciences. Reprinted by permission.

**Figure 8.17:** From Korn, H. and H. Martin, "Gravity Tectonics in the Naukluft Mountains of South West Africa" in *Geological Society of America Bulletin, 70,* pp. 1047–1078, 1959.

**Figure 8.18:** From Cloos, E., "Boudinage" in *American Geophysical Union Transactions, 28,* pp. 626–632, 1947. Copyright American Geophysical Union.

**Figure 8.20:** From Ramberg, H., "Evolution of Drag Folds" in *Geological Magazine, 100,* pp. 97–106, 1963. Copyright 1963 Cambridge University Press. Reprinted by permission.

**Figure 8.21:** From Hoeppener, "Das tektonische Inventar eines Aufschlusses in den Orthoceras-Schiefern bie Dillenberg" in *Geologische Rundschau, 44,* pp. 93–98, 1955. Copyright © 1955 Geologische Rundschau. Reprinted by permission.

**Figure 8.25:** Reproduced with permission of the Geological Association of Canada from Windley, B. F., "Evolution of the Early Precambrian Complex of Southern West Greenland" in *Geological Association of Canada Special Paper 5,* pp. 155–161, 1969. Copyright © 1969 Geological Association of Canada.

**Figure 8.26:** From Carey, S. W., "Folding" in *Journal of Alberta Society of Petroleum Geologists, 10,* pp. 95–144, 1962. Copyright © 1962 Canadian Society of Petroleum Geologists. Reprinted by permission.

**Figure 8.27:** From Terra, H. de, "Structural Features in Gliding Strata" in *American Journal of Science, Ser. 5, 21,* pp. 204–213, 1931. Copyright © 1931 American Journal of Science. Reprinted by permission.

**Figure 8.29:** From Funk, H., T. Labhart, A. G. Milnes, A. O. Pfiffner, U. Schaltegger, C. Schindler, S. M. Schmid, and R. Trümpy, Bericht uber die Jubiläumsex kursion "Mechanismus der Gebirgsbildung" der Schweizerischen Geologischen Gesellschaft in das ost-und zentralschweizerische Helvetikum und in das nördliche Aarmassiv vom 12. bis 17. September 1982: *Eclogae Geol. Helvetiae,* v. 76, pp. 91–123.

**Figure 8.30:** From Stearns, D. W., "Faulting and Folding in the Rocky Mountain Forelands" in V. Matthews, (ed.), *Laramide Folding Associated with Faulting in the Western United States. Geological Society of America Memoir 151,* pp. 1–37, 1978. Copyright © 1978 D. W. Stearns. Reprinted by permission.

## Chapter 9

**Figure 9.3:** From Oertel, Gerhard, "Extrapolation in Geologic Fabrics" in *Geological Society of America Bulletin, 73,* pp. 325–342, 1962. Copyright © 1962 Gerhard Oertel.

**Figure 9.5:** From Paterson, M. S. and L. E. Weiss, "Symmetry Concepts in the Structural Analysis of Deformed Rocks" in *Geological Society of America Bulletin, 72,* pp. 822–841, 1961.

**Figure 9.6:** From Weber, Klaus, "Guide to Excursions: Rheinisches Schiefergebirge" in *International Conference on the Effects of Deformation in Rocks,* Gottingen, 1980, p. 65. Reprinted by permission.

**Figure 9.7:** From Dennis, John G., *International Tectonic Dictionary, English Terminology, Memoir 7.* Copyright © 1967 American Association of Petroleum Geologists. Reprinted by permission.

**Figure 9.8:** From Wilson, Gilbert, *Introduction to Small-Scale Geological Structures.* Copyright 1982 Allen & Unwin, London. Reprinted by permission.

**Figures 9.10, 9.11, and 9.12:** From Friedman, M. and F. B. Conger, "Dynamic Interpretation of Calcite Twin Lamallae in a Naturally Deformed Fossil" in *Journal of Geology, 72,* pp. 361–368, 1964. Copyright © 1964 University of Chicago. Reprinted by permission.

**Figures 9.13 and 9.14:** From Sylvester, A. G. and J. M. Christie, "The Origin of Crossed Girdle Orientations of Optic Axes in Deformed Quartzites" in *Journal of Geology, 76,* 571–579, 1968. Copyright © 1968 University of Chicago. Reprinted by permission.

## Chapter 10

**Figure 10.8:** From Plessman, Werner, "Gestienslosung, ein Hauptfaktor beim Schieferungsprozess" in *Geol. Mitteil 4,* pp. 69–82, 1965.

**Figure 10.12:** From Turner, F. and L. E. Weiss, *Structural Analysis of Metamorphic Tectonites.* Copyright © 1963 McGraw-Hill Book Company, New York. Reprinted by permission.

**Figure 10.15:** From Wright, T. O. and L. B. Platt, "Pressure and Dissolution and Cleavage in the Martisburg Shale" in *American Journal of Science, 282,* pp. 122–135, 1982. Copyright © 1982 American Journal of Science. Reprinted by permission.

**Figures 10.17 and 10.18:** From Dieterich, James H., "Origin of Cleavage in Folded Rocks" in *American Journal of Science, 267,* pp. 155–165, 1969. Copyright © 1969 American Journal of Science. Reprinted by permission.

**Figure 10.22:** From Anderson, T. Bernard, "Kink-Bands and Related Geological Structures" in *Nature,* Vol. No. 202, pp. 272–274, 1964. Copyright 1964 Macmillan Journals Limited.

**Figure 10.25:** From Zwart, H. J., "The Chronological Succession of Folding and Metamorphism in the Central Pyrenees" in *Geologische Rundschau, 50,* pp. 203–218, 1961. Copyright © 1961 Geologische Rundschau. Reprinted by permission.

**Figures 10.27 and 10.28:** From Cloos, E., "Oolite Deformation in the South Mountain Fold, Maryland" in *Geological Society of America Bulletin, 58,* pp. 843–918, 1947.

**Figure 10.31:** From Cloos, E., "Lineation" in *Geological Society of America Memoir 18,* p. 122, 1946.

**Figure 10.33:** From Wilson, Gilbert, *Introduction to Small-Scale Geological Structures.* Copyright 1982 Allen & Unwin, London. Reprinted by permission.

**Figure 10.34:** From Weiss, L. W., "Geometry of Superposed Folding" in *Geological Society of America Bulletin, 70,* pp. 91–106, 1959.

## Chapter 11

**Figure 11.5:** From Bock, H., "Das Fundamentale Kluftsystem" in *Zeitschrift der Deutschen Geologischen Gesellschaft, 131,* pp. 627–650, 1980. Reprinted by permission.

**Figure 11.7:** From Wilson, Gilbert, *Introduction to Small-Scale Geological Structures.* Copyright 1982 Allen & Unwin, London. Reprinted by permission.

**Figure 11.8:** From Cloos, H., "Gang und Gehierk einer Falte" in *Zeitschrift der Deutschen Geologischen Gesellschaft, 100,* pp. 290–303, 1948. Reprinted by permission.

**Figure 11.12:** From Illies, H., "Graben Tectonics as Related to Crust-Mantle Interaction" in J. S. Illies and St. Mueller, (eds.), *Graben Problem: International Upper Mantle Project, Scientific Report No. 27,* pp. 4–27, 1970. Stuttgart, Schweitzerbart.

**Figure 11.14:** From Bankwitz, Peter, "Uber Klufte, II" in *Geologie, 15,* pp. 896–941, 1966. Reprinted by permission.

**Figure 11.17:** From Bannert, D., "Luftbildkartierung des Lineationsnetzes vom Ries und seiner Umgebung" in *Geologica Bavarica, 61,* 379–384, 1969. Reprinted by permission.

**Figure 11.18:** From Bolsenkotter, H., "Feintektonische Untersuchungen an Schlechten und Kluften in Steinkohlenflozen des Ruhrgebietes" in *Geologische Rundschau, 44,* pp. 443–472, 1955. Reprinted by permission.

**Figure 11.19:** From Schneider, H. J., "Der Bau des Arnspitzstockes und seine tektonische Stellung zwischen Wetterstein-und Karwendelgebirge" in *Geologica Bavaria, 17,* pp. 17–55, 1953. Reprinted by permission.

## Chapter 12

**Figure 12.2:** From Cloos, H., *Einfuhrung in die Geologie.* Berlin: Borntraeger, p. 503, 1936. Reprinted by permission.

**Figure 12.5:** From Dennis, J. G., *International Tectonic Dictionary, English Terminology, Memoir 7,* p. 196, 1967. Copyright © 1967 American Association of Petroleum Geologists. Reprinted by permission.

**Figure 12.11:** From Gill, J. E., "Normal and Reverse Faults" in *Journal of Geology, 43,* pp. 1071–1079, 1935. © 1935 University of Chicago. Reprinted by permission.

**Figure 12.15:** From King, G. C. P., "Geological Faults: Fracture, Creep and Strain" in *Royal Society of London Philosophical Transactions, A288,* pp. 197–212, 1978. Copyright © 1978 Royal Society of London. Reprinted by permission.

**Figure 12.16:** From Rickard, M. J., "Fault Classification-discussion" in *Geological Society of America Bulletin, 83,* pp. 2545–2546, 1972. Copyright © 1972 M. J. Rickard.

**Table 12.1:** From Dennis, J. G., *International Tectonic Dictionary, English Terminology, Memoir 7.* © 1967 American Association of Petroleum Geologists. Reprinted by permission.

## Chapter 13

**Figure 13.1:** From Secor, Donald T., "Role of Fluid Pressure in Jointing" in *American Journal of Science, 263,* pp. 633–646, 1965. Copyright © 1965 American Journal of Science. Reprinted by permission.

**Figure 13.3:** From Bruhn, R. L., M. R. Yusas, and F. Huertas, "Mechanics of Low-angle Normal Faulting: An Example from Roosevelt Hot Springs Geothermal Area, Utah" in *Tectonophysics, 86,* pp. 343–361, 1982. Copyright © 1982 Elsevier Scientific Publishing, Amsterdam. Reprinted by permission.

**Figure 13.4:** From Ode, H., "A Note Concerning the Mechanism of Artificial and Natural Hydraulic Fracture Systems" in *Colorado School of Mines Quarterly, 51,* pp. 19–29, 1956. Copyright 1956 Colorado School of Mines Press. Reprinted by permission.

**Figure 13.5:** From Ziony, Joseph I., "The Analysis of Systematic Jointing in Part of Monument Upwarp, S. E. Utah". Ph.D. Thesis, 1966, University of California, Los Angeles, p. 112. Reprinted by permission of Joseph I. Ziony.

**Figure 13.7:** From Bock, H., "Das Fundamentale Kluftsystem" in *ZeitschriftDeutsche Geologische Geselleschaft, 131,* pp. 627–650, 1980. Reprinted by permission.

**Figure 13.8:** From Price, N. J., "Mechanics of Jointing in Rocks" in *Geological Magazine, 96,* pp. 146–147, 1959. Copyright 1959 Cambridge University Press. Reprinted by permission.

**Figure 13.12:** From Raleigh, C. B., J. B. Healy, and J. C. Brodehoeft, "Faulting and Crustal Stress at Rangely, Colorado" in *Flow and Fracture of Rocks, American Geophysical Union Monograph, 16* (Grigg's Volume), pp. 275–284, 1972. Copyright American Geophysical Union.

**Figure 13.13:** From King, S. C. P., "Geological Faults: Fracture, Creep, and Strain" in *Royal Society of London Philosophical Transactions, A288,* pp. 197–212, 1978. Copyright © 1978 Royal Society of London. Reprinted by permission.

**Figure 13.15:** From Hubbert, M. K., "Mechanical Basis for Certain Familiar Geologic Structures" in *Geological Society of America Bulletin, 62,* pp. 355–372, 1951.

**Figure 13.16:** From Griggs, D. T. and J. Handin, "Observations of Fracture and Hypothesis of Earthquakes" in *Rock Deformation: A Symposium. Geological Society of America Memoir 79,* pp. 347–364, 1960.

**Figure 13.18:** From Rutter, E. H., "On the Nomenclature of Mode of Failure Transitions in Rocks" in *Tectonophysics, 122,* pp. 381–387, 1986. Copyright © 1986 Elsevier Scientific Publishing, Amsterdam. Reprinted by permission.

**Figure 13.19:** From Sibson, R. H., "Fault Rocks and Fault Mechanisms" in *Geological Society of London Quarterly Journal, 133,* pp. 191–214, 1977. Copyright © 1977 Geological Society of London. Reprinted by permission.

**Figure 13.20:** From Paterson, *Experimental Rock Deformation—The Brittle Field,* 1978. Berlin: Springer-Verlag. Reprinted by permission.

**Figure 13.21:** From Bankwitz, P., "Uber Klufte III & IV" in *Z. Geol. Wiss,* Bd. 6, 1978.

**Figure 13.22:** From Sales, Reno., "Ore Deposits of Butte, Montana" in *American Institute of Mining & Metallurgical Engineers Transactions, 46,* pp. 11–30, 1914.

**Figure 13.24:** From Balk, R., "Structural Behavior of Igneous Rocks" in *Geological Society of America Memoir 5,* p. 177, 1937.

**Figure 13.25:** From Murawski, H. and D. Merkel, "Cleat Examinations in the Ishikari Coal Region, Hokkaido/Japan: Some Remarks about the Connection Between Micro- and Macrotectonic Pattern" in *Neues Jahrbuch der Geologie & Palaeontologie Monatsch* (1983) pp. 181–192. Reprinted by permission.

**Figure 13.26:** From Stearns, D. W., "Structural Interpretation of the Fractures Associated with the Bonita Fault" in V. C. Kelly and F. D. Trauger (eds.), *Guidebook of East-Central New Mexico.* Copyright 1972 New Mexico Geological Society. Reprinted by permission.

**Figure 13.28:** From Cloos, E., "Experimental Analysis of Fracture Patterns" in *Geological Society of America Bulletin, 66,* pp. 241–256, 1955.

## Chapter 14

**Figure 14.2:** From Ter-Stapanian, G., "Klassifizierung der Erdrutschrisse" in *Geologie und Bauwesen, 28,* pp. 53–54, 1962. Reprinted by permission of Springer-Verlag, New York.

**Figure 14.3:** From Harrison, J. V. and N. L. Falcon, "Collapse Structures" in *Geological Magazine, 71,* pp. 529–539, 1934. Copyright 1934 Cambridge University Press. Reprinted by permission.

**Figure 14.5:** From Pridmore, C. L. and C. Craig, "Upper-plate Structure and Sedimentation of the Baker Peaks Area, Yuma County, Arizona" in E. G. Frost and D. L. Martin, (eds.), *Mesozoic-Cenozoic Tectonic Evolution of the Colorado River Region, California, Arizona, and Nevada.* Copyright © 1982 Cordilleran Publishing, San Diego. Reprinted by permission.

**Figure 14.6:** From Murawski, H., "Raumproblem und Bewegunglablauf an listrischen Flachen, inshesondere bei Tiefenstorungen" in *Neues Jahrbuch der Geologie & Palaeontologie Monatsch* (1976), pp. 209–220. Reprinted by permission.

**Figure 14.7:** From Erxleben, A. W. and G. Carnahan, "Slick Ranch Area, Starr County, Texas Gulf Coast" in A. W. Bally (ed.), *Seismic Expression of Structural Styles: AAPG Studies in Geology, Series #15, 2,* 2.3.1–23. Copyright © 1983 American Association of Petroleum Geologists. Reprinted by permission.

**Figure 14.8:** From Bruce, C. H., "Pressured Shale and Related Sediment Deformation: Mechanism for Development of Regional Contemporaneous Faults" in *AAPG Bulletin, 57,* pp. 878, 886, 1973.

Copyright 1973 American Association of Petroleum Geologists. Reprinted by permission.

**Figure 14.9:** From Dennis, John G. and V. C. Kelley, "Antithetic and Homothetic Faults" in *Geologische Rundschau, 69,* pp. 186–193, 1980. Reprinted by permission.

**Figure 14.10:** From Behrman, R. B., "Geologie und Lagerstatten des Oelfeldes Reitbrook bei Hamberg" in *Hannover Amt fur Bodenforschung,* pp. 190–221, 1949. Reprinted by permission.

**Figure 14.12 (bottom):** From Cloos, H., "Kunstliche Geberge" in *Natur und Museum, 59,* pp. 225–272, 1929.

**Figure 14.13:** From Mueller, S., E. Peterschmitt, K. Fuchs, and J. Ansorge, "Crustal Structure Beneath the Rhinegraben from Seismic Refraction and Reflection Measurements" in *Tectonophysics, 8,* pp. 529–542, 1969. Copyright © 1969 Elsevier Scientific Publishing, Amsterdam. Reprinted by permission.

**Figure 14.14:** From Zieglar, P. A., "Faulting and Graben Formation in Western and Central Europe" in *Royal Society of London Philosophical Transactions, A305,* pp. 113–143, 1982. Copyright © 1982 Royal Society of London. Reprinted by permission.

**Figures 14.16 and 14.18:** From Holmes, A., *Principles of Physical Geology, 3d ed.* © 1978 Halsted Press, New York. Reprinted by permission.

**Figure 14.19:** From Illies, H., "Bauplan und Baugeschichte des Oberr Heingrabens" in OBERRHEI NISCHE GEOLOGISCHE ABHANDLUNGEN, *14,* pp. 1–54, 1965. Reprinted by permission.

**Figure 14.20:** From Robinson, B. A., "Low-angle Normal Faulting, Mary's River Valley, Nevada" in A. W. Bally (ed.), *Seismic Expression of Structural Styles, 2: AAPG Studies in Geology Series #15,* 2.2.2–12. Copyright © 1983 American Association of Petroleum Geologists. Reprinted by permission.

**Figure 14.21:** From Proffett, John M., Jr., "Cenezoic Geology of the Yerington District, Nevada, and its Implications for the Nature and Origin of Basin and Range Faulting" in *Geological Society of America Bulletin, 88,* pp. 247–266, 1977. Copyright © 1977 John M. Proffett, Jr. Reprinted by permission.

**Figure 14.22:** From Seager, W. R., "Low-angle Gravity Glide Structures in the Northern Virgin Mountains, Nevada and Arizona" in *Geological Society of America Bulletin, 81,* pp. 1517–1538, 1970. Copyright © 1970 W. R. Seager. Reprinted by permission.

**Figure 14.23:** From Gross, W. W. and F. L. Hillemeyer, "Geometric Analysis of Upper-plate Fault Patterns in the Whipple-Buckskin Detachment Torrance, California, and Arizona" in E. G. Frost and D. L. Martin, (eds.), *Mesozoic-Cenezoic Tectonic Evolution of the Colorado River Region, California, Arizona, and Nevada.* Copyright © 1982 Cordilleran Publishing, San Diego. Reprinted by permission.

**Figure 14.24:** From Montadert, L., O. DeCharpel, D. Roberts, P. Guennoc, and J. C. Sibuet, "Northeast Atlantic Passive Continental Margins: Rifting and Subsidence Processes" in *American Geophysical Union Maurice Ewing Series No. 3,* pp. 154–186, 1979. Copyright American Geophysical Union.

**Figure 14.26:** From Tapponnier, Paul, G. Peltzer, A. Y. LeDain, and R. Armijo, "Propagating Extrusion Tectonics in Asia: New Insights from Simple Experiments with Plasticine" in *Geology, 10,* pp. 611–616, 1982. Copyright © 1982 Paul Tapponnier, G. Peltzer, A. Y. LeDain, and R. Armijo. Reprinted by permission.

**Figure 14.27:** From Sengor, A. M. C., "The North Anatolia Transform Fault: Its Age, Offset, and Tectonic Significance" in *Geological Society of London Quarterly Journal, 136,* pp. 269–282, 1979. Copyright © 1979 Geological Society of London. Reprinted by permission.

**Figure 14.28:** From Kennedy, W. Q., "The Great Glen Wrench Fault" in *Geological Society of London Quarterly Journal, 136,* pp. 41–76, 1946. Copyright © 1946 Geological Society of London. Reprinted by permission.

**Figure 14.30:** From Crowell, J. C., "Origin of the Late Cenozoic Basins in Southern California" in *Tectonics and Sedimentation, SEPM Special Publication No. 22,* pp. 189–204, 1974. Copyright © 1974 Society of Economic Paleontologists and Mineralogists. Reprinted by permission.

**Figures 14.33 and 14.35:** From Sylvester, A. G. and R. R. Smith, "Tectonic Transpression and Basement-controlled Deformation in San Andreas Fault Zone, Salton Trough, California" in *AAPG Bulletin, 60,* pp. 2081–2102, 1976. Copyright 1976 American Association of Petroleum Geologists. Reprinted by permission.

**Figure 14.34:** From Harding, T. P. and J. D. Lowell, "Structural Styles, Their Plate-tectonic Habitats, and Hydrocarbon Traps in Petroleum Provinces" in *AAPG Bulletin, 63,* pp. 1016–1058, 1979. Copyright 1979 American Association of Petroleum Geologists. Reprinted by permission.

**Figure 14.38:** From Pierce, W. G., "Jura Tectonics as a Decollement" in *Geological Society of America Bulletin, 77,* pp. 1265–1276, 1966. Copyright © 1966 W. G. Pierce. Reprinted by permission.

**Figure 14.39:** From Kerr, J. W. and R. L. Christie, "Tectonic History of the Boothia Uplift and Cornwallis Fold Belt, Arctic Canada" in *AAPG Bulletin, 49,* pp. 905–926, 1965. Copyright 1965 American Association of Petroleum Geologists. Reprinted by permission.

## Chapter 15

**Figure 15.1:** From Bailey, E. B., *Tectonic Essays: Mostly Alpine.* Copyright 1935 Clarendon Press, Oxford, England. Reprinted by permission.

**Figure 15.2:** From Butler, R. W. H., "The Terminology of Structures in Thrust Belts" in *Journal of Structural Biology, 4,* pp. 239–245, 1982. Copyright 1982 Pergamon Press, Ltd. Reprinted by permission.

**Figure 15.4:** From Woodward, H. P. "The Appalachian Region". 5th World Petroleum Congress. Paper 1–59, p. 19, 1959.

**Figure 15.5:** From Boyer, S. E. and D. Elliott, "Thrust Systems" in *AAPG Bulletin, 66,* pp. 1196–1230, 1982. Copyright 1982 American Association of Petroleum Geologists. Reprinted by permission.

**Figure 15.6:** From Fermor, E. M. R. and R. A. Price, "Imbricate Structures in the Lewis Thrust Sheet around the Cate Creek and Haig Brook Windows, Southeastern British Columbia" in *Geological Survey of Canada, Paper 76–1B,* pp. 7–10, 1976. Copyright Geological Survey of Canada. Reprinted by permission.

**Figure 15.7:** From Berthensen, A., "Recumbent Folds and Boudinage Structures Formed by Subglacial Shear: An Example of Gravity Tectonics" in *Geologic en Mijnbouw, 58,* pp. 253–260, 1979. Reprinted by permission.

**Figure 15.8:** From Dahlstrom, C. D. A., "Balanced Cross Sections" in *Canadian Journal of Earth Sciences, 6,* pp. 743–757, 1969. Reprinted by permission.

**Figure 15.9:** From Dahlstrom, C. D. A., "Balanced Cross Sections" in *Canadian Journal of Earth Sciences, 6,* pp. 743–757, 1969, (after Gallup, 1951). Reprinted by permission.

**Figure 15.10:** From Price, R. A., "The Cordilleran Foreland Thrust and Fold Belt in the Southern Canadian Rockies", a poster presentation, Conference on Thrust and Nappe Tectonics, Imperial College, London, April, 1979. Reprinted by permission of R. A. Price.

**Figure 15.11:** From Trumpy, R., "An Outline of the Geology of Switzerland" in *An Outline of the Geology of Switzerland: A Guidebook.* Copyright 1980 Wepg & Co., Basel, and Schweizerische Geologische Kommission. Reprinted by permission.

**Figure 15.14:** From DeJohn, K. A. and R. Scholten, *Gravity and Tectonics.* © 1973 John Wiley & Sons, New York. Reprinted by permission.

**Figure 15.15:** From Page, Ben M., "Gravity Tectonics Near Passo Della Cisa" in *Geological Society of America Bulletin, 74,* pp. 655–672, 1963. Copyright © 1963 Ben M. Page. Reprinted by permission.

**Figure 15.16:** Source: Scrope, G. P., *Considerations on Volcanoes.* London: W. Phillips, p. 270, 1825.

**Figure 15.17:** From Funk, V. H., T. Labhart, A. G. Milnes, O. A. Pfiffner, U. Schaltegger, C. Schindler, S. M. Schmid, and R. Trumpy, Bericht uber die Jubilaumsexkursion <<Mechanismus der Gebirgsbildung>> der Schweizerishen Geologischen Geselleschaft in das ost- und in das nordliche Aarmassiv vom 12, bis 17. September 1982: *Eclogae Geol Helvetiae,* v. 76, pp. 91–123.

**Figure 15.18:** From Price, R. A., "The distinction between displacement and distortion in flow and the origin of diachronism in tectonic overprinting in orogenic belts". International Geologic Congress, 24th (Montreal) Sect. 3, pp. 545–551, 1972.

**Figure 15.19:** From Elliott, David, "The Motion of Thrust Sheets" in *Journal of Geophysical Research, 81,* pp. 949–963, 1976. Copyright American Geophysical Union.

**Figure 15.20:** From Ramberg, Hans, "Some Remarks on the Mechanisms of Nappe Movement" in *Geologica Foreningen (Stockholm) Forhandl, 99,* pp. 110–117, 1977. Reprinted by permission.

**Figure 15.21:** From Engelen, "Gravity Tectonics in the Northwestern Dolomites, N. Italy" in *Geologica Ultraiectina, 13,* p. 92, 1963. Reprinted by permission.

**Figure 15.22:** From Korn, H. and H. Martin, "Gravity Tectonics in the Naukluft Mountains of South West Africa" in *Geological Society of America Bulletin, 70,* pp. 1047–1078, 1959.

**Figure 15.23:** From Pederson, S. A. S., "Regional Geology and Thrust Fault Tectonics in the Southern Part of the North Greenland Fold Belt, North Peary Land" in *Geological Survey of Greenland, 99,* pp. 79–87, 1980. Copyright 1980 Greenland Geological Survey. Reprinted by permission.

**Figure 15.24:** From Shiki, T. and Y. Misawa, "Forearc Geological Structure of the Japanese Islands" in J. K. Leggett, (ed.), *Trench-Forearc Geology.* Geological Society of London Special Publication 10, pp. 63–73, 1982.

**Figure 15.25:** From Boyer, S. E. and D. Elliott, "Thrust Systems" in *AAPG Bulletin 66,* pp. 1196–1230, 1982. Copyright 1982 American Association of Petroleum Engineers. Reprinted by permission.

**Figure 15.26:** From Cook, Frederick A., D. S. Albaugh, L. D. Brown, S. Kaufman, J. E. Oliver, and R. D. Hatcher, Jr., "Thin-skinned Tectonics in the Crystalline Southern Appalachians" in *Geology 7,* pp. 563–567, 1979. Copyright © 1979 Frederick A. Cook, D. S. Albaugh, L. D. Brown, S. Kaufman, J. E. Oliver, and R. D. Hatcher, Jr. Reprinted by permission.

**Figure 15.27:** From Ramsay, J. H. and Y. Huber, *The Techniques of Modern Structural Geology, Vol. 1.* © 1983 Academic Press, Orlando, Florida. Reprinted by permission.

## Chapter 16

**Figure 16.1** From Bischoff, G., "Statische Gesetzmassigkeiten des basischen Deckenvulkanismus und deren Hinweise auf Vorgange im oberen Erdmantel; mit Beispielen von Analogien aus Sudamerika und Afrika" in *Zeitschrift der Deutschen Geologischen Gesellschaft, 116,* pp. 813–831, 1966. Reprinted by permission.

**Figure 16.3:** From Cloos, H., "Bau und Tatigkeit von Tuffschlaten" in *Geologische Rundschau, 32,* pp. 711–800, 1941. Reprinted by permission.

**Figure 16.4:** From Hans Cloos, *Gesprach mit der Erde, (1954 edition).* © R. Piper & Co Verlag, Munchen, 1947.

**Figure 16.6:** From Cloos, H., *Einfuhrung in die Geologie,* p. 503, 1936. Berlin: Borntraeger. Reprinted by permission.

**Figure 16.7:** From Escher, A., E. Escher, and J. Watterston, "The Reorientation of the Kangemiut Dike Swarm, West Greenland" in *Canadian Journal of Earth Sciences, 12,* pp. 158–173, 1975. Reprinted by permission.

**Figure 16.9:** From Groschopf, P. and W. Reiff, "Das Stinheimer Becken. Ein Verleich mit dem Ries" in *Geologica Bavarica, 61,* pp. 400–412, 1969. Reprinted by permission.

**Figure 16.10:** From Armstutz, G. C., "Polygonal and Ring Tectonic Patterns in the Precambrian and Paleozoic of Missouri, USA" in *Ecologie Geologica Helvetiae, 52,* pp. 904–913, 1959. Reprinted by permission.

**Figures 16.11 and 16.12:** From Wegmann, C. E., "Tectonic Patterns at Different Levels" in *Geological Society of South Africa Proceedings, 66,* pp. 1–78, 1965. Copyright 1965 Geological Society of South Africa. Reprinted by permission.

**Figure 16.14:** From Eskola, P. E., "The Problem of Mantled Gneiss Domes" in *Geological Society of London Quarterly Journal, 104,* pp. 461–476, 1949. Copyright © 1949 Geological Society of London. Reprinted by permission.

**Figure 16.15:** Based on an unpublished drawing by C. A. Hopson. Redrawn by permission.

**Figure 16.16:** From Oertel, G., "Der Pluton von Loch Doon in Sud Schottland" in *Geotektonische Forschungen, 11,* p. 83, 1955. Reprinted by permission.

**Figures 16.19 and 16.26:** From Sylvester, A. G., "The Precambrian Rocks of the Telemark Area in South Central Norway III" in *Norsk Geolisk Tidsskrift, 44,* pp. 445–482, 1964. Copyright 1964 Universitetsforlaget, Oslo. By permission.

**Figure 16.22:** From Pitcher, W. S., "The Nature, Ascent, and Emplacement of Granitic Magmas" in *Geological Society of London Quarterly Journal, 136,* 627, 662, 1979. Copyright © 1979 Geological Society of London. Reprinted by permission.

**Figure 16.23:** From Cloos, H., "Der Sierra-Nevada-Pluton in Californien" in *Neues Jharbach der Geologie & Paleantologie (1936), 76,* (Beilage Bd), pp. 356–450. Reprinted by permission.

**Figure 16.24:** From McGregor, A. M., "Some Milestones in the Precambrian of Southern Rhodesia" in *Geological Society of South Africa Proceedings, 54,* pp. 27–71, 1951. Copyright 1951 Geological Society of South Africa. Reprinted by permission.

**Figure 16.25:** From Balk, R., "Structural Behavior of Igneous Rocks" in *Geological Society of America Memoir 5,* p. 177, 1937.

**Figures 16.27 and 16.28:** From Cloos, H. and E. Cloos, "Die Quellkuppe des Drachenfels am Rhein; ihre Tektonik und Bildungsweise" in *Zeitschrift Vulkanologie, 11,* pp. 33–40, 1927.

**Figure 16.30:** From Holder, M. T., "An Emplacement Mechanism for Post-tectonic Granites and its Implications for their Geochemical Features" in M. P. Atherton and J. Tarney, (eds.), *Origin of Granite Batholiths, Geochemical Evidence,* pp. 116–128. Copyright 1979 Shiva Publications Ltd., Natwich, Cheshire, England.

# Index

## G

Gangel, L., 248
Geological maps, 352
Geological Society of America, 212
Geometric analysis, xv–xvi
Geosynclines, 23
Ghosh, S. K., 133
Gibson, I. L., 19
Gilbert, G. K., 33
Gill, J. E., 216, 217, 218, 220
Girdle, 156, 157, 188
    crossed, 163
Gjar, 39
Glacier, Malaspina, 103
Glacier foliation, 104
Glarus double fold, 284
Glass, I. G., 19
Gneiss
    augen, 228
    mylonite, 228
Gneiss domes, 317
Gouge, fault, 226, 228
Graben, 260–65
    Rhine, 260, 261, 264, 265
    Viking, 262
Graded bedding, 56, 62
Gradient, 338, 355
Grain fabrics, 102, 155–59
Granites
    bedding, 328
    I-type, 324, 325
    rift, 328
    S-type, 324, 325
Gravity, 254–57
Gravity faults, 254
Gravity sliding, 295
Gravity spreading, 295, 296, 298
Gray, D. R., 182
Great Glen fault, 271, 272
Greene, G. W., 35
Gretener, P. E., 61
Griffith, A. A., 234
Griffith cracks, 234
Griffith criterion of failure, 234
Griggs, D. T., 90, 92, 94, 102, 242

Groschopf, P., 313
Gross, W. W., 267
Growth faults, 257–59
Gutenberg, B., 7, 87

## H

Hackle marks, 205, 206
Halbouty, M. T., 86
Hall, Sir James, 82
Haller, J., 317
Halokinesis, 106
Ham, W. E., 22
Hamberg, A., 104
Hammond, R. H., 48
Handin, J., 77, 78, 89, 94, 242
Hanging wall, 212, 213
Hardin, G. C., Jr., 86
Harding, T. P., 275
Harris, L. D., 285
Harris, R. L., Jr., 171
Harrison, J. V., 255
Hartmann, L., 238
Hartmann's rule, 237, 238, 240
Heard, H. C., 91, 92, 93
Heave, 219
Heezen, B. C., 9, 18
Heim, A., 82, 154, 179, 284
Heirtzler, J. R., 12
Helvetic Alps, 296
Helvetic nappes, 149, 301
Hercynotype, 323, 325
Herschel, Sir John, 31
Hess, H. H., 9
Heterolithic unconformity, 57–58
Hiatus, 57
Hill, M. L., 20, 220, 272
Hillemeyer, F. L., 267
Hinge (fold), 116, 124
Histogram, 207
    polar, 207
Hodgson, R. A., 205
Hoeppener, R., 71, 144, 251
Hofmann, R. B., 30
Holder, M. T., 328
Hollingworth, S. E., 108

Hollister (California), 37
Holmes, A., 8, 262
Homoaxial (folding), 128
Homocline, 122, 123
Homogeneity, 155
Homogeneous strain, 69
    finite, 70
Homogeneous stress field, 68
Hooke body, 74
Hopson, C. A., 318
Horizontal separation, 216, 217
Horse, 286
Horsetail veins, 246
Hossack, J. R., 188, 288
Hot spots, 20, 21
Hubbert, M. K., 83, 84, 85, 87, 236, 240,
    241, 289, 291
Huber, M. I., 72, 136, 301
Hunt, C. B., 35
Hutton, C., 6
Hutton, J., 294, 328
Hydraulic fracturing, 236

## I

Ignimbrites, 322, 323
Illies, H., 204, 264, 265
Imbricate, 224, 286
Incompetent, 136
Indicator minerals, 155–56
Infinitesimal strain, 70
Inflation, 328
Infrastructure, 315–18, 320
Initial dip, 55
Inoue, E. 28, 29
Interlimb angle, 117
Intrafolial folds, 176
Inverted limb, 117
Irvine, T. N., 63
Isacks, B., 4, 12, 16
Island arc, 324
Isoclinal folds, 117
Isogons, dip, 120, 121
Isostasy, 5–7
Isotropic fabric, 156
Isotropic stress, 66